LAST CALL

Humanity Hanging from a Cross of Iron and
Our Escape to Another Planet

"El sueño de la razón produce monstruos". (The dream of reason produces monsters.) Etching from Los Caprichos de Francisco de Goya (1799).

LAST CALL

Humanity Hanging from a Cross of Iron and Our Escape to Another Planet

Daniel R. Altschuler
University of Puerto Rico, USA

World Scientific

NEW JERSEY · LONDON · SINGAPORE · BEIJING · SHANGHAI · HONG KONG · TAIPEI · CHENNAI · TOKYO

Published by

World Scientific Publishing Co. Pte. Ltd.
5 Toh Tuck Link, Singapore 596224
USA office: 27 Warren Street, Suite 401-402, Hackensack, NJ 07601
UK office: 57 Shelton Street, Covent Garden, London WC2H 9HE

Library of Congress Cataloging-in-Publication Data
Names: Altschuler, Daniel R., author.
Title: Last call : humanity hanging from a cross of iron and our escape to another planet /
    Daniel R. Altschuler.
Description: New Jersey : World Scientific, [2022] | Includes bibliographical references and index.
Identifiers: LCCN 2022008889 | ISBN 9789811253614 (hardcover) |
    ISBN 9789811254383 (paperback) | ISBN 9789811253621 (ebook for institutions) |
    ISBN 9789811253638 (ebook for individuals)
Subjects: LCSH: Science--Philosophy--History. | Nature--Effect of human beings on.
Classification: LCC Q174.8 .A48 2022 | DDC 306.4/5--dc23/eng20220503
LC record available at https://lccn.loc.gov/2022008889

British Library Cataloguing-in-Publication Data
A catalogue record for this book is available from the British Library.

For any available supplementary material, please visit
https://www.worldscientific.com/worldscibooks/10.1142/12760#t=suppl

The pub was silent for a moment longer, and then, embarrassingly enough, the man with the raucous laugh did it again. The girl he had dragged along to the pub with him had grown to loathe him dearly over the last hour or so, and it would probably have been a great satisfaction to her to know that in a minute and a half or so he would suddenly evaporate into a whiff of hydrogen, ozone and carbon monoxide. However, when the moment came she would be too busy evaporating herself to notice it. The barman cleared his throat. He heard himself say: "Last orders, please."

— Douglas Adams, *The Hitchhiker's Guide to the Galaxy*

Thanks

I thank my wife, Celia, who taught me many things and encouraged me when I wanted to give up. Then I thank all those who, over many years, helped me understand. You know who you are. Finally, all those who wrote books that taught me many things I could not have learned otherwise. Special thanks to Grammarly for a review of my writing.

Contents

List of Figures

Chapter 7

Chapter 8

Chapter 9

Chapter 10

Chapter 11

Chapter 12

Chapter 13

Chapter 17

List of Tables

Chapter 1

Introduction

Man is a rational animal: so at least I have been told. Throughout a long life, I have looked diligently for evidence in favour of this statement, but so far, I have not had the good fortune to come across it, though I have searched in many countries spread over three continents. On the contrary, I have seen the world plunging continually further into madness. I have seen great nations, formerly leaders of civilization, led astray by preachers of bombastic nonsense. I have seen cruelty, persecution, and superstition increasing by leaps and bounds, until we have almost reached the point where praise of rationality is held to mark a man as an old fogey regrettably surviving from a bygone age.

— Bertrand Russell[1] (Endorsed by Daniel Altschuler)

To sin by silence, when we should protest,
Makes cowards out of men. The human race
Has climbed on protest. Had no voice been raised
Against injustice, ignorance, and lust,
The inquisition yet would serve the law,
And guillotines decide our least disputes.
The few who dare, must speak and speak again
To right the wrongs of many. Speech, thank God,
No vested power in this great day and land
Can gag or throttle. Press and voice may cry
Loud disapproval of existing ills.
May criticize oppression and condemn
The lawlessness of wealth-protecting laws
That let the children and child bearers toil
To purchase ease for idle millionaires.

— Ella Wheeler Wilcox (1850–1919)[2]

I look at the scene through my window, sitting on a comfortable chair. I see the Caribbean Sea reflecting different shades of blue. White waves break on shallow mangrove islets, and the bright sunlight reflects off dark green leaves buffeted by the wind. From this point of view, the world looks beautiful, more so because through science and scholarship, we mostly understand how it came to be. It is an astonishing story, better than all fiction you might have read, but it might end quite differently from what you may think.

The curved horizon I see hints at the spherical Earth, which, as far as I can see, looks vast, and in contrast, I am but a small ant. A simple calculation tells me you could place all humans on the island of Hawaii (standing), so how can *we* be a problem for the planet, as you may have heard?

It starts when a human, who perhaps weighs 150 pounds, uses a vehicle that weighs about twenty times as much to move around. And that is just one example of thousands you can think about. But, then, what matters is our *ecological footprint*, as I will discuss below. And then, there is all the rest which is what this book is about. The bottom line is that we have changed the world in the last 50 years more than during all of human existence, and those changes push us to an abyss.

Perhaps I need to look at it from a different point of view than from my comfortable chair to understand a few things that escape my personal experience. Shifting one's point of view is always an excellent tactic to learn something new. And what I learn is not so comfortable.

In 1609, a man equipped with a tube and a couple of lenses, later called "telescope," opened a new window to the skies, a window through which, with the advance of science and new technologies, we have come to grasp our place in the Universe ("Mostly harmless"). Aside from amazement, the main message is humility: we are here on a minuscule microscopic dot of an inconceivable vast Universe, both in time and space, probably and hopefully not the only sentient beings, but we do not yet know.

Here I want to open a different window, one that looks down upon our planet and our humanity, telling you what I see, trying to comprehend this unhinged world we live in. I will look at it from an "astronomical" perspective, a view of humanity from as far away as I can, a view from nowhere, considering our place in the cosmos, our astonishing pigheadedness, and the realities discovered by science. These have forced on us a different and less comfortable metaphysical view that upset many (and still does), that sprouted from the minds of Galileo di Vincenzo Bonaulti de Galilei (1564–1642), Charles Robert Darwin (1809–1882) and Santiago Ramón y Cajal (1852–1934), (whom you will soon meet). We learned how the world is and not how we wish it were from them and their followers. A view that goes against deep-rooted myths that still obscure many minds. These myths infuse our thoughts, and we should at least understand them to have a chance to enlighten our minds. (Other cultures have different myths, equally hazardous.)

This book considers the potential extinction of a species, a frequent event over the deep history of life on Earth. A maladapted species, or one whose habitat changes radically, does not survive leaving no descendants, although some are fossilized for us to learn about their unfortunate departure. It is happening right now, and it is no longer big news when the last one of a species dies, be it plant or animal. Gorillas, elephants, blue whales, tigers, orangutans, and many lesser-known forms of life are on the verge, and the primary reason for this is *us*. We did follow the instructions: *and let them have dominion over the fish of the sea, and over the fowl of the air, and over the cattle, and over all the earth, and over every creeping thing that creepeth upon the earth* (Genesis 1:26). The newsworthy species I will tell you about is *Homo sapiens*, which we call ourselves with a bit of exaggeration.

True, bad news is never welcome. We see enough of that every time we look at a screen or read a paper (never mind books which are also an endangered species). It is well known that in some cases, bad news is simply not believed, denied so as not to confront an unacceptable and challenging fact, known to psychologists as a "cognitive dissonance." Others choose to shoot or fire the messenger, as if that settled anything, or might resort to "the power of positive thinking," but it will not get them far.

The future will not be as great as it is portrayed in many movies and novels, suggesting that we shall conquer space and colonize other planets, even if we are turned into silicon chips.

But just look around you far and wide, not with your prejudiced eyes, not through pink glasses, just with your plain eyes, whatever color they may be. It is mostly not a pretty sight. You imagine that we will cross a new frontier, that progress will make life better for all. But imagination is often very far from reality. It cannot avoid some facts for which there are no alternatives, including this one: sooner or later, you will go back to where you came from and stay there for eternity, a natural event, unavoidable.

Suicide, for instance, can be avoided if someone heeds the signs of a hopelessly depressed friend and calls for help (unless you are in jail) but depending on circumstances, help might arrive too late. Unfortunately, every year, about 45,000 persons commit suicide in the US (12 per 100,000), and it is the tenth leading cause of death (the global rate is about 11 per 100,000). Thus, as a global civilization, we face what seems unavoidable, with nobody to call for help, neither mythical gods nor illusory extraterrestrials, who might save us from mass suicide.

I, at least, prefer to live with both feet on the ground. And that ground is provided by science and scholarship, a knowledge that can give us some perspective like the fact that the Earth is about 4 billion years old and that *Homo sapiens* is about 150,000 years young. Think about that! (Imagine that you want to walk from San Francisco to Buenos Aires (although I do not recommend it). If you want the journey to last the age of the Earth, then you should take one step every 1,000 years or so.

Perhaps "metaphysical" makes it sound too deep. As far as I am concerned, all it means is a "worldview," which implies a certain way of thinking and the acceptance of particular narratives. If you hold on to obsolete metaphysics, you live in a different world from the real one, which can cause a lot of personal and collective grief. Therefore, I will discuss what changed because of the individuals mentioned above. *Without realizing it, we have poisoned the Earth by our emissions of greenhouse gases and weakened it by taking for farmland and housing the land that once was the home of ecosystems that sustained the environment. We have driven the Earth to a crisis state from which it may never, on a human time scale, return to the lush and comfortable world we love and in which we grew up*, wrote distinguished British environmentalist James Lovelock[3] in 2007 (who in 2019 turned 100). This concisely describes part of the view through this window (but there is more!).

Throughout history, many predicted the world's end, time's end (although it means the end of humans). Some have even wished for it expecting a better life (or death), generally related to some religion or supposed secret discovered in some ancient book or calendar. Despite all these predictions, the fulfillment of which many anticipated with joy, we are still here, on what is becoming an increasingly deeper "shithole" to use the word of a shameless president. This book is not about another eschatological prophecy. It does not give a date to lament or celebrate the long-awaited end and the entrance to a new kingdom. Still, it does paint what can be expected from a scientific point of view — a less abrupt, less glorious, and sadder ending of what could be considered a miracle.

The famous Isaac Newton wasted most of his time trying to determine a date for the event, which for him at least, came on March 31, 1727, when he died. His search for answers to a myth was bound to be fruitless. Still, this prodigy also left us with ideas that continued and consolidated those of Galileo Galilei and Johannes Kepler, who preceded him. Physics and modern science continued at an accelerated pace, spawning all types of technologies which to them would have looked like magic, and we reached the Moon.

The gap between our technological capabilities and our mental attitudes has widened significantly in the last 100 years. Part of the problem arises because some of our mental attitudes are hard-wired and very difficult to modify. We are the way we are. As distinguished sociobiologist Edward O. Wilson (born 1929) wrote[4]: *The real problem of humanity is the following: we have paleolithic emotions; medieval institutions; and god-like technology. And it is terrifically dangerous, and it is now approaching a point of crisis overall.*

Let us not forget that *Homo sapiens* is an animal (an unsettling fact for some). But, unfortunately, the gap between our scientific understanding of the world and what people think about it has become an almost insuperable chasm; witness all the hot air and nonsense about our hotter air which I will address in detail below.

Predictions are difficult since many unknown and unforeseen events might happen to spoil them. This is what happened to the dinosaurs (and many other species) about 66 million years ago when a 10 km diameter uninvited asteroid collided with the Earth changing the course of life. On the other hand, we would not be here if it were not for that accident.

Maybe, it will not be *that* bad (unless another asteroid hits or some idiot hits the red button). Perhaps part of humanity (the wealthy and powerful) might find a way to squeeze through the bottleneck we face and survive after the nightmare, but it will hurt.

We are entering a new stage of the Holocene ("Recent Whole"), which is how we call our current geologic epoch, which began after the last ice age about 12,000 years ago with a major shift in the climate. It is not the age of Aquarius; it is the *Anthropocene Transition*,[a] characterized by a **catastrophic convergence**[5] of serious problems: *climate and environmental crisis caused by our activities, the waging of wars, the existence of nuclear weapons* with the horrifying idea that some could be used, and *increasing inequality* at all levels (promoted by unbridled capitalism), problems that do have a profound ethical character and threaten to end forever with our misery. In addition, the gray eminence of *population growth* (which is no longer news) drives the whole thing. Of course, these are not the only ones (pandemics are in everyone's mind these days), but this book is long enough as it is.

Add to this the *idiotization* of the masses, including the delegitimization of scientific knowledge and the adherence to obsolete ideas and childish myths, including the idea that problems are solved with violence, and we are done.[6] And this idiotization is not due to some fault in our educational system. Still, it is built into the system so that those in control can easily indoctrinate the young about almost anything since they were not educated to *think critically* — to acquire a *scientific temper*. And some fools go to court to block the teaching of science (evolution) to uphold a myth.

---

[a] Although the term "Anthropocene" is fashionable I argue that it does not represent a new geologic epoch, like the one's geologist have identified in the past, separated one from the other by a drastic change in climate and life. So, we identify the Cretaceous and the Paleogene (Tertiary) as different geologic eras, mainly because at the end of the Cretaceous there occurred one of the five major extinctions of life on Earth. A thin layer of material separates sediments of these two eras, characterized by an unusually high content of Iridium. That actual layer of clay a few inches thick is not called by any name, it just marks the transition. The same can be said of the transition between the Permian and the Triassic, 252 million years ago, which wiped out most of life on Earth. So, I will use the term, "Anthropocene", in this restricted sense, meaning a transition to a new state of the world, and it will not much matter how that will be called.

My goal is to help you understand these problems, tell you as best as I can what the facts are and why I think they steer us to Dante's final gate.[b] I hope that after reading this book, you will be better prepared to evaluate the claims you hear and thus safer navigate through the treacherous waters of the sea of ambivalent thinking, "fake news," and "alternative facts." Perhaps you do not need persuasion, but I give you some tools to make others understand since otherwise, they will kill us. Thus, you will be able to avoid the currents that flow through the planet's fiber optics networks from transporting you to a torrid zone in which frightful storms of confusion and hatred abound. From there, the firm ground of reason is barely noticeable as if it were a vision in the mist. Reason, as Aristotle already thought, is the only thing that distinguishes us from all other living beings on the planet. It is at the same time the cause of the serious global problems that we face, two good incentives to try to understand why we are where we are, why we act like perfect idiots, if possible, without causing *metaphysical horror*.[7]

I will take you on tour and review a few crucial facts discovered by science to give you some background and perspective. I will do this because we live in a scientific age and because science is telling us some "inconvenient truths." I will tell you what science is and how it differs from technology. Scientists cannot detach themselves from the rest of society, working comfortably in government, industrial, or university laboratories, precisely because they are part of society and carry the burden of knowledge, an excellent reason to write this book.

Scientific knowledge clashes with certain widely held political or religious beliefs, and in their name, not a few bad things happen. This is not meant to be a political book. Many are being published in these turbulent times, but politics is unavoidable even if I write about often misunderstood science. So, I will not shy away from telling you what I think about politicians, any politicians, despite the good things that might be said in their favor, since that is irrelevant. Hitler loved his dog. I dislike most of the new generation of Teflon-clad stone-faced politicians of stunning mediocrity, who often embarrass our humanity.

---

[b] Dante Alighieri (born c. 1265, in Florence, Italy — died 1321, in Ravenna), Italian poet, prose writer, literary theorist, moral philosopher, and political thinker is the author of the Divine Comedy.

This is particularly important when science is being sidelined by those who make decisions, leading us to disaster (a word originated when losing the view of stars on a sailboat spelled trouble). You have seen this happening in real-time as you have heard what government officials (of the Trump administration) were babbling about what was happening to us due to the SARS-CoV-2 virus.

I will tell you what I think of miracles, souls, and the supernatural. Of the many myths which infect our minds, these are the worst and deeply entrenched. You cannot solve any problem if you start from falsehoods. Although we read in many places that "in God we trust," it can quickly transform into a copout or something even worse, as shown by what is happening in the region we associate with the "cradle of civilization." It seems that killing others for thinking differently or because their skin is different is how we are. It is a bloody thread that runs through the tapestry of our nightmarish history.

I will tell you about the always fashionable and appealing conspiracy theories because they can lead to tragedies under certain circumstances, as illustrated by one of the worst in recent history (the Shoah[c]), which we should remember since forgetting can lead us to repeat. When reason is lost, terrible things happen. When truth succumbs to myth, terrible things happen. When money is all that counts, nothing counts. When you believe nonsense, you place yourself in danger. And ethics should not be just a course you took in college.

And suppose you agree that humanity's future is bleak, and you have bought into the idea of a technological solution, a "fix," consisting of abandoning ship and traveling onboard another vessel to colonize a nearby planet (nearby in this case might mean 1,000 light-years). In that case, I shall review the exciting recent discoveries of extrasolar planets. This, aside from a fascinating story, will convince you that it is either here or nowhere; much of the rest of the book should convince you that it is also now or never.

---

[c] Commonly known as the "Holocaust", but a holocaust was an ancient offering to the gods by burning an animal, so the Hebrew Shoah, meaning "destruction" is preferable.

I will also tell you how I view the empire, that is, the United States. Probably most of its citizens do not think about this matter, and perhaps are not even aware of this or do not care, or worse, might even be proud of it uncritically accepting that they are exceptional. This, however, robs them of a good chunk of their hard-earned dollars that go to support a costly and massive military engaged in constant and futile wars. Or perhaps they swallowed the narrative that we are a "good empire" with good intentions (under the euphemism of "global leadership" or "leader of the free world"). Instead, we create havoc and bomb and kill people in distant lands in a quest for human rights, freedom, and democracy, to defend the Nation. History tells us that there never were good empires, only bad ones, some worse than others.

Official narratives (as said by George Orwell: *he who controls the present controls the past*) are often quite different from reality, established by schooling, and all sorts of media, think tanks, "public relations" efforts, and government propaganda. *Plus, it's a lot easier to believe that your country is a benevolent and liberating force for good in the world, rather than as an imperial nation that enters (and wages) war to further its strategic and economic interests*, write Sirvent and Haiphong.[8] It very much boils down to finding the narrative that best fits the real world and will help us get out of the mess we are in. In the words of critic Caitlin Johnstone[9]: *The real underlying currency of our world is not gold, nor bureaucratic fiat, not even military might. The real underlying currency of our world is narrative and the ability to control it. Everything always comes down to this one real currency..*

Philosopher Jason Stanley writes[10]: *If one can convince a population that they are rightfully exceptional, that they are destined by nature or by religious faith to rule other populations, one has already convinced them of a monstrous lie.* Journalist Max Blumenthal[11] has a different narrative: *For most Americans, the digital abstraction of the war and the dual-layer patina of patriotic hoopla and humanitarian goodwill overwhelmed their critical faculties and ensured their consent. The stage was set for the era of drone warfare that saw the United States carrying our robotic assassinations from Yemen to the Philippines with little political backlash at home.*

The US invaded Iraq to help them get rid of an evil dictator (he was) who had weapons of mass destruction (he did not), but look at the wreckage left behind. Think that we are best friends of Saudi Arabia and sell them armaments used to liquidate people in Yemen. The US is also allied (married might be the word) to Israel, a country that seems to have forgotten its roots and now treats Palestinians as if they were not human.

We were also good friends of Cuban dictator Fulgencio Batista (1901–1973), the New York and Las Vegas mafia chum. Later the US invaded Cuba (bay of pigs) to get rid of the one who got rid of our friend, (Fidel Castro Ruz (1926–2016)), simply because he did not play along (I could go on and on, but one example is enough).

For 60 years, the US government has kept the pressure on Cuba with an economic embargo that hurts its citizens because it just does not like the way Cuba manages its affairs (or perhaps more to the point, to keep Cuban immigrant support in a critical state). It is then absurd to claim that this as proof that socialism (the dreaded s-word) does not work. As told by William Blum[12] in a must-read book, it started under the Eisenhower administration, with a memo by then deputy assistant secretary of state for inter-American affairs (Lester D, Mallory (1904–1994)) that stated in part: *Most Cubans support Castro... There is no effective political opposition (...) The only possible way to make the government lose domestic support is by provoking disappointment and discouragement through economic dissatisfaction and hardships (...) Every possible means should be immediately used to weaken the economic life (...) denying Cuba funds and supplies to reduce nominal and real salaries with the objective of provoking hunger, desperation and the overthrow of the government.* So much for human rights, so much for good intentions. It might be understandable in the context of the Cold War and the (exaggerated) fear of communism, which was utilized to justify capitalism as the only and best economic system, but it is reprehensible to continue. The same reasoning operates against Venezuela, Iran, and any other countries that do not play along.

At the same time, we are best friends with Saudi Arabia, which manages its affairs in repugnant ways (sometimes with a chain saw). Corporations can set up shops in communist China, but not in communist Cuba. It is difficult to accept and respect this double standard, and it was a stupid move for the US in the first place, as I write on a computer made in China.

As Michael Parenti explains in a book[13] that would open your mind: *The overall aim is to promote a global order dedicated to private ownership of the world's financial and industrial wealth, expropriation of its natural resources, and advantageous control of its consumer and labor markets. This is a world where the gap between the wealthy few and the many poor grows ever greater, where the masses are experiencing a drastic decline in living standards. The goal is a world composed totally of exploitative, repressive, free-market countries like Indonesia, Nigeria, and Haiti rather than prosperous social democracies like Finland, Sweden, or Denmark (whatever their respective flaws). Thus far, the empire builders have been quite successful.*

Over history, many nations became empires and for a while determined the course of events, but just as is the case for species, empires do not last forever (neither do diamonds, by the way), and what they leave behind is not a pretty sight. The Roman Empire and the British one, among others, eventually collapsed, leaving behind offending statues and a good number of corpses. The present empire maintains itself by having by far the mightiest military in the world, squandered effort (in the words of General Dwight Eisenhower), which in the end will not help because the game is changing. Neither a pandemic nor a hurricane can be stopped with nuclear weapons or a wall, and the wounds we have inflicted on our planet, are not healed with aircraft carriers. Current events also hint that the empire is crumbling from within.

The ever more strident calls for urgent change and global crisis management are ignored by the Nation, which, if not the leader, should be one of the major players in this live-or-die game. Instead, it is immersed in thinking about solving its sad internal problems and achieving Martin Luther King Jr.'s dream of 60 years ago. But the dream should cross US borders and engulf the world, but it will not be achieved by dropping bombs. *I can't breathe*, implored George Floyd (1973–2020) as a policeman's knee pressing on his neck took his life. Racism, one of the terrible ills of humanity, thrives in the US and other countries. Do you remember Rodney King? That's the problem.

Imagine an international football tournament: the US team arriving with glittering helmets and coaches and players psyched-up to bring the trophy home. But upon entering the field, they see the other team: they wear shorts and no helmets, and it is a *round* ball. Wrong game!

A "sleeping giant" has woken (and let us hope we do not fall into Thucydides' Trap (see below). It uses its enormous resources on more productive pursuits than just modernizing weapons that cannot be used, outsmarting those with the oval ball. It is rapidly becoming the world's largest economy (and polluter), and every year has a more substantial presence in world affairs. Moreover, China and Russia are engaged in a different kind of war, a shadow war[14] waged by attacking the vulnerabilities of the US as an open society. We need brains instead of muscles to face these challenges.

How dare you! said the 16-year-old Swedish schoolgirl in a short emotional address to those gathered at the United Nations Climate Action Summit on September 23, 2019. Greta Thunberg[15] said: *People are suffering. People are dying. Entire ecosystems are collapsing. We are at the beginning of mass extinction. And all you can talk about is money and fairy tales of eternal economic growth. How dare you.* Indeed, how dare we!

You might question what a scientist has to say about things that are beyond science. How can I talk about economics, ethics, war, and peace, things that do not belong to the scientific domain? My formal training is not in those areas. However, the scientific way of thinking, which I will present later as the *scientific temper,* is not exclusive to science. On the contrary, it is relevant to all other aspects of our society because the world is one, composed of scientific facts and human action. As sung by Paul Simon, I also wonder what's gone wrong. This book is my attempt to find an answer. Furthermore, you do not have to be an ichthyologist to know when fish stink. What we do or not can and needs to be illuminated by the scientific temper, and that is what I intend to do, and you will see what comes out of it and be the judge.

1.1 A Note on Quotations and Notes

> In quoting others, we cite ourselves.
>
> — Julio Cortazar[16]

You will find the text infused with many quotations by others because I could not say it better and illustrate that many of my concerns were already expressed years if not centuries ago. They will also familiarize you with some you might not have heard of.

Notes are not there to appear authoritative but document what I say and provide references to books for further reading. I have not read them all, but at least browsed them or read a few chapters to recommend them. There are many, and you might not have time, but try, books are still the best source of knowledge (if you know which to read), and I hope those I mention will provide. The only way to escape from the narrative matrix is to search for the real world described in serious, well-researched books. An appendix will give a selection (See the note.[d]) By the way, the epigraphs at the beginning of the chapters or sections are not there for decoration. They are relevant, things said in the past by wise people and tried to warn us, sadly with little effect.

1.2 Things I Need to Tell You

What a chimera then is man, what a novelty, what a monster, what chaos, what a subject of contradiction, what a prodigy! Judge of all things, yet an imbecile earthworm; depository of truth, yet a sewer of uncertainty and error; pride and refuse of the universe. Who shall resolve this tangle?

— Blaise Pascal[17]

We, humans, are creatures of paradox. When our talents and capabilities are put to positive use, we compose music that makes the spirit soar, create works of art and architecture of timeless beauty, and design and build devices that multiply our physical capabilities, extend the reach of our unaided senses and connect us to the wider world. Yet we can and do also use those remarkable talents and capabilities to very different effects, despoiling our environment, debasing and degrading each other, and creating products and processes so destructive that we may yet become the first species responsible for its own extinction.

— Lloyd Dumas[18]

---

[d] I have tried to document everything said as it should be, indicating the original sources, and the books that treat the different topics I write about in more detail. I have preferred, if there are alternatives, not to cite cybernetic sources because of their volatility. For each book quoted there are dozens that deal approximately with the same topic at different levels and styles. Beyond a documentation of sources, I consider it a list of readings that I recommend to those who wish to dig deeper. A selected list is given in the Appendix.

I began writing this well before the COVID-19 pandemic (BC). You will read this after COVID-19 (AC), after the deaths of millions and more infected. Thanks to science, we have developed vaccines that are slowly pushing back on this scourge. We would be more successful if people understood that vaccines save lives and did not believe weird things. But although many hope (and we must) that we will radically change our way of life in AC, I am afraid that AC will resemble BC regarding the vital issues I will tell you about. People yearn to go back to "normal" when they should not. I also began in BT (before Trump), and you will read in AT. Significant changes, perhaps, but what I must tell you does not change much.

And I think this way because history informs me that we have not altered our behavior very much despite the warnings, the calamities we have been through, and the hallowed progress we have made despite all the bloodshed. I wrote during the miserable Trump presidency, a "splendid marriage between ignorance and arrogance."[e] I hope that President Biden's administration will work toward internal peace, listen closely to the experts, and change things for the better. But he must also work toward world peace, which is uncertain unless he agrees (and manages) to dismantle the empire. Some time ago, Chalmers Johnson[19] (1931–2010), professor of political science at the University of California, San Diego (and a consultant for the CIA from 1967 to 1973), suggested this as "our last hope." We must join forces with all nations to change course, for which energetic revolutionary action is needed. The course we are on is a curse.

This book is relevant to that, especially since Biden is a creature of the establishment, and we do not need more of the same. Just rejoining the Paris accords and the WHO (World Health Organization) will not change much. Much more needs to be done; much needs to change, especially after Trump's destruction. If the idea is to return to "traditional American global leadership," then it is a bad idea, as you will understand as you read on. The way to win the war on terror is to stop being a terrorist and bullying others, just as we could win the war on drugs by decriminalizing

---

[e] As expressed by CNN's Chris Cuomo.

all of them. Do you want to kill yourself with drugs? Be my guest. After all, there is no war on suicides, and people do kill themselves.

I cannot and do not pretend to be neutral; nobody can. I suppose an extraterrestrial could (those who neither visit us, despite the current Pentagon hype nor send telepathic messages). I am a human being with a mind-forged over eons by evolutionary processes and influenced by the culture that nourished it (just like yours). A mind that was born a few months before the Soviets arrived at Auschwitz to find piles of dead or dying humans abandoned by the Nazis. Primo Levi[20] survived, as did a few others to tell the harrowing story. My mind spoke German with my parents, who had fled, but they never mentioned that, and I only found out later. They never discussed much about being Jewish, never went to synagogue so that I was able to keep my atheism (oh my god, why is that so terrible?) with which we are all born.

Scrawled on the walls of my city, I often saw *Yankee go home*, but in my youth, I did not understand. Hadn't the US and its allies defeated those evil men my family fled from? Were they not the best in science and technology? Wasn't there something called the "American Dream" that everyone desired? Didn't the US have the best universities, which indeed I was lucky to attend?

Years have passed, and my mind learned many things, some related to my profession, and others simply because of the accumulation of experiences, lived or studied. I confess that I do not like the civilization we have created. Sure, it has many good and positive things, but the bad ones overwhelm it. This is not because of anybody, not because of a conspiracy. Still, there are plenty who push in the wrong direction, fooled by a false narrative facilitated by ignorance. I fight with these apathetic times simultaneously horrified and fascinated.

Everything indicates that we may be reaching a breaking point, a point of no return. A consequence of the accumulation of deeds perpetrated on humanity and the planet by dark minds, such as that of the current president of Brazil, Jair Messias! Bolsonaro, who in December 2008, said[21]: *The error of the* (Brazilian) *dictatorship was that it tortured, but did not kill.* (It *did* kill.)[22] He is not the only one.

Each one of us can make a long list of shameful past and present humans who disgrace us. Probably the most despised one (except for some lunatics) is Adolf Hitler (1880–1945), and his henchmen, including Hermann Göring (1932–1945), who was sentenced to death by hanging at the Nuremberg trials (but committed suicide the night before) as did the other beast, Heinrich Himmler (1900–1945). The list of German criminals is long, but other nations are not exempt, featuring the ruthless Russian Joseph Vissarionovich Dzhugashvili (Stalin) (1878–1953), the clownish Italian Benito Mussolini (1883–1945), the founding father of fascism, and Spanish dictator Francisco Franco (1892–1975).

Augusto Pinochet (1915–2006) can be added to the list (including many other Latin American dictators) being responsible for the death and suffering of tens of thousands (with the support and knowledge of the US government). Or, if you are interested in Africa, I offer ruthless Idi Amin Dada (1925–2003) "the Butcher of Uganda," responsible for the murder of hundreds of thousands and a dozen more African murderers. Or perhaps you have heard of Pol Pot (1925–1998), who ruled Cambodia for 4 years and in that time managed to kill about two million Cambodians, a quarter of its population. Today you can pick among many other heads of state and politicians who many wish were never born (you will have your candidates depending on your knowledge of history and where you live). You could include the charming prince of Saudi Arabia and a long list of mostly men (would a world ruled by females be different?) who think that their position of power (democratically obtained or not) gives them the right to abuse it.

Although we could wish to erase them from history, they will remain forever, vile figures with boorish minds that if it were not for the support from the rest of the tribe would not be where they were (or are). The fault is communal; the fault is that we are not what we think we are. We are a chimera, as observed by Pascal, our reason and our ethical values are often the prisoners of our passions. The sleep of reason creates monsters, as renowned Spanish painter Francisco José de Goya y Lucientes (1746–1828) illustrated.

Once an idea is implanted (Jews killed God, other "races" are inferior, capitalism is the only way) or a technology is implemented (combustion engine, WWW, nuclear weapons), it acquires its own dynamics, as does the broom in Johann Wolfgang von Goethe's (1749–1832) fable: *Der Zauberlehrling* — The Sorcerer's Apprentice. The difference is that there is no higher sorcerer to stop the vortex in which we find ourselves.

The most I can hope for is to be objective (although nobody is perfect) and try to explain the reasons and show the evidence that led me to believe one thing and not another. But I am no exception when it comes to biases, we all are biased one way or another, and the worst bias is to think one is not. Furthermore, biases are often unconscious until you are forced to face them and hopefully understand them, admit them, and take corrective action. It is a lot to ask. Just imagine the different world view of someone who only watches Fox instead of CNN, not to mention Al Jazeera, reads the *The Wall Street Journal* instead of *The Washington Post* or subscribes to *The Nation* (as I do). And the "bubbles" I will tell you about later are worse.

I will briefly tell you my biases: First and foremost, I was trained as a scientist, and my bias is that if you claim something as true, you must show *me* the *evidence,* and it is not my job to show that you are wrong (although it is a strong temptation). Related to this is Brandolini's law[23] which states that, *The amount of energy needed to refute bullshit is an order of magnitude bigger than that to produce it.* As a scientist, "alternative fact" just makes me laugh. (There is a place for an alternative *interpretation* of the facts, but that is perfectly normal.) I believe that without a *scientific temper* held by all citizens in a society which through technology, is dominated by the results of scientific research, we will be in trouble. We already are when leaders deny climate change, perceive conspiracies all over the place, and want your insides illuminated with UV light to kill a virus. My deepest bias is that I am right, and "they" are wrong, and this is the main problem with the world because "they" have the same bias. At least I am not willing to kill for my bias, and that is in my favor. I also like people who have nothing to say and therefore shut up. Reading this, you might also think I am biased against what the US government does in our name. Still, although other countries have their own faults, my bias arises from my experience, which would be different if I lived elsewhere. Those others have their own problems, but none have been more important than the US regarding its power and its claims to leadership.[24]

It is reckless to misuse science to push pseudoscience or reject scientific results in biology, environmental science, or fundamental physics. Furthermore, not believing that something is true does not make it untrue. So, I will devote some time to help you get other people's minds out of the Middle Ages by explaining the most essential features of science, and under that light, look at our beliefs.

I agree with what the 1776 US declaration of independence states[25] (with some caveats that will become evident as you read on): *We hold these truths to be self-evident, that all men are created equal and endowed by their creator with certain unalienable rights including life, liberty, and the pursuit of happiness.* (We would revise this today from "all men" to "all persons" I suppose.) Perhaps they were "all created equal," but it is evident that they were and are not treated as equals, and worse if they belong to other countries. Further down in the document, you can read as one of the grievances against King George III that, *He has excited domestic insurrections among us and has endeavored to bring on the inhabitants of our frontiers, the merciless Indian Savages, whose known rule of warfare, is an undistinguished destruction of all ages, sexes, and conditions.* Could the Taliban of Afghanistan not say the same about US presidents?

I also believe that these inalienable rights cannot be boundless, that the pursuit of happiness does not mean you can have as much as makes you happy whereas others remain destitute, (Zygmunt Bauman, a Polish-born sociologist (1925–2017) had this to say: *in today's world all ideas of happiness end up in a store.* Let us not forget that the "Founding Fathers" acquired their wealth from slavery and the theft of indigenous lands, so that the "American Revolution" had its very downsides.

If life is an unalienable right (meaning: not capable of being taken away), then nobody and no collective can take it: so, then what are we doing killing people all over the planet or executing them in our prisons?[f] If to be happy means a healthy, well-fed, and educated meaningful life,

---

[f] The national death-row population is roughly 42% black — nearly three times the proportion in the general population. Ford, M. Racism and the execution chamber. *The Atlantic*, June 2014.

then all should have equal access to sustenance, health care, and education regardless of their economic circumstances or "ethnicity," and this is clearly not the case.

The preamble to the US constitution reads (emulated by many other countries): *We the People of the United States, in order to form a more perfect Union, establish Justice, insure domestic Tranquility, provide for the common defense, promote the general Welfare, and secure the Blessings of Liberty to ourselves and our Posterity, do ordain and establish this Constitution for the United States of America.*

I observe a very imperfect union. People on "one side of the aisle" look at the others as enemies instead of as fellow citizens who happen to think differently. And when it comes to race or ethnic background, what I observe is disunion. If you look at the US from afar, you see a mighty nation but not a great nation — might does not make right.

Justice seems to be a lost memory, liberty is curtailed by state power, domestic tranquility is in turmoil, and the general welfare is quite restricted. The "land of the free" becomes the home of the world's largest prison system, disproportionately populated by non-white persons. The common defense means going to battle (as if it were a crusade) in distant places where few welcome us. I do not believe that "we the people" want this (with some notorious exceptions). The tragedy is that although we have elections every few years and choose our leaders (from a very short list), they then go and do whatever the wealthy oligarchy[g] wishes, shielded by secrecy if convenient. And as already noted by US writer/critic Dwight Macdonald in 1945[26]: *It is a terrible fact, but it is a fact, that few people have the imagination or the moral sensitivity to get very excited about actions which they don't participate in themselves (and hence about which they feel no personal responsibility). The scale and complexity of modern Governmental organization and the concentration of political power at the top are such that the vast majority of people are excluded from this participation.* That was then; imagine now.

---

[g] Aristotle said that, *oligarchy is when men of property have the government in their hands... wherever men rule by reason of their wealth, whether they be few or many, that is an oligarchy.*

I also do not believe in the existence of any god (who lovingly gave us free will to slaughter each other). There is again no evidence to support this belief well summarized by the German theologian and ex-catholic priest Eugen Drewermann (born 1940): *A God who can do anything and yet does nothing when he looks on so much calamity, deserves, not to be considered good, or conversely: if he were good, but even he could not stop it, he would be not almighty: both qualities do not reconcile with each other as long as the world is as it is: a valley of tears. Both attributes, omnipotence, and goodness, are indispensable to the divine, according to Christian theology. It is the world itself, which refutes the Christian God as its creator, or in other words, the moral claim embodied in the Christian idea of a deity is led to absurdum by the reality of the world itself* (no wonder he is an ex-priest).

It is difficult to understand how God allowed the butcheries associated with Verdun, Somme, Stalingrad, Auschwitz, Nagasaki, Diem Bien Puh, Khe Sanh, Basra, Cambodia, Rwanda, Iraq, and many others.[28] If God knew about the horrors of Treblinka, Buchenwald, and many other torture and extermination camps of the Nazis and the Gulag of the Soviets, he (she, it?) is the first one who should have been prosecuted by the Nuremberg tribunal as the mastermind, or perhaps for desertion, or malicious abandonment. And do not give me the evasion of "inscrutable ways." Think about that.

There is no point in pretending to understand something about our origin without considering what biological evolution tells us, to argue without a clue about logic, or understand how we think without knowing what the neurology of the brain and psychology tells us. A philosophy that does not consider what we know about stars, planets, life, and the brain will be empty. But also, a science that ignores some philosophical questions will turn into mere technique, and that is, unfortunately, happening at universities that increasingly, with a mercantile vision, graduate employees instead of thinkers. Artificial academic disciplinary barriers are the first enemy of knowledge. There is nothing more transdisciplinary than the world.

Some will say that I have simplified things, and it is true. Simplify to see what is primary, and possibly understand the cause of some phenomenon, however complex it may be, and in this way, approach understanding. That is to say that the parallelepiped cow can be a handy concept when considering certain agricultural matters.[h] In the words of the Uruguayan journalist and author Eduardo Galeano (1940–2015)[29]: *Initially one feels that intellectual work consists in making complex what is simple, and then one discovers that intellectual work consists in making simple what is complex. And a case of simplification is not a matter of dumbing down. It is not about simplifying to lower the intellectual level, nor to deny the complexity of life and literature as an expression of life. On the contrary, it is about achieving a language that is capable of transmitting life's electricity, suppressing everything that is not worthy of existence.* An ancient proverb says: *say only that which is better than silence.* Furthermore, if I did not simplify, it would not fit in one book.

On the other hand, it is possible to hide reality behind a curtain of complications, arguing that everything is relative and uncertain, that it all depends on the point of view and the context, that we do not know everything, and that the data is precarious. It is a tactic widely used by demagogues of darkness, either when they try to argue against biological evolution, vaccines, or against the climate crisis we are causing. One ought not to confuse darkness with depth, which is only valid in the ocean, but false in life. Darkness is merely a lack of light.

We have awakened to a morning of serious problems. For many years we have been hearing about the contamination of water, land, and air, have been concerned about desertification and deforestation, have worried about relentless population growth and nuclear proliferation, a novel discourse just 50 years ago.

Perhaps you have heard it so many times that you no longer pay attention to it, so many that it is already dull, it is normal. But we must

---

[h] As my friend Gonzalo Carbonell taught me many years ago.

repeat so that it does not get lost in the din of the virtual world, and remains in everyone's subconsciousness so that it is not forgotten, as Hiroshima and Nagasaki; Auschwitz, and Dachau slowly evaporate from collective memory.

If we forget, it is possible and probable that facing a crisis, we will return to something even worse, and the crisis is knocking on our door. As famously said by philosopher George Santayana (1863–1952)[30]: *Those who cannot remember the past are condemned to repeat it.* (Although as well argued by David Rieff,[31] sometimes collective memory can be toxic, and it might be better to forget.) "Millennials" (and you might be one of them) have little direct knowledge of fascism (no matter if you saw Schindler's List or others of that genre) because all this happened during the Second World War, so they cannot *remember.* A few are still alive who might have some (bad) memories of this period (nuclear bombs and concentration camps), while others ask why it is called "second."

I write, neither from the left nor right, not very useful concepts in this tangled and crazy world — where even the US looks like a funny farm. Many people think "socialist" and "intellectual" are insults ignoring that our "founding fathers" were intellectuals and that even in the capitalist US, many things are socialized, and that communist China is capitalist. As already noted over 80 years ago by Spanish philosopher Ortega y Gasset: *Being on the left is, like being on the right, one of the infinite ways that man can choose to be an imbecile: both, in effect, are forms of moral paralysis. Moreover, the persistence of these qualifiers contributes not a little to further falsifying the "reality" of the present, which is already false, because the complexity of the political experiences to which they respond has been distorted, as can be seen by the fact that today the right promises revolutions and the lefts proposes tyrannies.* Still valid.

Nor do I pretend a better future for me since I will not be here to know whether I was right for purely biological reasons related to the maximum age reached by an average human being. But I had to write this because scientists and scholars should do all they can to get people to understand. As was said by long-forgotten author and poet Ella Wheeler

Wilcox (1850–1919): *To sin by silence, when we should protest, makes cowards out of men.*

As you read on, I will mention some numbers and statistics that illustrate certain things. Much has been said about "how to lie with statistics,"[32] and we know how that goes, but it is also true that you cannot get at the truth without them. The problem often has to do with the availability of uncertain numbers often kept secret for some reason (usually not the real reason). So, when you read them here (for instance, the number of nuclear weapons in a country), be aware that these are estimates and not exact numbers, but it does not matter for the discussion. It is the idea behind that number that counts.

I hear people argue about the number of victims of Hitler's henchmen. Still, it does not matter if it was four million or six million Jews, three million Soviet prisoners, and two million Polish civilians, among many others; one is too many. As sociologist William Bruce Cameron[33] (not Einstein as is often misquoted[34]) wrote: *not everything that can be counted counts, and not everything that counts can be counted.*

Since one of our worst faults is to generalize, let me state that my criticism of the US is not meant for all those living there. I know very well that the US is a nation of great diversity and currently starkly divided, with a population that also suffers the consequences of the actions of those in power who in their name, but mostly not for their benefit, make decisions, nationally and internationally.

It will be evident as you read that I do not think much of Mr. Trump. Instead of "cleaning the Swamp," Trump filled it with crocodiles. One positive consequence of the Trump presidency, I must admit, is that he encouraged all the racists and other misfits to come out of the woods for all to see. But even after learning the nature of the beast, more still voted for him in 2020, which is more than worrisome. Were they all gaslighted, or are they all that ignorant, that devoid of understanding, that devoid of humanity? If the latter is true, the future looks grim, and fascism will keep knocking on the door. According to Robert Graef[35]: *The Big Ignoramus includes global warming deniers, the Lord-will-provide Christians, fossil fuel advocates, destructive exploiters of land and sea, uber-consumers, and that great segment of society that goes along because it can't be bothered with the discomfort of knowing and changing.*

I know very well that the US is not the only one in the game, Russia's Putin (who is currently trying to beat Hitler and Stalin in inhumanity), China's Xi Jinping, Brazil's Bolsonaro, and an increasing number of autocrats come to mind, but it remains for others to criticize and solve their problems (some risking their lives by so doing) and try to better their societies. Larry Diamond, a political scientist at Stanford, writes[36]: *After three decades in which democracy was spreading and another in which it was stagnating and slowly eroding, we are now witnessing a global retreat from freedom.* But we should start at home, and it is time to lead by example and not by raw might. What right does the US have to impose ways of thinking and behavior on others, accusing them of actions we also engage in?

As quoted by Rob Riemen,[37] the Italian Federico Fellini (1929–1993), an outstanding filmmaker (*La Dolce Vita, La Strada, Amarcord*), said it well: *Fascism always arises from a provincial spirit, a lack of knowledge of the real problem, and people's refusal — through laziness, prejudice, greed or arrogance- to give their lives deeper meaning. Worse, they boast of their ignorance and pursue success for themselves or their group, through bragging, unsubstantiated claims, and a false display of good characteristics, instead of drawing from true ability, experience, or cultural reflection. Fascism cannot be fought if we do not recognize that it is nothing more than the stupid, pathetic, frustrated side of ourselves, of which we should be ashamed. To curb that part of ourselves., we need more than activism or an antifascist part because latent fascism is hidden in all of us. It once gained a voice, authority, and trust, and it can do it again.*

And this is happening all over our planet in a dangerous global gathering storm. And that is the crucial bit. A growing number of the problems we face are global, so we need global solutions where everyone agrees and contributes; that is the rub. Increasing polarization, different worldviews between and within the east and west, north and south, make any globalized vision unlikely. Even within "one Nation under God, indivisible," the division is approaching civil war. But the Earth's atmosphere belongs to everyone, as do the oceans, which cover 70% of our planet's surface. It seems evident that a nation cannot unilaterally wreck them. And although huge forests cover about 30% of land area and are the "lungs of the planet," they *are* in some countries. It is less obvious how to maintain them and encourage national governments to act for the common good. Do we bomb Brazil to halt deforestation of the Amazon (estimated at 1.4 million hectares per year and increasing) threatening global welfare and US security and is more important than middle eastern oil? It is worrisome that the president of Brazil has promised to remove all environmental and pro-indigenous legislation that protects the Amazon and its people. Is this acceptable? But it is Brazilian citizens who must decide.

But I am American, and that is what I know best and can talk about. *America* (named after the Italian explorer Amerigo Vespucci (1454–1512)) is a continent with 35 *sovereign* states with altogether one billion people, the United States (henceforth the US) with 330 million inhabitants, Brazil with 210 million, and Mexico with 120 million being the most populous. As noted by historian Daniel Immerwahr,[38] before 1898, "America" was seldom used; it was the United States. The US can be proud of some of its people's achievements (although I've always have had a bit of a problem with being proud of what others had achieved, happy, yes, proud no). Many looked up to the US for its sacrifice helping defeat what was an "axis of evil" during the Second World War. But it slowly lost the moral high ground and has reached an appalling state of moral turpitude after events in Indochina, South and Central America, the Middle East, and elsewhere, which brought "shock and awe" to innocent people changing governments (regime change) to its liking. It is telling that those who defend US militarism need to go back to WWII to justify it. Why not aspire to be a nation that others in the world could look up to, setting an example of democracy, tolerance, intellectual achievements, and decency in action? *National pride is to countries what self-respect is to individuals: A necessary condition for self-improvement. Too much national pride can produce bellicosity and imperialism just as excessive self-respect can produce arrogance*, writes Richard Rorty.[39]

Washington DC is no longer a "City upon a Hill" if it ever was. "City upon a Hill" as explained by Sarah Churchwell,[40] is a phrase coined by John Winthrop (1587–1649; third governor of the Massachusetts Bay Colony) in 1630: *We must consider that we shall be as a City upon a Hill, the eyes of all people are upon us; so that if we shall deal falsely with our god in the work we have undertaken and so cause Him to withdraw His present help from us, we shall be made a story and a byword through the world, we shall open the mouths of enemies to speak evil of the ways of God and all professors of God's sake.* Indeed.

In his farewell address to the Nation in 1989, President Reagan said[41]: *I've spoken of the shining city all my political life, but I don't know if I ever quite communicated what I saw when I said it. But in my mind, it was a tall, proud city built on rocks stronger than oceans, wind-swept, God-blessed, and teeming with people of all kinds living in harmony and peace; a city with free ports that hummed with commerce and creativity. And if there had to be city walls, the walls had doors, and the doors were open to anyone with the will and the heart to get here. That's how I saw it and see it still.*

And so, it was, up to a point. Unfortunately, the torch on the statue of liberty does not shine brightly these days, and it might as well be "sailing away to sea" in the words of Paul Simon. I will take tons of burnishing solution to restore some brightness to the murky US image. After enduring the fool on the hill (to mention another song), we have a new, hopefully, wiser government to rebuild what has been destroyed. But for this to happen, real and profound changes are needed.

Over modern history, the US (let us be clear that when I mention the US, I mean those who are in power, be it public or private, (the rest, only matter every couple of years) has increasingly acted as the world's bully and is slowly paying the price for not practicing what it preaches, not even in its own territory. We are pushing on some natural and social variables that keep an increasingly complex and entangled system in a fragile equilibrium. There is a delicate balance between people of different interests and beliefs and of all people with the planet, which, when broken, usually leads to bloodshed, so that one must tread carefully.

The current pandemic helps those with anti-democratic agendas. Recently a US president tried to steal an election. Under the cloak of an emergency, laws are passed, money is provided with little oversight, fear drives authoritarian measures, and somebody is always ready to take advantage of a chaotic situation. An old Spanish proverb says: *A río revuelto, ganancia de pescadores*, or "it's good fishing in troubled waters." Private military companies (PMC) profit handily and represent a potential threat to a weak democracy.[42]

We may have saved democracy *for a while*, but this does not change what I have to say about the US and its role in the world (which is remarkably different from the official narrative). To heal a complex system and change direction, you need to change much more than who inhabits the White House. If we adhere to the current system as if there were no alternatives when it has been clear for quite some time that it is no good (except for a few), we are heading the wrong way.

Finally, it is not my intention to ruin your day with a depressing tale, but it can destroy the lives of your children and grandchildren if you do not consider some of the things that I have to say. I suspect that you will not like what I will tell you, probably because you still believe that the US is a great country and in the "American dream," which for many has turned into a nightmare, and other narratives you were fed as a child. We are known for not letting go of our beliefs, no matter the evidence.

If our leaders do not do something for our agonizing world, we will be left without a world. Be that as it may, give yourself a chance, you might learn more by reading something you might disagree with, something that goes against your cherished beliefs, than something about what you already know, and I apologize for the inconvenience.

The words of French Catholic priest at Étrépigny, in Champagne, France, Jean Meslier (1664–1729) the first modern atheist come to my mind[43]: *Let the priest preachers, scholars, and all the instigators of such lies, errors, and impostures be scandalized and angered as much as they want after my death; let them treat me, if they want, like an impious apostate like a blasphemer an atheist, let them insult me and curse me as they want. I do not really care since it will not bother me in the least. Likewise let them do what they want with my body; let them tear it apart cut it to pieces, roasted or fricassee it and then eat it, if they want, in whatever sauce they want, it will not trouble me at all. I will be entirely out of their reach; nothing will be able to frighten me.* He did worry and kept his writing secret while alive since he knew that in those days, they would have burned him at the stake. In some places on this wretched planet, we still kill people who think differently.

Chapter 2

The Fleeting Visit of the Enlightenment

Men fear thought as they fear nothing else on Earth — more than ruin, more even than death. Thought is subversive and revolutionary, destructive and terrible, thought is merciless to privilege, established institutions, and comfortable habits; thought is anarchic and lawless, indifferent to authority, careless of the well-tried wisdom of the ages. Thought looks into the pit of hell and is not afraid ... Thought is great and swift and free, the light of the world, and the chief glory of man.

— Bertrand Russell[1]

We seem to be urged to share any knowledge that can be expected to help cure the body politic of its disease and return it to the healthy state to which we, as children of the Enlightenment, think our fellow-citizens have the right and duty to aspire at the end of this blood-drenched century: a state that is rational, progressive, anti-superstitious, pro-science, and free of the medieval curses of folk magic, miracle, mystery, false authority, and mindless iconoclasm.

— Gerald Holton[2]

Just for being human, we have rights (as do animals). Steven Bronner writes[3]: *Without the idea that individuals are ends unto themselves, rather than instrumental means for achieving other aims, rights and laws are stripped of purpose [...] Human dignity as a universal quality was anathema to the aristocracy and the true believer. And for good reason. Women, slaves, Jews, and those without property soon recognized the use of "rights" and natural law in pressing their claims for inclusion into society.* Perhaps the best-known document in this respect is the French Declaration of the Rights of Man and of the Citizen written in 1789 (Déclaration des droits de l'homme et du citoyen) during the French revolution. Thomas Jefferson

(1743–1826), the third president of the US (1801–1809), had a hand in writing it. This and similar proclamations — the 1776 US Declaration of Independence, and the 1789 US Bill of Rights — (which comprises the first ten amendments to the US Constitution), inspired the 1948 Universal Declaration of Human Rights of the United Nations.[4]

Peace, justice, and freedom are included in the first sentence of the preamble, which states that human rights are *the basic rights and freedoms to which all humans are entitled*. The first article says: *All human beings are born free and equal in dignity and rights. They are endowed with reason and conscience and should act towards one another in a spirit of brotherhood*. The fifth article says: *No one shall be subjected to torture or cruel, inhuman or degrading treatment or punishment*. These two of the thirty articles are enough to realize that there is a long way to go for all humans to have these rights. In too many places, they do not even know they have them.

What happened to the Enlightenment's ideas that sought to promote a world of knowledge and justice and led to a radical change in the vision that until then, humans had of themselves and their relationship with others and the world? The program of the Enlightenment can be summarized as the attempt to remove the superstitious fear humans felt so that they could better deal with reality.

Renowned British philosopher Sir Isaiah Berlin (1909–1997) tells us[5]: *Dark mysteries and grotesque fairy tales which went by the names of theology, metaphysics, and other brands concealed dogma or superstition with which unscrupulous knaves had for so long befuddled the stupid and benighted multitudes, whom they murdered, enslaved, oppressed, and exploited*.

With its faith in rationality — the ability to recognize reasons and arguments that are objectively valid and general — and its struggle against that which has no foundation and corrupts the mind, Enlightenment proposes a common basis for understanding between humans. It rejects the authority (of kings, presidents, or gods), recognizing that authority must be guided by a future vision and not by outdated past rules.

The Enlightenment affirms the universality and equality of human beings ("all men are created equal") and embodies the idea that knowledge to guide humanity is obtained from reason supported by experience (science). This was already understood by Galileo, for which he had to suffer. Galileo wrote[6]: *I do not feel obliged to believe that the same God who has endowed us with senses, reason, and intellect has intended us to forego their use and by some other means to give us knowledge which we can attain by them. He would not require us to deny sense and reason in physical matters which are set before our eyes and minds by direct experience or necessary demonstrations. This must be especially true in those sciences of which but the faintest trace (and that consisting of conclusions) is to be found in the Bible.*

Science was born with him, and slowly and against powerful currents, the secularization of western society proceeded. But we are in danger of going back to superstitious obscurantism.

It is essential to follow scientific understanding and procedures to contribute to the resolution not only of scientific questions but also of social and political questions. It is necessary that all (not only scientists) understand something of the way science proceeds, as I will explain later. I say "contribute" because other factors are not accessible to a scientific analysis that are of the utmost importance when solving our increasingly overwhelming problems. Above all are ethical questions that need to be considered alongside our supposedly rational choices so that we do not become mere machines. An ethical guide is to be found in our humanity, not in outdated dogmatic ancient books, although they contain some worthy ideas. Of greater importance is that ethics must be understood by all and must be heeded. For example, when someone, calling himself a Christian, commits murder, he is not heeding the sixth commandment. If you do not love your neighbor, you are not Christian (Leviticus 19:9–18: ...*you shall love your neighbor as yourself*). This is a complex topic mainly outside the scope of this book, so I recommend Johnathan Haidt's excellent book: *The Righteous Mind.*[7]

History is not a straight line of progress as many believe. After that flash of light, we saw the slaughters of the last century, the genocides and massacres that eclipsed that light. There is no law stating that things will get better with time, and in any case, "better" and "progress" are slippery concepts. Our lack of wisdom, together with widespread ignorance, and our deeply ingrained but false beliefs, lead us, as if we were a bunch of idiots, to never-ending cycles of troubled times. Thomas Homer-Dixon writes[8]: *Within the larger society, stock markets crash, revolutions break out, and floods devastate communities. The simple mental models in our heads, the models that guide our daily behavior, are built around assumptions of regularity, repetition of past patterns, and extrapolation into the future of slow, incremental change. These mental models are the autopilots of our daily lives. But no matter how much we plan, build buffering institutions and technologies, buy insurance, and develop forecasts and predictions, reality constantly surprises us.*

Think about the destruction under Roman Christian rule of what was then the greatest center of learning and classical knowledge in Alexandria, Egypt (The Great Library). The Musaeoum — a shrine to the muses — a forerunner of modern universities) and the horrible murder by Christian zealots of the great philosopher Hypatia in the year 415.[9] Whatever it was, it was not progress, and today we have zealots of all types.

We owe this and many other tragedies to the course of events that began with Emperor Constantine's (272–337) conversion to Christianity, which led the world into 1,000 years of darkness.[10] As Catherine Nixey writes: *But there is another side to this Christian story, one that is worlds away from the bookish monks and the careful copyist of legend. It is a far less glorious tale of how some philosophers were beaten, tortured, interrogated, and exiled and their beliefs forbidden; it is a story of how intellectuals set light to their own libraries in fear. And it is above all a story that is told by absences: of how literature lost it's liberty; how certain topics dropped from philosophical debate — and then started to vanish from the pages of history. It is a story of silence.* We owe a great debt of thanks to those Arabs who translated and preserved major Greek works destroyed by Christians; book-burning was not invented by the Nazis.

Goethe had this to say in a letter to his friend and composer Carl Friedrich Zelter in 1812 (but it could be today)[11]: *When one sees how the world in general, and especially the young ones, not only give themselves up to their lusts and passions but how at the same time the higher and better part of them are perverted and disfigured by the somber follies of the age so that everything that should lead them to be blessed, lead them to damnation, so one is not surprised by the misdeeds by which man rages against himself and others.*

We can attribute the abolition of slavery, the universality of voting rights, the struggle for human (and women) rights and civil liberties, and the rejection of authoritarian regimes to Enlightenment thought. But we seem to be slipping backward in many places where we thought we had overcome the worst. But if my team loses 3 to 1 and at the last minute scores a second goal, it will have progressed, but it will nonetheless have lost the game. (I mean the one with the round ball). We have not built on the Enlightenment ideas of Hume (1711–1776), Voltaire (1694–1778), Diderot (1713–1784), Holbach (1723–1789), Clifford (1845–1879), Helvétius (1715–1771), and many others who long ago understood much that we (meaning the majority) do not seem to understand, all of them worthy intellectual heirs of Galileo, René Descartes (1596–1650) and Francis Bacon (1561–1626).

Those who cautiously met at 10 rue des Moulins, in Paris at Holbach's "salon," understood[12,13] that knowledge was better than ignorance, that a reasonable action based on research and analysis was preferable to waiting for God, and that the future should be a human project and not one governed by authoritarian rules of the past. They understood that debate was superior to fanaticism and that the barriers to an inquiry whether religious or secular, were pernicious to understanding.

Isaiah Berlin[14] writes about two factors that have shaped our world: *There are in my view two factors that, above all others, have shaped human history in the twentieth century. One is the development of the natural sciences and technology, certainly, the greatest success story of our time — to this, great and mounting attention has been paid from all quarters. The other, without doubt, consists in the great ideological storms that have altered the lives of virtually all mankind: the Russian Revolution and its aftermath — totalitarian tyrannies of both right and left and the explosions of nationalism, racism, and, in places, religious bigotry [...] When our descendants, in two or three centuries' time (if mankind survives until then), come to look at our age, it is these two phenomena that will, I think, be held to be the outstanding characteristics of our century — the most demanding of explanation and analysis.* To be clear "if mankind survives until then," are his words, not mine.

The romantic reaction to the Enlightenment and well-established scientific ideas is not something new and as the eminent historian of science and physicist Gerald Holton[15] (born 1922) pointed out, it served to inspire Hitler, Himmler, Goebbels, and Rosenberg, the main leaders of German National Socialism (which had nothing to do with socialism).

We have believed the narrative that we are heirs of the Enlightenment, and it is true if we stay within the walls of a university (and beware!). However, it is an academic heritage in every sense of the word; the Enlightenment never transcended those walls. We think that the citizenry has forgotten it, and this is false since it is not possible to forget something that was never known. That is the explanation that Isaiah Berlin was looking for. Voltaire knew this and wrote in a letter to a friend[16]: *Enlightenment times will enlighten only a small number of honest people. The vulgar masses will always be fanatics.*

For me and others,[17] "preaching" the ideas of the Enlightenment is a question related not only to our survival but also a quest that it be not merely survival but provide for a full and dignified life. For this, it is necessary to transcend the walls within which we comfortably discuss its merits, while outside, they kill for a god or die for a dogma.

Of course, we live in a very different world today, so we need to update a few things, as written by renowned French-Bulgarian historian Tzvetan Todorov (1939–2017)[18]: *When we look at today to Enlightenment thinking for support in dealing with our current difficulties, we cannot adopt unaltered the propositions formulated in the eighteenth century, not only because the world has changed, but also because the thinking was multiple, not one. Instead, what we need today is to re-establish Enlightenment thinking in a way that preserves the past heritage while subjecting it to a critical examination, lucidly assessing it in light of its wanted and unwanted consequences.*

As well argued by philosopher Charles W. Mills Enlightenment liberalism, had triumphed over the divine right of kings, but only then to codify a social right of white people over non-white people.

There is no need to look very far to see misery, hunger, and violence, a consequence of millenarian tribalism, modern triviality, and unrestrained capitalism. A proof that the schemes we have adopted to order our societies clearly do not work for most people and that the "invisible hand" of Adam Smith (1723–1790) does not push in the right direction. My concern arises from the many years I have heard and read different arguments, studies, and considerations regarding the problems we face, accompanied by marches and demonstrations demanding a change leading to a better future, together with a desperate lack of progress. This sets the tone of this book.

We live in a world wounded by the violence that splashes blood on white walls with intelligent bombs or bomb intelligences. A world saddened by the misery of the destitute while those who have everything are even more miserable.

We face countless problems caused by the increasing impact of our actions on the environment that in modern times has grown exponentially. That last word is already a problem because many do not understand what "exponentially" means and visualize growth as a linear process. We can imagine the result of adding one mile of highway per day but have no clue what exponential growth would mean with its explosive development. Currently, a good example is to observe the growth of the pandemic caused by the SARS-CoV-2 virus. People have trouble seeing it coming, and those who sound the alarm are dismissed as alarmists, and politicians do not understand.

Our knowledge of the world has also snowballed, but the great mass of humanity lacks the background and time to analyze, understand, and imagine. They have been left behind, powerless to do much about their plight, and militarized police will be waiting if they manage to organize in protest. What some of us know collides head-on with the most cherished beliefs of the majority who cannot revise them for fear of being left suspended in the air. As the German philosopher Günther Anders (1902–1992) observed some time ago, we have become obsolete. He meant that our ways of thinking, in the face of the accelerated knowledge of the world, and our technological capacities to communicate and act, have become obsolete. This has caused what Thomas Homer-Dixon[19] called an "ingenuity gap," meaning a gap between our need for practical and innovative ideas to solve our complex problems and our actual supply of those ideas.

The accelerated technological change has made us capable of exterminating ourselves with our weapons. I would add, eliminate ourselves by our ignorance of the consequences of our actions because if you push hard on a part of a complex system, you can destroy it. (The case of stratospheric ozone is a good example that I will present later.) We are not fit to inhabit this globalized, computerized, and complex world with our outdated ethical, social, and economic norms. We are at the mercy of currents that we do not comprehend nor control, a huge, massive ship drifting with a broken rudder without knowing where we are headed.

The "terrible" question: *Is there hope for man?* with which economist Robert Heilbroner (1919–2005) began his 1973 book[20]: *The Human Prospect* (revised in 1991), continues to be the mother of all questions. I am afraid that the answer, as he also thought, is closer to no than to yes, and not because I am a pessimist by nature, quite the opposite, but the facts cannot be hidden or altered.

The mere fact of the question (we are not concerned with the inescapable death of the Sun in a future too distant to consider) points to a new consciousness that includes the possibility of apocalyptic suicide, either sudden by nuclear confrontation or by the slow agony of global cancer. A bleeding moon is perceived through dark clouds. Instead of fearing the power of the mythical gods, we must now fear the real power of humans.

But this awareness is ambiguous because we cannot really understand the meaning of a possible apocalypse; we do not think it can happen outside the myth because somehow, a solution will be found. There will be problems with no solution, and we shall suffer for it. It is simply not possible to climb Mount Everest in 10 minutes or stop an earthquake. And yes, failure *is* an option. The terrible question also points to a fundamental ethical dilemma: Do we really care about humanity's future beyond that of our own grandchildren (if at all)?

For some time, this has been a concern to such scholars as Paul R. Ehrlich and Anne H. Ehrlich, who have dedicated their lives to trying to make us understand. They say[21]: *Sadly, much of the progress that has been made in defining, understanding and seeking solutions to the human predicament over the past 30 years is now being undermined by an environmental backlash, fueled by ideas and arguments provided by the brownlash.*[a] *While it assumes a variety of forms, the brownlash appears most clearly as an outpouring of seemingly authoritative opinions in books, articles, and media appearances that greatly distort what is or isn't known by environmental scientists. Taken together, despite the variety of its forms, sources, and issues addressed, the brownlash has produced what amounts to a body of anti-science — a twisting of the findings of empirical science — to bolster a predetermined worldview and to support a political agenda. By virtue of relentless repetition this flood of anti-environmental sentiment has acquired an unfortunate aura of credibility.* That was in 1996, over 25 years ago, and these days we see it in action.

Let us briefly review some of today's answers and how to understand them within a new metaphysical framework that removes the myths that for millennia have formed a web of chains that immobilized the mind and the human spirit and did not let it soar. Of course, as the saying goes: It is hard to soar like an eagle when you are surrounded by turkeys.

---

[a] By this they mean the efforts being made to minimize the seriousness of environmental problems.

Chapter 3

Perspective

You, then, who are shut-in and prisoned in this merest fraction of a point's space, do ye take thought for the blazoning of your fame, for the spreading abroad of your renown? Why what amplitude or magnificence has glory when confined to such narrow and petty limits?

— Boethius[1]

Let us anchor our ideas to the foundation of everything we have learned about the world and about ourselves, even though the foundation is a bit slippery, and the anchor might not hold as firmly as we think. Therefore, I begin by summarizing what we know about us and our place in the universe in space and time. This will provide us with a bit of perspective and much-needed humility.

What follows must be taken for what it is: the *best explanation* we have, provided by science, about the world and its inhabitants, always tentative, always improvable but much better than any baseless idea propagated on the WWW and other illusions.

The new perspective allows us to know ourselves better and get closer to the perennial questions: Where do we come from? Where are we going? Who are we? These are questions that fill us with anguish. The problem is that the answers I have suggested elsewhere: *from and to nowhere in particular* to the first two and *an unfortunate accident* to the third, are not to anyone's liking. Still, I reaffirm that they are better adjusted to the truth than all the myths that have been invented to answer them more reassuringly.

To gain some perspective, consider first the following image of a tiny part of the sky taken by the Hubble Space Telescope. It is called "The Hubble

Figure 3.1. The Hubble eXtreme Deep Field (HXDF) NASA; ESA; G. Illingworth, D. Magee, and P. Oesch, University of California, Santa Cruz; R. Bouwens, Leiden University; and the HUDF09 Team.

eXtreme Deep Field (HXDF)" (Figure 3.1). It represents one 32 millionth of the sky. It is the size that a square piece of paper 1-mm (0.04 inches) per side would cover when held about arm's length! It shows about 5,500 galaxies (each point of light is a galaxy), the oldest of which was 13.2 billion light-years away. It was obtained by adding images taken over 10 years, corresponding to approximately 23 days of exposure time. Extrapolated to the entire sky, it means that you would get 5,500 × 32 million or about 176 billion galaxies, each with billions of stars. If this will not humble you, nothing will. It was already clear to the eminent Bertrand Russell almost 100 years ago[2]: *In the visible world, the Milky Way is a tiny fragment; within this fragment, the solar system is an infinitesimal speck, and of this speck, our planet is a microscopic dot. On this dot, tiny lumps of impure carbon and water, of complicated structure, with somewhat unusual physical and chemical properties, crawl about for a few years, until they are dissolved*

Figure 3.2.   NASA's Cassini spacecraft Saturn's rings and our planet Earth (arrow) and Moon.

again into the elements of which they are compounded. They divide their time between labour designed to postpone the moment of dissolution for themselves and frantic struggles to hasten it for others of their kind. Natural convulsions periodically destroy some thousands or millions of them, and disease prematurely sweeps away many more. These events are considered to be misfortunes; but when men succeed in inflicting similar destruction by their own efforts, they rejoice, and give thanks to God. In the life of the solar system, the period during which the existence of man will have been physically possible is a minute portion of the whole; but there is some reason to hope that even before this period is ended man will have set a term to his own existence by his efforts at mutual annihilation. Such is man's life viewed from the outside. Perspective indeed!

If this is too mindboggling (as it is for me), I offer you another image that might help. On July 19, 2013, the wide-angle camera on NASA's Cassini spacecraft captured Saturn's rings and the Earth and Moon in the same frame (Figure 3.2). In this image, Earth, which was 900 million miles (1,440 million km) away, appears as a blue dot marked by the arrow. The following image shows a detail (Figure 3.3).

These are the results of a long intellectual journey. To summarize[3]:

Figure 3.3. Detail of the above image showing Earth and Moon.

3.1 Laws

We have discovered that nature operates with certain regularities that we call *natural laws*. These determine what is possible and what is not. They represent the best summary of what the world is like, and our desire that it be different is useless. They have led to the development of powerful technologies that have transformed the world, for better or for worse. It is, for example, a natural law that nothing can exceed the speed of light (300,000 km/s in a vacuum); it is not a technological limit. While mentioning this, let me clarify an often-misunderstood term: "Light-year." It is not a measure of time; it is a measure of *distance* such that one light-year is equal to the distance traveled by light in 1 year at its enormous speed, a vast distance. So, if we observe something 1,000 light-years away (not far), we see it as it *was* 1,000 years ago. For all you know, it does no longer exist.

We have discovered that everything, including you and me (forgive me if I call you a thing), what we eat and drink, and the bewildering variety of phenomena we observe, are the complex manifestation of a system built with different combinations of just three particles: neutrons, protons, and electrons that form atoms. These atoms (in most cases, only a few types) can form molecules that contain carbon and are the components of all living things or form the minerals that make rocks. The bonds that combine atoms into different compounds result from electrical forces between their electrons.

The extraordinary diversity of phenomena we observe results from chemical reactions between these atoms and compounds, with rare but important exceptions (nuclear reactions), which combine or divide to form new compounds. Although there are other elementary particles (such as neutrinos and pions), our everyday world manifests the interaction between atoms, molecules, and photons (particles of electromagnetic radiation), of which light is a part.

There are other laws of nature without which fundamental processes cannot be understood: I present the first and second laws of thermodynamics. The first one is not difficult to understand, except that "energy" is not precisely what most people think: *Energy is conserved in an isolated system* (completely isolated systems are difficult to obtain in practice). In any process, the initial amount of energy (a number obtained *via* a formula) precisely equals the final amount (if you are careful not to miss anything in your accounting, especially heat that might dissipate and is a form of energy).

This leads to the second law, which states that the quality of the energy involved is degraded; some of it cannot be recovered. This is characterized by another less familiar quantity called entropy. The lower the entropy, the higher the energy's quality (the ability to do some work). The second law says that: *In an isolated system, entropy increases.* At a basic level, entropy is a measure of disorder; higher entropy means greater disorder. (My desk is a place of high entropy, and I need to intervene occasionally to lower it.)

For example, the air in your room does not all spontaneously go to the opposite corner of where you are, which would mean greater order and a decrease in entropy, and by the way, that you would die. You, as an organism, are very ordered (low entropy), but this is possible only because you are not an isolated system, continuously absorbing energy from food (ultimately from the Sun). When you stop doing this, you die, your constituent molecules and atoms disperse, and what was your ordered body becomes disordered, and entropy increases. The same can be applied to economic systems that feed on energy and need ever more of it to grow, but, as we shall see, this has created a very critical problem.

These laws are potent, pertain to any process, and explain why a hot cup of coffee with milk left to itself cools down (until it reaches the temperature of its surroundings), and why you never see the milk separating spontaneously from the coffee, you poured it into, among many other things. Combustion generates energy (part of which is lost as heat into the environment) to move an engine, a tool in a factory, or a turbine to generate electricity. The ordered atoms (carbon and hydrogen), which compose hydrocarbon molecules that burn as they combine with oxygen (that is what "burning" means), are transformed into a more disordered arrangement of smaller molecules which disperse, such as CO_2 (more on that later) and water (H_2O). Entropy increases.

3.2 Big Bang

The observation of the expansion of the universe (distant galaxies recede from us, the furthest ones moving faster; known as Hubble's law) and the microwave background radiation, among other astronomical measurements, indicate that about *14 billion[a] years* ago, the "Big Bang" which gave rise to our universe occurred. From that beginning, having formed protons (hydrogen nuclei) and electrons in a sea of electromagnetic radiation, the structures we observe were formed: galaxies containing stars (the Sun and its planetary system being merely one of them), gas, and interstellar dust. Within the stars, through a well-studied process known as *nucleosynthesis*, throughout millions, even billions of years, all the atoms of the periodic table were formed, the heaviest (such as iron, gold, or silver) in the explosive final moments of the life of a high mass star. These explosions (supernovae) "fertilized" interstellar space with the resultant atoms, thus allowing the formation of new stars and planetary systems (which we are finding, as I will explain below) that incorporated the newly formed chemical elements. It is a fascinating story.[4,5]

---

[a] One billion is 1,000 million.

3.3 Life

On our planet, formed with the entire solar system, life arose in its early history. It *evolved*, perhaps starting with replicating molecules (such as RNA), inhabitants of the primitive oceans, perhaps around deep hydrothermal vents. A few multiplied, and many others disappeared because of local environmental conditions. The algorithmic process (which I will explain later) of *natural selection* began to operate at that time, and here we are after *4 billion years*! This unimaginable long age has been obtained from the study of certain radioactive atoms found in some minerals.

The fossil record shows us that in the past, there were life forms that do not exist in the present and that present forms of life did not exist then. There are no longer dinosaurs, and in their time, there were no humans. However, some falsely indicate (for religious reasons) that there were. A high percentage of living matter is composed of hydrogen, nitrogen, oxygen, and carbon, not coincidentally the most abundant elements in the universe on average (along with helium which, however, does not produce compounds being a "noble" gas). An organism's blueprint is encoded in a long macromolecule found in the nucleus of nearly all cells (deoxyribonucleic acid — DNA), one of the most significant discoveries of the last century.

A human is built of about 30 trillion (30,000,000,000,000) cells of about 200 different types. Their DNA consists of about three billion base pairs which, if they were placed along a line, would be about 2 m (6.5 feet) long. DNA is deciphered with a unique genetic code. The nucleic acids that constitute these molecules (DNA and RNA) use the same five chemical bases in all organisms (Adenine, Thymine, Guanine, Cytosine, and, Uracil which substitutes for thymine in the structure of RNA). This is astonishing; think that you could fit about 200 cells on the dot of this i.

All organisms have a similar biochemistry. All the complexity of the living world arises from the different sequences that can be formed by only four bases in DNA that serve to specify the amino acids that form the proteins of the organism. They also regulate the biological processes that determine if an organism becomes a banana or a chimpanzee.[b]

There *is great unity* in the beautiful diversity that we observe. Everything suggests a single origin. Everything indicates that we do not need special laws nor "life energies" for life. The fossil record also teaches us that obsolete species go extinct. There were five cases in which a high percentage of organisms disappeared relatively fast — a traumatic mass extinction. The last one occurred 66 million years ago, caused by the impact of an asteroid. At the same time, a niche was opened for the development of mammals. It is in that sense that above I wrote "an unfortunate accident."

3.4 Homo Sapiens

It is estimated that *Homo sapiens* in its modern anatomical form has existed for some 200,000 years, a long time for us but the blink of an eye in Earth's history. The ancestors of *Homo sapiens*, other animals belonging to the genus *Homo,* emerged in Africa and used stone tools about 2 million years ago. For most of that stone age, these tools did not change much. Circumstances were very different then. Small bands inhabited large spaces with virtually unlimited resources, and any contamination was minimal and easily absorbed by natural processes. The problems between them were sometimes settled with violence, but with the available technique — a stick or a stone — the damage was limited. At some point, we invented weapons that acted at a distance (the spear, the bow and arrow), and we went from being preyed to being predators, the first step to total domination, promoted by technique and more recently by technology, which today allows drone assassination at a distance.

---

[b] The publication of the discovery of the structure of DNA in 1953 by James Watson and Francis Crick (1918–2004), culminated the search for the foundation of inheritance, demonstrating how a molecule in the nucleus of cells could be the carrier of replicable genetic information. The short publication (Molecular Structure of Nucleic Acids: A Structure for Deoxyribose Nucleic Acid, *Nature*, 171, 4356), contains one of the most famous litotes (look for it in the dictionary) in the history of science: *It has not escaped our notice that the specific pairing we have postulated immediately suggests a possible copying mechanism for the genetic material.*

The paleontological record shows many innovations, including the use of fire by Neanderthals 100,000 years ago (*Homo neanderthalensis* — Man from the Neander Valley), the beginning of the controlled use of fuels, until attaining the use of iron 3,000 years ago. The rest is history, culminating with the technological explosion of the last century. Today, we continue with the tradition of solving our problems between people and nations with violence, but now it is possible to eliminate millions by simply pressing a button. Unfortunately, we carry an old mortgage of aggressiveness without which *Homo sapiens*, with its frail body, would not have survived the predators of the African savannas. Canceling that mortgage is not easy, although at least some of us have come to understand that we must overcome our nature if we wish to survive as a civilization. You can read Yuval Harari's excellent book *Sapiens*,[6] for all the details.

3.5 Viruses and Bacteria

We have discovered, beginning with the pioneering observations of the Dutchman Anton van Leeuwenhoek (1632–1723), inventor of the microscope (although there were precursors), a vast population of microscopic beings, alive (bacteria) and others less alive (viruses). They are descendants of the first inhabitants of our planet, and without some of them, we would not survive. But a few are dangerous and can make us sick and even kill us. We have developed defenses against them, antibiotics, and vaccines through scientific research, which allowed us to overcome many afflictions that would have been fatal only 100 years ago.

Think of the polio outbreaks (poliomyelitis — also known as infantile paralysis) of the first half of the last century, which affected tens of thousands for the rest of their lives or killed them. The development of vaccines and vaccination enforcement has mostly eradicated this disease (but the virus is still there). The current race to beat COVID-19 with vaccines that have been developed thanks to scientific advances is in peril because of groundless resistance by people who seem to prefer the disease, which is *not* just another flu.

3.6 Imagination

The fossil record documents a gradual increase in the cranial capacity of our evolutionary ancestors. Lucy, our umptieth grandmother (so-called because her discoverers listened to the Beatles' "Lucy in the Sky with Diamonds"), lived about 3 million years ago in what is now poor Ethiopia. She belonged to *Australopithecus* (southern ape) *africanus*, was about four feet tall, and had a cranial capacity that was one-third of ours. Compared to other animals, our brain requires a large amount of energy, and no other being in the long history of life developed such an organ. It seems to be an aberration.

Cranial capacity alone does not necessarily correlate with intellectual development. What seems to matter is the density of neurons in certain parts of the brain and how they are interconnected (the *connectome*). We tend to think that our more recent ancestors were "primitive," but everything indicates that if we could bring a Mesopotamian baby and raise it in our time, it would have no difficulty.

On the other hand, our mental structures and cognitive faculties were formed in the deep past, responding to the demands to survive in the frigid Pleistocene. (The geological epoch that began approximately 2.6 million years ago and ended about 12,000 years ago.) They are not optimal for our present world. The ability of symbolic communication emerged at some point, an essential feature of modern *Homo sapiens* and the basis of our development, our power over other creatures, and social life. It allowed us to leave the prison of immediate experiences and reactions to become aware of our being and to represent how the world is. But as "collateral damage," it also allowed us to invent gods and money.

The lion-man (*Löwenmensch*) was found in the cave of Hohlenstein-Stadel, in Germany in 1939 (Figure 3.4). It is a 30-cm-high mammoth ivory sculpture with a human body and a lion's head. The age obtained by carbon-14 dating is about 40,000 years! Thus, it is the oldest sculpture we know and shows that our ancestors already had the faculty of imagining fictitious beings.

Figure 3.4.   Löwenmensch.

3.7 The Holocene

The Holocene interglacial geological period, which placed the Earth in a new warm state (the "long summer," according to Fagan[7]), marks a crucial geophysical event that follows the Pleistocene at the end of the last ice age (when the estimated population of humans was about a million). From a climatological perspective, the Holocene has been a remarkably stable period compared to previous interglacial periods. This is why agriculture emerged, and various civilizations were established, an important example of the interaction between humans and the environment. At the beginning of the Holocene, the planet warmed globally (mainly related to the orbital configuration of the earth) by about 5°C and the consequences were colossal. It is estimated that 14,000 cubic km of ice melt every year (imagine a cube of water 25 km (15 miles) per side), and sea level rose one meter every 20 years.

An increasing population and the new fertility of the land as the ice retreated led to agriculture and the domestication of animals. The selection of better varieties of plants and animals produced a revolution in the conformation of societies until then nomadic. Fundamental socioeconomic changes occurred with the notion of private property, not very useful for the previous hunter-gatherer societies. Thus, the seeds for power, war, inequality, and poverty were planted.

Although hunter-gatherer people had few things, they were not poor. Poverty is an invention of civilization, as is the associated concept of work. It is not that the individuals of primitive societies did not work, but it did not imply a relationship of subjugation. This relationship arose after inventing private property so that it became possible to differentiate between the "have" and the "have-nots," who had to live off their work. By releasing some individuals for other tasks, since not all hands were needed to obtain food, a remarkable social differentiation began with the possibility of maintaining armed forces dedicated to the defense of a society that had ceased to be nomadic. As populations grew and formed states with people bound by tradition, religion, language, ethnicity, and other common traits, the need for an organizational structure with leaders and bureaucracies to manage and control society also grew.

Juan Grompone writes[8]: *Human history develops between the dialectic of fraternity and domination. The appearance of the tool-making man led some human beings to dominate others and lose the original fraternity. The rest of human history is the attempt by diverse ways almost all failed, to recover the lost fraternity.*

3.8 Energy

About 250 years ago (when the human population was about 800 million — ten times less than at present), the industrial revolution, ushered by the invention of the steam engine by James Watt (1736–1819), converting heat into mechanical work, changed the character of life (and war). Machines before this operated with the energy of wind, water, and animals.

The primary source of energy went to the subsoil, through fossil fuels — coal, natural gas, and oil, and as we will see, this has led to a challenging problem. (The first oil well in the US was drilled in Pennsylvania in 1859 by Edwin Drake.) These fuels have been crucial for our accelerated development and supply most of our growing energy needs, without which we would collapse. They were formed during the Earth's long history by a process that transformed the remains of organisms that died hundreds of millions of years ago (that is why they are called "fossil"). In that sense, your car runs on solar energy stored in the distant past. Energy is distinguished by its source, but it is always the same. Thus, we distinguish between nuclear energy (related to changes that occur in the atomic nucleus), chemical energy (related to changes that occur in the chemical bonds between atoms), and kinetic energy (related to the movement of an object), among others.

A high fraction of living beings draws energy directly from sunlight. Photosynthesis is undoubtedly the most important biological process on Earth, used directly by plants or animals that eat plants. In this process, water, atmospheric carbon dioxide (CO_2), and light energy synthesize glucose (hence its name). Because it absorbs atmospheric CO_2, deforestation is worrisome.

3.9 Unknowns

It is also healthy to reflect on what we do not know (if we know), what (for now) are mysteries. We do not understand how animated matter arose from unanimated matter (using anachronistic words) — how life arose. We have discovered that a high percentage of the Universe is composed of "dark matter" and "dark energy," but we do not know what they are. We also do not know what existed before the Big Bang (if the idea of "before" makes any sense in this context).

At least for the last question, uniting philosophy and science, there is nothing contradictory or difficult with the idea of a "multiverse" that exists eternally without beginning nor end. Ours would be one of the countless possibilities, a gigantic and growing bubble in which conditions have been such that we can exist to think about it (see the anthropic fallacy below). And having mentioned thinking, we do not know what consciousness is, although we suspect that it emerges from the complexity of the brain's connectome (the web of connections between all neurons), with its 86 billion neurons (not 100 billion as is often stated). While we are on this subject, it is not true that we only use 10% of our brains; we use all of it, except those who proclaim that we only use 10%. But perhaps 86 billion neurons are not enough to understand.

It will take time to transform these and other "mysteries" into problems that we can solve. For example, it took over 50 years to go from the mathematical calculations made by the British theoretical physicist Peter Higgs (born 1929) (together with others) in the 1960s to the discovery of the Higgs boson (to explain the mass of elementary particles) at the particle accelerator at CERN. The experimental work was the product of an international effort of thousands of scientists, engineers, and technicians.[c] The Nobel Prize for Physics for 2013 was awarded to Higgs and the Belgian François Englert (born 1932). Another Nobel Prize awaits those who will discover the nature of dark matter.

The previous summary (which has no pretense of being complete) is the result of the commitment of a handful of men and women, some more recognized than others, who in recent centuries have dedicated their lives to lifting the veil behind which nature hides and to approach the truth of the world. They devoted themselves to reading the book of nature, as proposed by Galileo, and fostered a radical change in our conception of the physical world.

The above summary describes our state of knowledge, always tentative and subject to change, but it is the best model we have until further notice

---

[c] I recommend the film *Particle Fever*.

of what the world is like. These are the facts, and there is no alternative, like it or not. Ignoring them leads to a fictional world, perhaps more in keeping with some of the beliefs that we have maintained for millennia. Still, if we do not accept that the world is the way it is and not as we would like it to be, we will go the wrong way (and I would say we are), and it will all end in tears.

In terms of the planet's history, we do not realize the suddenness of the recent changes. If *Homo sapiens* is about 200,000 years old, this corresponds to about 7,000 generations (of 30 years). The great majority of these generations inhabited caves and used stone tools. The most recent 150 generations used metals. The last 20 generations knew the printed word. The combustion engine and electricity began to be used in the previous four generations. Most of the technologies that are familiar to us are products of the last two generations. The material and technological progress has been dizzying. We handed over control to the machines and presently and unnoticed also to artificial intelligence.

But much of our hallowed progress has been obtained by mortgaging the future of humanity. We must return to nature, but not in the sense of repudiating science and technology, as some wish. On the contrary, we need to think with great care more than ever to obtain even more profound knowledge of nature and our relationship with it. Let us not lose sight of the fact that the sometimes-idealized life of our ancestors was, in the words of Thomas Hobbes (1588–1679): *solitary, poor, nasty, brutish and short.*

Chapter 4

A New Era and the Catastrophic Convergence

We stand now where two roads diverge. But unlike the roads in Robert Frost's familiar poem, they are not equally fair. The road we have long been traveling is deceptively easy, a smooth superhighway on which we progress with great speed, but at its end lies disaster. The other fork of the road — the one less traveled by — offers our last, our only chance to reach a destination that assures the preservation of the earth.

— Rachel Carson, *Silent Spring*[1]

Two roads diverged in a yellow wood,
And sorry I could not travel both
And be one traveler, long I stood
And looked down one as far as I could
To where it bent in the undergrowth.

Then took the other, as just as fair,
And having perhaps the better claim,
Because it was grassy and wanted wear;
Though as for that the passing there
Had worn them really about the same,

Yet knowing how way leads on to way,
I doubted if I should ever come back.
And both that morning equally lay
In leaves no step had trodden black.
Oh, I kept the first for another day!

I shall be telling this with a sigh
Somewhere ages and ages hence:
Two roads diverged in a wood, and I —
I took the one less traveled by,
And that has made all the difference.

— Robert Frost (1916)

Several milestones mark our collective awakening to a new reality that requires a radical change in our ideas about the future of humanity and a new global ethic. Unfortunately, current events show me that we have a long way to go and have learned little form history. This demands a new meta-education, that is, a reconsideration of the reasons we educate. Even deeper, a reconsideration of the significance of our lives. It will also be necessary to accept a new metaphysics, a new understanding of humanity's place in the world, generated by several scientific developments. The milestones:

Figure 4.1. Image of Hiroshima after the bomb.

- On the morning of August 6, 1945, in Hiroshima, a city in southwestern Japan, and nearby Nagasaki 3 days later, nuclear bombs were detonated, killing about 200,000 in an instant (see Figure 4.1). It ushered in a new era ("the nuclear age") in which we acquired the ability to eradicate ourselves, an era in which the possibility of the world's end became real without the need for an apocalyptic divine intervention. As described by Toyofumi Ogura,[2] a survivor: *I couldn't believe it. All around me was a vast sea of smoking rubble and debris with a few concrete buildings rising here and there like pale tombstones, many of them shrouded in smoke. That's all there was as far as the eye could see.* I wish all who think of nuclear war would read it.

  This genocide connects with the one committed by the Nazis and collaborators on the Jews and others, symbolized by Auschwitz, an almost unspeakable act of human cruelty, which marked one of Europe's darkest times. As noted by Sven Lundqvist: *Auschwitz was the modern industrial application of a policy of extermination on which European world domination had long since rested.* This happened when Earth's human population was a mere 2.4 billion (with about 80 million killed during World War II).

- Rachel Carson's (1907–1964) groundbreaking 1962 book: *Silent Spring* provided the spark that ignited ecological thinking.[3] In it, we read: *Along with the possibility of the extinction of mankind by nuclear war, the central problem of our age has therefore become the contamination of man's total environment with such substances of incredible potential for harm — substances that accumulate in the tissues of plants and animals and even penetrate the germ cells to shatter or alter the very material of heredity upon which the shape of the future depends.* That was over 50 years ago! Earth's human population had grown to 3.1 billion. Since then, an increasing and unstoppable environmental poisoning of land, air, and water has threatened us all.

- The photo obtained by the Apollo 8 astronauts (Commander Frank Borman, Command Module Pilot James Lovell, and Lunar Module Pilot William Anders) in orbit around the Moon in December 1968, shown by media worldwide (we did not have WWW then). The awareness of the finitude and solitude of "Spaceship Earth," then inhabited by about 3.6 billion human beings, was implanted in our minds. And it will remain solitary, no matter how many extrasolar planets astronomers are discovering (see Figure 4.2). (It was an actual "photograph" taken on film with a Hasselblad camera). (Unfortunately, it was also the year of the assassinations of Martin Luther King (1929–1968) and Robert Kennedy (1925–1968).)

- The 1972 report[4] sponsored by the Club of Rome (founded by the Italian Aurelio Peccei (1908–1984) warning about the limits to growth on a finite planet, warning about the consequences of uncontrolled demographic growth combined with the voracity of the industrial system, which destroyed the capitalist myth of unlimited growth, the conclusion being that without significant changes humanity was in danger of growing beyond the physical limits of the planet. Those changes have not occurred, and we have already passed those limits (see below). By then, Earth's population had increased to 3.8 billion.

Figure 4.2. Photo obtained by astronaut Bill Anders from Apollo 8 on December 24, 1968 (NASA).

- The 1985 discovery of the ozone hole by a group of British researchers[5] (more below) who worked in Antarctica and demonstrated beyond any doubt the global impact produced by the 4.9 billion humans then living.
- The September 11, 2001 tragedy, when some 3,000 people were murdered by a gang of fanatics, began a profound transformation of US society, not used to being attacked since Pearl Harbor (It was a real war then). It embarked the US on a disastrous and permanent "War on Terror," which cost the lives of millions abroad, started wars in the middle east that devastated entire countries, in the process creating more terrorists.[6] We were 6.2 billion by then.

- The revolution started in 2007 and is ongoing when many internet social technologies took off. (The I-phone, Facebook, Twitter, Android, and much else[7].) It all started with Sir Tim Berners Lee (born in 1955). In 1989, working at CERN (The European Organization for Nuclear Research), he invented the WWW. By then, we were 6.6 billion. Soon we will be 8 billion.

You might wonder why I did not include the Apollo 11 Moon landing on July 16, 1969. Our first visit to another celestial body as astronauts Neil Armstrong, Michael Collins, and Edwin "Buzz" Aldrin got to the Moon. It was a remarkable, heroic feat, a testimony of our technical prowess and the science behind it. Many considered it a first step to conquer the Cosmos, a "giant leap for mankind." But did it change anything fundamental in how we view ourselves and our relationship to the world, in the same sense as the Apollo 8 photo? I do not think so.

On the contrary, for a while, it made us think that if we could do that, we could do anything, and that is not so. It backfired with many comments of the sort: "we went to the Moon, but we can't solve (choose your problem)." The much-talked-about fall of the Berlin wall and the end of the Cold War (if indeed it did end), although important political events are also not earthshaking.

We have determined that *the terrestrial system moves far outside the range of natural variability exhibited during the last million years.* The nature of the changes *occurring simultaneously* in the terrestrial system, their magnitudes, and the rate of change are extraordinary. We are at a unique moment in the history of *Homo sapiens*, nurtured by the accelerated growth of the human population and its activities. We have become a geologic force leading us to the *Anthropocene transition.*

In an important book, Christian Parenti writes[8]: *Climate change arrives in a world primed for a crisis. The current and impending dislocations of climate change intersect with the already-existing crises of poverty and violence. I call this collision of political, economic, and environmental disasters the* **catastrophic convergence**. *By catastrophic convergence, I do not merely mean that several disasters happen simultaneously, one problem* *atop another. Rather, I argue that problems compound and amplify each other, one expressing itself through another.*

Paul and Anne Ehrlich write in an essay[9]: *In the face of an absolutely unprecedented emergency, the world community has no choice but to take dramatic action to avert a collapse of civilization. Either humanity will change its ways, or they will be changed for us.*

I suspect the first option will not happen. We are not prepared to leave behind the mythical world and avoid entering a virtual one. The heavy baggage of a metaphysical misunderstanding of reality does not permit us to think freely and move fast. Instead, we hold beliefs consonant with Middle Age thought and move as if we were in a muddy bottomed shallow swamp.

Chapter 5

A New Physical and Social Setting

[Earth] is a living system, an immense organism, still developing, regulating itself, making its own oxygen, maintaining its own temperature, keeping all its infinite living parts connected and interdependent, including us. It is the strangest of all places, and there is everything in the world to learn about it. It can keep us awake and jubilant with questions for millennia ahead if we can learn not to meddle and not to destroy. Our great hope is in being such a young species, thinking in language only a short while, still learning, still growing up.

— Lewis Thomas[1]

The new vision of reality we have been talking about is based on awareness of the essential inter-relatedness and interdependence of all phenomena — physical, biological, psychological, social, and cultural. It transcends current disciplinarian conceptual boundaries and will be pursued within new institutions.

— Fritjof Capra[2]

Aside from the milestones I discussed above, we must also realize that we live in a rapidly changing world and that past ideas and formulas might need substantial revision. Therefore, I will set aside the byzantine complexities to which many debates sometimes lead, perhaps interesting for a rainy Sunday afternoon, but which in general terms do not contribute to the objectives I have set for this book. Instead, I aim to remain pragmatic by appealing to informed reasoning and considering the reality of *this* world so that we may at least distinguish between gymnasium and magnesium, both important for a healthy body.

Now it is a question of life or death. In the words of Günther Anders[4]: *It is not enough to change the world, we do this anyway, and it mostly happens without our efforts, regardless. What we must do is interpret these changes. So that we, in turn, can change them. So that the world doesn't go on changing without us — and not ultimately become a world without us.*

The new setting also includes a political-cultural shift that is turning against the spirit of the Enlightenment. Things in the US (and elsewhere) have changed quite markedly (reaching rock bottom under POTUS 45th) since Hermann Melville (1819–1891) wrote[5]: *Let us waive that agitated national topic, as to whether such multitudes of foreign poor should be landed on our American shores; let us waive it, with the one only thought, that if they can get here, they have God's right to come; though they bring all Ireland and her miseries with them. For the whole world is the patrimony of the whole world; there is no telling who does not own a stone in the Great Wall of China.*

President Trump made a mockery of ... *Give me your tired, your poor, your huddled masses yearning to breathe free, The wretched refuse of your teeming shore. Send these, the homeless, tempest-tost to me, I lift my lamp beside the golden door!.*[6] That did indeed represent the greatness of the US and not what he proposed. In a meeting on immigration, Trump reportedly said: *Haiti? Why do we want people from Haiti here? Then they got Africa. Why do we want these people from all these shithole countries here? We should have more people from places like Norway.* (Never occurred to him that people in Norway might have no interest.)

There is a worldwide tendency to turn to authoritarianism in various forms, basically driven by fear, distrust of political leaders (well deserved), irrationalism, conspiracy theories, anger about the present, and understandable anxiety about the future. This is an undercurrent that runs deep in the US population, and is worrisome because of all that I will discuss, and was called *the paranoid style* by historian Richard Hofstadter[7] a style held by *angry minds* which: *has to do with the way in which ideas are believed and advocated rather than with the truth or falsity of their content.* In his brilliant essay, Hofstadter writes: *The distinguishing thing about the paranoid style is not that its exponents see conspiracies or plots here and there in history, but that they regard a "vast" or "gigantic" conspiracy* **as the motive force** *in historical events. History is a conspiracy set in motion by demonic forces of almost transcendent power, and what is felt to be needed to defeat it is not the usual methods of political give and take but an all-out crusade. The paranoid spokesman sees the fate of this conspiracy in apocalyptic terms — he traffics in the birth and death of whole worlds, whole political orders, whole systems of human values. He is always manning the barricades of civilization.* Mind you, this was written in 1963.

We have seen it taken to extremes during the Trump administration, but it has encompassed social and political thought in the US (changing the bad guys over time). However, it is not exclusively a US phenomenon. The signposts are clearly visible in many nations for those who know some history, but that is the catch.

Historian Anne Applebaum[8] writes: *Polarization, conspiracy theories, attacks on the free press, an obsession with loyalty. Recent events in the United States follow a pattern Europeans know all too well,* and as said by Mark Twain: *History doesn't repeat but sometimes rhymes.*

Times have changed significantly since the last wave of fascism took over and dehumanized a part of humanity. Bertram Gross[9] writes: *Faceless oligarchs sit at command posts of a corporate-government complex that has been slowly evolving over many decades. In efforts to enlarge their own powers and privileges, they are willing to have others suffer the intended or unintended consequences of their institutional or personal greed. For Americans, these consequences include chronic inflation, recurring recession, open and hidden unemployment, the poisoning of air, water, soil, and bodies, and more important, the subversion of our constitution. More broadly, consequences include widespread intervention in international politics through economic manipulation, covert action, or military invasion.* (Currently it is the Ukraine that must suffer the consequences.)

Philosopher and Nobel laureate Bertrand Russell (1872–1970) once said[10]: *The fundamental cause of the trouble is that in the modern world the stupid are cocksure while the intelligent are full of doubt.* There is a well-researched phenomenon, a cognitive bias called the "Dunning Kruger effect,"[11] in which people of low ability have imagined superiority and mistakenly assess their cognitive ability as greater than it is. Put less academically; the more stupid you are, the more you think you are not.

This is how we end up with confident clowns in government and insecure intellectuals. It is our nature to pay more attention to the former.

Back in 1940, renowned philosopher Hannah (palindrome!) Arendt (1906–1975)[12] wrote: *Intellectual, spiritual, and artistic initiative is as dangerous to totalitarianism as the gangster initiative of the mob, and both are more dangerous than mere political opposition. The consistent persecution of every higher form of intellectual activity by the new mass leaders springs from more than their natural resentment against everything they cannot understand. Total domination does not allow for free initiative in any field of life, for any activity that is not entirely predictable. Totalitarianism in power invariably replaces all first-rate talents, regardless of their sympathies, with those crackpots and fools whose lack of intelligence and creativity is still the best guarantee of their loyalty.* In the words of Elton John: "Nine times out of ten, well I see the storm approaching, long before the rain starts falling." Nine times out of ten.

All societies have contradictory norms, but it strikes me as bizarre that flying the Nazi flag with chants of "Jews out!" is accepted, invoking freedom of expression. But, if a female in Charlottesville showed her breasts in public, she would be arrested for indecent exposure. Meanwhile, hundreds have been killed and injured in churches, schools, and clubs in recent years by some madman carrying a semiautomatic weapon and inspired by the vile idiocy of white supremacy.

With the poor logic of "guns don't kill people, people kill people" (I would say people *with* guns kill people) and citing the second amendment (1791): *A well-regulated Militia, being necessary to the security of a free State, the right of the people to keep and bear Arms, shall not be infringed.*), a right granted when the weapons then available consisted of revolvers or simple rifles, gun control in the US is vehemently opposed. Perhaps it would be wise to revise and update a nation's constitution every 50 years or so. After all, there are many things happening today that the "founding fathers" could not even have dreamed or having nightmares about, even such things as a president pardoning himself. It is also a fact that more people are killed (some say lynched) by police every year[13] (usually African Americans[14]) than are killed in mass shootings which are also becoming more frequent). Be that as it may, "well regulated" seems to have been overlooked. It is surreal. The use of "homeland" also bothers me; it reminds me of *Vaterland*. I do not see why a perfect word with no connotation is not used: "The United States."

What is right and what is wrong, issues of ethics (or moral philosophy) and justice, pose difficult questions whose answers we will not find in ancient books that some describe as "sacred" (nor in a modern physics book or laboratory) and for which many have killed in abundance, which is proof that they are not adequate.

The idea that there are alternative facts (a euphemism for telling lies) and a world in which truth is relative, that there is "post-truth," is simultaneously perverse and inept and is pure demagoguery. It will sink us with certainty and will entangle everyone, even the crooks who propose it. Timothy Snyder writes[15]: *post-truth is pre-fascism*. According to the Oxford Dictionary, which chose post-truth as "word of the year 2016," it is *relating to or denoting circumstances in which objective facts are less influential in shaping public opinion than appeals to emotion and personal belief*. But the truth is all-important, and there are good ways to establish it. Hannah Arendt had this to say: *The ideal subject of totalitarian rule is not the convinced Nazi or the convinced Communist, but the people for whom the distinction between fact and fiction (i.e., the reality of experience) and the distinction between true and false (i.e., the standards of thought) no longer exist.*

One thing is a fact, and another is our knowledge and interpretation of that fact, which is open to alternatives. Take, for example, the current sometimes bitter discussion about the origin of the SARS-Cov-2 virus. It is a fact that it either escaped from the Wuhan Laboratory (for whatever reason) or that it was a case of zoonosis (jumping from an animal to humans). One of these two alternatives is a fact, it is true (unless it was brought by aliens), but we do not (yet) know.

There is a new industry of "fact-checking," but who checks the fact-checkers? The issue is a difficult one, and as mentioned by Noam Chomsky[16]: *The power of the government propaganda apparatus is such that the citizen who does not undertake a research project on the subject can hardly hope to confront government pronouncement with fact.*

It is easy to distort the facts or invent them if you appeal to one of the most intense human emotions: Fear. Psychologists understand that we are tribal descendants, and our brains are that way.[17] Humans who are different are to be feared (even if they just think differently), they are suspect and menacing, and this made sense eons ago. Today it does not, but it is hard-wired into our brains to benefit all demagogues who play us like a violin.

The demonization of "others" (defined by ethnic, religious, or racial features) is the standard operating procedure of fascist thinking, which resort to these tactics to scare people into supporting an authoritarian leader.

People who do not know how to think are confused by a bunch of preachers and grifters who speak from ignorance (or worse). So, they try persuading the credulous about miraculous cures, the easy solution of a complex problem, or what awaits them in the "hereafter" (after the failure of the miraculous cures).

Governments are indeed dominated by persons with degrees in law and related disciplines; after all, they are called legislatures. To get a law degree might require basic physics, but some seem to have squeezed by with a "D." But the actual legislature consists of natural laws that no legislator can change, even if they do not like them. As an advance: inject carbon dioxide into the atmosphere of any planet, and its surface will be heated — basic physics — that we ignore at our own peril. Shawn Otto states[18]: *Without a better way of incorporating science into our policymaking, democracy may ultimately fail its promise. We now have a population that we cannot support without destroying our environment — and the developing world is advancing by using the same model of unsustainable development. We are 100% dependent on science and technology to find a solution.*

The setting has also changed due to the increasing complexity of our societies and the fact that 50 years ago, the human population was half (3,851,650,000) of what it is today. Complex modern societies must be governed with knowledge of the behavior of complex systems, and for this, those who govern them are ill-prepared. (And perhaps nobody is prepared, but that is no excuse to put ignorance in charge.) The planet we live on can be divided into various "spheres" (lithosphere, cryosphere, hydrosphere, atmosphere, biosphere, etc.). Some are less complex than others but interact in ways we do not fully understand, producing a very complex system of systems. It can also be divided into different cultures, primarily defined by religious beliefs, as argued in a famous book[19] by political scientist Samuel P. Huntington (1927–2008), that make global understanding very difficult. This demands special precautions when one system component is pushed to extreme values (such as temperature), lest we get into total failure. Complex systems are finicky, press the wrong button, and all hell breaks loose because their response is nonlinear, and small changes can have significant effects, described as tipping points. Modern societies and economies are also complex systems, sometimes showing novel behaviors difficult to control or predict, leading to collapse. Our brain, a complex system of neurons, demonstrates this novel behavior as thought and consciousness, "emergent" properties.

A simple example illustrates this idea. Water is a wet liquid, a combination of Oxygen and Hydrogen that are not wet liquids (at room temperatures). Wetness is a new property that *emerges* from the combination of these two atoms. Neurons do not think; it is the complex system of billions of neurons that do. This difficulty has been discussed by Canadian political scientist and University Research Chair at the University of Waterloo, professor Homer-Dixon in his excellent book that tells us of significant "tectonic stresses" which interact — differential population growth between rich and poor countries (predominantly African), energy scarcity (as we must eliminate fossil fuels), environmental degradation (on land and water), climate change, and economic instability (especially the widening wealth gap) — and writes[20]: *The stresses and multipliers are a lethal mixture that sharply boosts the risk of collapse of the political, social, and economic order to individual countries and globally*, in other words, a catastrophic convergence.

This is quite different from the situation for complex systems designed and built by us. For these at least, we have the blueprints and understand the whole from its constituent parts, be it a bicycle or the space shuttle composed of millions of interacting components. We have all the information necessary to understand the system's response to a change in one of its parts. Should there be a system failure, we can investigate and understand what happened. This was the case for the space shuttle Challenger, where a simple O-ring that lost its elasticity in the cold weather provoked the fatal accident.

But we have nothing even close to complete and detailed knowledge of "System Earth." There are no blueprints, and only slowly and belatedly are we beginning to understand some of the complex interactions that affect the system. Only experts in relevant science can offer their best knowledge and advise those who are the leaders of society (if they are willing to listen).

The entropic character of every process (the second law of thermodynamics), including the economic one, cannot be ignored and is the reason why there are limits that paint a bleak future. In his valuable work, Romanian mathematician, and economist Georgescu-Roegen[21] (1906–1994), laid the foundation of an economic theory that would not ignore thermodynamic limits. A financial system, a complex society, a plant or animal, need the energy to function and more power to grow. If we do not constantly add energy, society will decay.[22] He wrote: *If we abstract from other causes that may knell the **death bell** of the human species, it is clear that natural resources represent the limiting factor as concerns the life span of that species. Man's existence is now irrevocably tied to the use of exosomatic* (outside the body) *instruments and hence to the use of natural resources just as it is tied to the use of his lungs and of air in breathing, for example. We need no elaborated argument to see that the maximum of life quantity requires the minimum rate of natural resources depletion. By using these resources too quickly, man throws away that part of solar energy that will still be reaching the Earth for a long time after he has departed. And everything man has done during the last two hundred years or so puts him in the position of a fantastic spendthrift. There can be no doubt about it: any use of natural resources for the satisfaction of nonvital needs means a smaller quantity of life in the future. If we understand well the problem, the best use of our iron resources is to produce plows or harrows as they are needed, not Rolls Royce's, not even agricultural tractors.*

I perceive the "death bell of the human species" in these turbulent times. Natural philosophy must return to be a fundamental pillar of education to remove us from the impasse generated by the increasingly ambitious goals of the *financial-industrial-religious-military* international complex (FIRM — expanding on Eisenhower), which opposes any change and is global.

Those that could encourage the necessary changes are acutely myopic, inept, criminal, or possibly all of the above, and perhaps worse. More likely, they do not want, do not care, or disagree with actions that are contrary to their material interests and ideological agendas. Thus, in many countries, democracy is replaced by *Cleptoineptocracy*. Kevin MacKay writes[23]: *The problem [...] is that a small minority of wealthy individuals benefit immensely from the current set-up and create a social system in which they maintain a near-total monopoly on decision-making power. This describes a system of class-based political oligarchy.*

As described by the 2001 Nobel laureate for economics, Joseph Stiglitz, the experts from international organizations that establish economic policies for poor countries (IMF, WTO, World Bank, controlled by the US), stay in five-star hotels when they visit. They join elegant meetings and sumptuous banquets to discuss issues that affect the welfare of the poor with members of the elites without them having neither voice nor vote.

They have not seen schools without books or chalk, hospitals without medicines in which the sick crowd together like merchandise in a warehouse. They have not visited neighborhoods where the only running water is the one that passes through a smelly and choleric trench in front of a wood and cardboard shack. And they argue about gender-neutral bathrooms, a stupid and trivial problem, utterly irrelevant for the world without toilets. Meanwhile, let us spend trillions on war.

Or perhaps they think like King Louis XV: *Après Moi le deluge!* Alternatively, it could be more sinister. I can imagine that the wealthy minority and a few fools might think the coming "tropic of chaos" is a Godsend. There is no need to worry about overpopulation, resource depletion, biodiversity, misery, genocides, and devastation if it happens in faraway places to "inferior" people. Anyway, biodiversity is for sissies; real men go hunting. All they must do is move to higher and safer ground, buy more powerful air conditioners, and be prepared with weapons behind fortifications to keep "the wretched refuse of their teeming shores" at bay and wait for the carnage which will relieve the pressure. If you think this is farfetched, think about Trump's insistence on a wall and his idea of sending troops to the southern border of the US to defend the country from an invading caravan of destitute people with sore feet.

In Stiglitz's words[24]: *The problem is not with globalization, but with how it has been managed. Part of the problem lies with the international economic institutions, with the IMF, World Bank, and WTO, which help set the rules of the game. They have done so in ways that, all too often, have served the interests of the more advanced industrialized countries — and particular interests within those countries — rather than those of the developing world. But it is not just that they have served those interests; too often, they have approached globalization from particular narrow mindsets, Shaped by a particular vision of the economy and society.*

I can see this at work as I write after Puerto Rico was hit by the most powerful hurricane ever to do so (Maria in September 2017), devastating this beautiful island, leaving it in precarious shape, thousands without electricity nor a roof on their homes. The island, a US territory with about 3.3 million *US citizens* (dwindling because of migration to the US), is also deeply in debt. Corrupt or inept local government in collusion with US banks continued to issue loans at rising interest rates even after it was clear that it would lead to a repayment crisis. (On the other hand, ask a bank for a mortgage loan, and they might even require a colonoscopy.)

And here we are with a "Financial Oversight and Management Board for Puerto Rico," established by the US congress. It constitutes a government over the elected government making decisions without regard for the damage their decisions will inflict on US citizens, whose median household income is about US$20,000 (representing 45% of persons in poverty). At the same time, the board's executive director earns a salary (paid for by the Puerto Rican people) of US$650,000. Nice, isn't it? She was the Ukrainian Finance Minister from December 2014 until April 2016; one wonders.

After the hurricane, many hoped that the mightiest military in the world would come to the rescue, ships, helicopters, and airplanes bringing emergency supplies and personnel to help those in need and rebuild what looked like the result of a nuclear strike. Indeed, if you could mount the logistics for an attack on Iraq, you could mount a humanitarian operation to help fellow citizens in peril. Instead, we waited in vain and were visited by Mr. Trump, who added insult to injury by throwing paper towels into the expectant crowd. (Since you might not believe this, just look it up on YouTube.) Vile.

A study published in the *New England Journal of Medicine*[25] estimates that about 5,000 persons died, stating that: *interruption of medical care was the primary cause of sustained high mortality rates in the months after the hurricane.*

Vigorous action by the US government could have avoided this, and I dare say that if the Puerto Rican population was composed of white, blond, and blue-eyed persons, the story would have been quite different. Then we had a series of earthquakes and now the virus. Paradise.

The rest are powerless to push the necessary changes to remove us from this incredible mess in which we find ourselves being oppressed and/or depressed. Changes will come only if those with the power to implement them realize that in the long run, it is for their good or that of their descendants, that what they gain in the long term is more than what they lose in the short term. Maybe it is too much to ask. Besides, we are all dead in the long term, which does not stimulate concern for it. That is how we are.

In democracies governed by majority votes (not so in the US), it is possible to ascend to positions of power by mere image and demagoguery (and a fortune) instead of by an enlightened attitude. Ignorance facilitates the return to the "throne and the altar."

We urgently need to find new ways of thinking, to solve the problems caused by the old ways of thinking. Maintaining the status quo is the currency of all those in power. The only tolerated changes are those necessary to maintain the structure or those that do less damage if the situation is critical. A dramatic change, a revolution, is resisted by force if necessary. In many countries (I would say the majority), the armed forces exist for this, and not to protect the nation from some external enemy, as alleged. This was the case with the armed forces of Latin American countries, which, starting in the 1960s, were transformed from a national defense mission against nobody to one of "internal security" to combat the enemies of the elites supported by the United States: the unions, the social workers, the sympathizers of left-wing ideas, those who cried out for social justice, all encompassed under the label of "communists" or "subversives." We have declared war on terror as if this were a nation, appealing to outdated ideas and creating hundreds of new terrorists in the process. A 2008 report by the RAND corporation states[26]: *Military force usually has the opposite effect from what is intended: It is often overused, alienates the local population by its heavy-handed nature, and provides a window of opportunity for terrorist-group recruitment.*

We know circumstances in which national security has been brandished to cause total personal insecurity. Thus, there is "Nacht und Nebel" in some unhappy countries, or as was the case in many countries in Latin America: *Desaparecidos.* But a nation is nothing more than the sum of its individuals. The previous situation is one of those absurdities demonstrating the fundamental divergence of objectives between the rulers and the ruled.

Noam Chomsky, Institute Professor in the Department of Linguistics and Philosophy at MIT, writes[27]: *What about the security of the population? It is easy to demonstrate that this is of marginal concern to policy planners. Take two prominent current examples: global warming and nuclear weapons. As any literate person is doubtless aware, these are dire threats to the security of the population. Turning to state policy, we find that it is committed to accelerating each of those threats in the interest of its primary concerns, protection of state power and all the concentrated private power that largely determines state policy.*

Secrecy is abused for personal protection, and anytime a government needs to cover up some wrongdoing, it can be stamped "Top Secret." *Authoritarian regimes, murderous military organizations, human rights breaching spy agencies, polluting or corrupt organization, mind control religious cults, and many more examples are available where their ability to continue with the illegal or illegitimate actions or to hide past events all must utilize secrecy and impose punishment on leakers to ensure that secrecy,* writes Australian journalist Greg Bean.[28] For example, the elimination of the Jews during Hitler's Third Reich was top secret, although the undertaking was of such a dimension that many knew and more suspected that this was happening.

For all secrecy, there will be leaks, and thus we occasionally learn what some people in government do in the name of "we the people" and all the lies they utter. Unfortunately, these "whistleblowers" are hounded as if they were traitors, and they are if you are one of the liars. Daniel Ellsberg notes[29] that *offers its readers a profound education in politics and morality: The reality unknown to the public and to most members of Congress and the press is that secrets that would be of the greatest import too many of them can be kept from them reliably for decades by the executive branch, even though they are known to thousands of insiders.*

You might think that I am flogging a dead horse since the "Pentagon Papers" leaked by Daniel Ellsberg in 1971, leading to the end of the tragic Vietnam war, is old history. It caused about three million deaths between military and civilians from 1955 to 1975 (about 60,000 US troops), but the apocalyptic horse is alive.

You may wonder why Bin Laden or Saddam Hussein were killed when they could have been (and Hussein was) taken alive. To my mind, the best explanation is that the dead do not speak (despite all those swindlers who, for a fee, claim they can talk to them) because if they could, they might have some distressing things to say.

By the way, the CIA is not what you might think. Carved in stone on the walls of the original CIA building, you could read the biblical motto,[a] *And ye shall know the truth, and the truth shall set you free.* How can an agency dedicated to undercover activities, lying, and deception, even think of this? Well, that is their mission. In one of the few books vetted by the agency, former agent Ralph W. McGehee tells us[30]: *The CIA is not now, nor has it ever been a central intelligence agency. It is the covert action arm of the President's foreign policy advisers. In that capacity, it overthrows or supports foreign governments while reporting "intelligence" justifying those activities. It shapes its intelligence, even in such critical areas as Soviet*

---

[a] From the Bible John 8:31–32 (KJV).

nuclear weapon capability, to support presidential policy. Disinformation is a large part of its covert action responsibility, and the American people are the primary target of its lies.

Over the years, independent of the party in power, the erratic and incoherent US foreign policy implemented including overt and covert military interventions, has placed everyone in danger and is, in the end, counterproductive. This is at the root of why "they hate us." You can read all about this in the revealing book by Max Blumenthal.[31]

"State secrets" serve to protect corrupt rulers and activities no decent person would accept. Thus, torture is instituted (under the euphemism of enhanced interrogation), a process that the majority (let us assume some level of decency) would oppose. I do not wish to belabor the point; there are abundant examples of things done in our name that we would condemn (perhaps you remember the infamous "torture memos," and you can read about the sordid US torture practices in the book by Stephen Gray[32]). For example, Robert Baer, former CIA agent, indicates[33]: *If you want a serious interrogation, you send a prisoner to Jordan. If you want them to be tortured, you send them to Syria. If you want someone to disappear — never to see them again — you send them to Egypt.*

These "disappearances" of so-called high-value detainees (HVD) within the context of the "War on Terror" continue and constitute acts of terror on individuals who have no recourse and are often innocent. It is a logical contradiction to fight terror with terror. You can read about these appalling torture and murder actions by the US in collaboration with the United Kingdom in Clara Usiskin's recent book,[34] which is: *A tough-minded investigation of how legal process and human rights have been ignored in the search for often non-existent terrorists in Africa.*

It has been revealed that since 1970 the encryption machines purchased from a Swiss firm, Crypto AG, used by many governments to communicate their dirty secrets, were rigged so that the CIA could decode all messages and gain knowledge of all that was planned. This included torture and assassinations by Latin American dictatorial regimes in the seventies. The firm was secretly owned by the CIA and its German counterpart, the BND (Bundesnachrichtendienst). It later expanded to spy on hundreds of countries, friends, or enemies. It continued until 2018 when Crypto AG was sold, or so they have stated.

So, it turns out that for a long time, some in the US government and I would think all Presidents knew about all the horrible things perpetrated by governments in other countries and simply stood by. Ethics?

David Stockman,[35] an experienced Washington, and Wall Street insider, sums it up as follows: *It's the power-hungry government officials and morally bankrupt politicians who continue to pillage our country and rob our citizens to satisfy their never-ending lust for power. It's the warmongers who run our national security apparatus who send our troops to die fighting wars in places we have no business being in. It's the insatiable greed on Wall Street and the Keynesian money printers running the Fed... who continue to inflate our markets, ruin our economy, and then siphon billions of dollars right out of our pockets, and the pockets of our children and grandchildren.*

It's the traitors in our intelligence agencies who undermine our democracy, and workday and night to violate our privacy and our constitutional rights in the name of "national security" It's the powerful elites who are unelected, unimpeachable, unaccountable, and virtually untouchable ... who act with impunity and complete disregard for what's good for this country.

Scientific studies show that we are drastically affecting the earth's climate. All those who ignore the red alert today and do not take and support the necessary measures to mitigate the damage will be the intellectual authors of the suffering and death generated. The problem as well stated by John Schumacher,[36] is that: *We are unquestionably a creature of genius with awesome intellectual ability. At the same time, we are akin to an earthworm engaging in actions of such stupidity that our own survival is jeopardized. We stand alone in the animal kingdom with our highly specialized logic and reasoning that summon us toward new truths. Yet we turn around and show that we are cesspools of irrationality ready to defend and even die for something in total defiance of that same logic and reason. We are capable of remarkably sensitive, higher-order information-processing on the one hand, and of the most stumfumbling distortions of bold-faced facts on the other. We glorious creatures can be indescribably tender and kind and risk our very lives for a single soul in distress. At another time or place, however, we are just as apt to play a part in equally unimaginable acts of savagery.*

In a speech to the public at the American Geophysical Union (AGU), the distinguished journalist Dan Rather expressed[37]: *Are we going to guide our future by science, reason, and knowledge, or are we going to succumb to superstition, ignorance, and propaganda?* That is the crucial question, and it looks like we are going for the second option.

Chapter 6

About Science

6.1 Science and Bullshit

There is a cult of ignorance in the United States, and there always has been. Anti-intellectualism has been a constant thread winding its way through our political and cultural life, nurtured by the false notion that democracy means that my ignorance is just as good as your knowledge.

— Isaac Asimov[1]

We have to do better as a nation of talking about science and denouncing fake news and sensationalism; otherwise, we let internet trolls win out on topics that determine the fate of our planet and our public health.

— Dan Rather

After reading all the above, I hope you will understand why urgent change guided by knowledge is needed, so it is time to look at how science fits (or not) into this messy world and how science and society interact. Most of the planet's population has (perhaps) an adequate education to look after themselves: reading, writing, and some knowledge of arithmetic (the proverbial three R's: Reading, "riting and rithmetic") and a minimal understanding of often distorted history, geography, and a tidbit of science. And that tidbit will kill us.

According to a United Nations[2] report, some one billion people in the world (one in eight) are illiterate, two-thirds of them women. On the other hand, as many optimists have told me, some things have improved according to available statistics: the percentage of illiteracy has dropped, as has extreme poverty, among other things. And no one can deny that the present is much better than 100, or even 50 years ago for many. But that is not the point; statistics only show one side, usually the side you want to highlight, and depend on your definitions of the variables in question (what is "better"?). Sometimes they are redefined to obtain a nicer result. Other times we do not know certain variables with sufficient precision to conclude anything, and as the saying goes: *Garbage in, garbage out.*

Furthermore, even with no illiterates, what matters is *what* they read. Much worse is scientific literacy, where poll after poll shows abysmal results. For example, the National Science Foundation (NSF) Science and Engineering Indicators report for 2018[3] determined that nearly half of those surveyed (49%) correctly indicated that *human beings, as we know them today, developed from earlier species of animals*, meaning that 51% got it wrong.

The idea is not that every citizen should understand the content of science or other scholarly endeavors. This goal would be impossible to achieve since even scientists know little about scientific areas that fall outside their respective specialties. But how can we allow citizens to be blissfully unaware of what the future might bring and let them believe childish things?

Is it too much to ask that any citizen understand why homeopathy is a lucrative scam? That he understands why you cannot "see" the future no matter how many crystal balls you gaze at? That there is no such thing as a "miracle"? That he realizes that unfettered capitalism will drive most to poverty? He grasps that there is no reason to think that vaccines cause autism and cell phone cancer and understands that if he does not know anyone with Polio, it is because of the past vaccination instituted by government health agencies. Recognize that we, with our activities, are affecting the Earth's health and are lacerating our planet? Is it that hard? Recognize that if shit happens, it does not necessarily mean it is the result of an evil conspiracy? Understand that if the eight wealthiest persons on the planet have as much wealth as the poorest half of humanity, something is wrong? And that this significant difference in wealth undermines democracy, which depends on the resources a candidate for public office has? Recognizes that bombing a country is not going to bring democracy to those left in the rubble? And finally, that dihydrogen monoxide is not dangerous? I do not doubt that you will have your favorites; the list is long.

Refusing to look at our real world through the lens of science and even showing contempt for science and misrepresenting its results concern me greatly. You have undoubtedly heard public figures (politicians) say: *Well, I am not a scientist*, and then continue with bullshit about the science in question. This is not just idle talk; think of the politicization of a very elementary fact: confronting a disease, transmitted by exhalation, it is basic science and logic that the disease would be controlled if everyone wore a mask for a while since the disease would not propagate as easily. It is appalling that many in the US do not understand this, and others selfishly protest that this infringes their "freedom." When science gets politicized, you get politics.

Philosopher Harry Frankfurt[4] explains: *Bullshit is unavoidable whenever circumstances require someone to talk without knowing what he is talking about. Thus, the production of bullshit is stimulated whenever a person's obligation or opportunities to speak about some topic exceed his knowledge of the facts that are relevant to that topic.* I have no firm statistics, but it seems to me that there has been an increase of bullshitters, or maybe it is that they now have found an outlet for their stupid opinions that they did not have before the internet revolution. Errol Morris[5] tells us: *Two hundred years ago, 99.999 percent of human idiocy went unrecorded. Now we have the Internet.*

The growing presence of bullshitters at a time when we need truth is dangerous, and according to Frankfurt, they are even worse than liars: *it is impossible for someone to lie unless he thinks he knows the truth. Producing bullshit requires no such conviction. A person who lies is therefore responding to the truth, and he is to that extent respectful of it. When an honest man speaks, he says only what he believes to be true; and for the liar, it is correspondingly indispensable that he considers his statements to be false. For the bullshitter, however, all these bets are off: he is neither on the side of the true nor on the side of the false. His eye is not on the facts at all, as the eyes of the honest man and of the liar are, except insofar as they may be pertinent to his interest in getting away with what he says*

Because scientific knowledge is essential for survival in modern times, its rejection or ignorance by important sectors of the citizenry, particularly by those who govern, poses a crisis that scientists, educators, and social leaders must confront.

I see it while I write: the fool on the hill (Mr. Trump) insisting that anthropogenic climate crisis is a hoax (while four violent hurricanes fed by hot ocean waters cause great destruction in Texas, Florida, North Carolina, devastating the Bahamas, and Puerto Rico). We also heard him recommending miracle cures for COVID-19 against all evidence. In general, he showed a disturbing contempt for the facts, reason, and science (which made "America great" to some extent) that led to the disaster caused by the mismanagement of the COVID-19 pandemic.

It is a tragedy that surveys about the public understanding of science in the US obtain worrisome results. The causes are multifactorial (educational level, age, ethnicity), but what is consistent is a widespread lack of understanding about what science is and why it is the best resource to alert us and help solve some of our most pressing problems.

There is an entire cottage industry of pseudoscience propagating pure, undiluted bullshit. It claims something supposedly obtained from scientific studies when there is no scientific basis for the claim, be it a perpetual motion machine, telepathy, studies of the curative effects of quartz or hydroxychloroquine, or homeopathic remedies. In some cases, it might be harmless (say, the "scientific" study of spontaneous human combustion). But in other cases, such as homeopathy, it can harm those who resort to this and therefore do not get the treatment they need and because, in this case, it is a multimillion-dollar scam. Similarly, the inability to use numbers, ignorance of statistics, and probabilities, lead to errors that sometimes cause horrors.

Let's look at homeopathy as an example of pseudo-medicine and how we know it cannot work. I do not know the credentials of those who work in the multi-million-dollar homeopathic industry, but they could be best described as crooks. For centuries so-called homeopathic remedies have been sold to those seeking help, first formulated by the German Samuel Hahnemann (1755–1843) (Figure 6.1).

He even has a monument in Washington that should, as a minimum, have an explanatory plaque affixed if not being dynamited outright. (I wrote this before the current discussions about statues.) These products are sold OTC in pharmacies, and, understandably, the public will think that they are a remedy for their ailments, but they are *demonstrably a fraud*.

Figure 6.1.   Hahnemann in Washington, DC. The George F. Landegger Collection of District of Columbia Photographs, in Carol M. Highsmith's America, Library of Congress, Prints and Photographs Division. Sculptor: Charles Henry.

Hahnemann postulated two enigmatic principles. The first one was: *equal cures equal (similia similibus curentur)*. This means that if you want to treat malaria — which causes intense fevers — you must administer as medicine a substance whose effect is to cause fever in a healthy person. The law was a relic of the magical ideas of the Middle Ages when it was thought that the similarity or "sympathy" between particular objects allowed to operate on one to affect the other. You would probably kill your patient if you did just that. So Hahnemann invented a second more enigmatic principle: *The smaller the dose of the curative substance, the more powerful its effect*, designated as the law of infinitesimals (if you see the logic, let me know). To prepare his medicines Hahnemann used a process of successive dilutions. Thus, a part of the extract of a medicinal plant was diluted with ten or one hundred parts of water. It was again diluted with water after mixing well (in their jargon called "succussion"). By repeating this step many times, he got to the point that there was nothing left, not even one molecule of the supposed active ingredient. For example, coffee can cause insomnia. Therefore, homeopathically diluted coffee can be used instead of sleeping pills, in this case: *coffea cruda* (Figure 6.2) (if you do

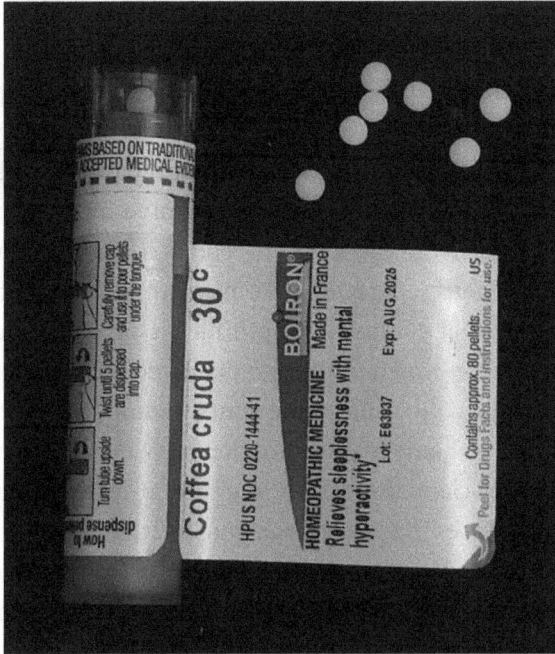

Figure 6.2. Coffea cruda.

not believe me, you can buy it online). At Amazon, it costs $6.99, and it says: *non-Drowsy, no known side effects, no known drug interactions*, a good description of the effects of pure water. Nor is there the danger of an overdose. Surprisingly, a high percentage of persons give it 4 or 5 stars. This means that these people get to sleep, but psychologists will tell you that it was not because of the product

Let me show you why I am so confident that it is a scam.

It is common for homeopaths to continue the dilution process 30 or 60 times, diluting by ten or 100. The result is a part of medicine and 10,000,000,000,000,000,000,000,000,000,000 parts of water (for the case of 30 dilutions by 10). You will remember (and if you do not remember, go back to an elementary physics or chemistry book or Wikipedia) that Avogadro's number ($N_A = 6.023 \times 10^{23}$ molecules/mol) is the number of molecules contained in one mole of a substance. One mole is a quantity in grams equal to the atomic or molecular weight of the substance. One mole of water, H_2O, composed of one oxygen and two hydrogen atoms, is a mass of 18 grams of water and contains 6.023×10^{23} molecules of water.

Suppose you start with one mole of the substance to prepare the remedy. When you reach a 24 times dilution, 0.6 molecules of the substance will remain (meaning no substance) in the preparation, less than one molecule of the "active" ingredient. If you continue, as usual, no molecule will remain. The absence of the active ingredient in homeopathic remedies makes it possible to fairly characterize homeopathy as a form of "medicine" without medicine that claims that nothing dissolved in water is more effective than water in which nothing is dissolved.

This lack of even a molecule of the remedy administered to the patient is openly admitted by homeopaths since they have no other remedy (never better said). So, they claim that the healing substances leave in the solution some "spiritual essences" that water "remembers" the active ingredient. It is then necessary to ask, as suggested by the Australian Council Against Health Fraud,[6] and quoted by Edzard Ernst, whether water also "remembers" all the other molecules with which it has been in contact throughout its history: *Strangely, the water offered as treatment does not remember the bladders it has been stored in, or the chemicals that may have come into contact with its molecules, or the other contents of the sewers it may have been in, or the cosmic radiation which has blasted through it.*

According to the "equal cures equal" principle, a homeopathic recipe for treating the irritation caused by the baby's diaper indicates that it be treated with *Rhus toxicodendron*, better known as "poison ivy." Luckily for the baby, the second dilution principle ensures that the child's buttocks do not feel anything other than the refreshing effects of a bit of water.

Oscillococcinum is a homeopathic product produced by the French company *Boiron* to prevent and relieve the symptoms of colds and flu, with annual sales exceeding 300 million euros. Its "active ingredient" is *Anas Barbariae Hepatis et Cordis Extractum* (extract of liver and heart of Barbary duck), HPUS 200CK (200 dilutions by the method of Mr. Korsakoff). However, the process of dilution does not matter since what is left is water. The inactive ingredients are sucrose and lactose. The name comes from the microbe "Oscillococcinum," which the French doctor

Joseph Roy (1891–1978) thought to have seen when examining victims of the influenza epidemic of 1918, which oscillated in his microscope. The microbe does not exist, and less is understood what the poor duck has to do with all this. The preparation method consists of the following steps:

- Fill a sterile bottle with one liter of water and glucose.
- Catch a mallard, cut off its head and extract its liver and its heart.
- Add to the bottle 35 g of the liver and 15 g of the heart.
- Leave the bottle at rest for 40 days; the liver and heart will dissolve.
- Empty the bottle without rinsing and fill it with pure water. Shake with energy (agitation or "succussion," for "dynamization").
- Repeat the previous steps (shaking, emptying, and filling) 200 times.
- Use the obtained water to impregnate 5 mg lactose tablets.

The little box of pills is sold at a good price, and a couple of doses a day are indicated. For CYA, the package also says: *If symptoms persist beyond three days, consult your doctor.* It's not just a joke; it's a multi-million-dollar scam, and, shockingly, the authorities (especially those related to public health) ignore it. On the Boiron website, I read: *We are interested in developing specific benefits for our employees to enrich their well-being. It is essential for us to link global economic performance with individual benefits.* Hmmm.

Accepting homeopathy implies discarding extensively corroborated medical, biological, physical, chemical, and pharmacological knowledge. Therefore, promoting it is highly irresponsible. It also means discarding the logical principle of contradiction (something cannot be X and not-X simultaneously). If a homeopathic medicine is pure water, it is contradictory to say that it has healing properties that pure water does not have. Edzard Ernst, referring to Prince Charles's support of homeopathy, said: *You can't have alternative medicine just because Prince Charles likes it, because that is not in the best interest of the patients. The quality of the research is not just bad, but dismal. It ignores harms. There is a whole shelf of rubbish being sold and that is simply unethical.*

In Hahnemann's times, medical practices were in many cases more harmful and painful than the alternative of doing nothing, which is exactly what homeopathic remedies did, which made it better in that context. With such a different view of science, it could be assumed that homeopathy practitioners would be eager to make novel predictions that could be empirically demonstrated.

This simple requirement is not met,[7,8] and it is usually the skeptics who have taken the initiative to direct the controlled studies that have been done, with results not favorable for homeopathy, which is not surprising.

The previous discussion might have been a bit complicated. Still, I wanted to show how to scientifically determine if something is true or false for a specific case. Homeopathy belongs to the pseudoscience garbage bin.

Let me also mention why the US FDA does nothing about it if they are marketed as medicines. In 1938, US Congress enacted the Federal Food, Drug and Cosmetic Act, which authorizes the FDA to assess the safety of food, drugs, and cosmetics. However, the then New York senator, and primary author and sponsor of the bill, Royal S. Copeland (1868–1938, senator from 1923 to 1938), a homeopathic physician, managed to exempt from these requirements all products listed in the *Homeopathic Pharmacopoeia* and so on up to the present. Corruption is nothing new.

6.2 The Scientific Temper

… without at least an elementary mathematical and scientific education a man remains a total stranger in the world in which he lives, a stranger in the civilization of the time that bears him. Whatever he meets in nature, or in the industrial world, either does not appeal to him at all, from his having neither eye or ear for it, or it speaks to him in a totally unintelligible language.

— Ernst Mach[9]

The only way to have real success in science, the field I'm familiar with, is to describe the evidence very carefully without regard to the way you feel it should be. If you have a theory, you must try to explain what's good and what's bad about it equally. In science, you learn a kind of standard integrity and honesty. In other fields such as business, it's different. For example, almost every advertisement you see is obviously designed, in some way or another, to fool the customer. The print that they don't want you to read is small; the statements are written in an obscure way. It is obvious to anybody that the product is not being presented in a scientific and balanced way. Therefore in the selling business, there is a lack of integrity.

— Richard Feynman[10]

We've arranged a global civilization in which the most crucial elements profoundly depend on science and technology. We have also arranged things so that almost no one understands science and technology. This is a prescription for disaster. We might get away with it for a while, but sooner or later, this combustible mixture of ignorance and power is going to blow up in our faces.

— Carl Sagan[11]

The "big" questions have not changed much from the time humans first asked them; we continue to question the origin of life, the universe, and everything (if they had one), what is good and what is not, and the nature of consciousness that allows us to ask these questions. There are questions concerning the *world's ontology* (of what there is), and *epistemology* (how we acquire knowledge about the world), how to know reality, accepting that there is such a thing. A philosophical stance, known as "solipsism," posits that your mind is all there is. I cite the eminent Bertrand Russell, who wrote[12]: *As against solipsism it is to be said, in the first place, that it is psychologically impossible to believe, and is rejected in fact even by those who mean to accept it. I once received a letter from an eminent logician, Mrs. Christine Ladd-Franklin, saying that she was a solipsist, and was surprised that there were no others. Coming from a logician and a solipsist, her surprise surprised me.* (Christine Ladd-Franklin studied at Vassar College and was accepted for graduate studies in 1878 at Johns Hopkins University, although women were not allowed; her application was for C. Ladd.) She completed the requirements for a Ph.D., but it was officially issued to her in 1927 (44 years after she had earned it) at seventy-eight.[13] In some countries, women are still not allowed to study.

However, the answers have changed, and the setting has changed. The accelerating epistemological advance of science has provided us with procedures that bring us closer to the truth. If it were not so, airplanes would not fly, antibiotics would not cure, GPS would not take to its destination, either you to the house of a friend or a missile to that of an enemy, and nuclear weapons would not explode, among many other things. Philosopher of science Peter Godfrey-Smith[14] writes: *Unless we have made some very surprising mistakes in our current science, the world we now live in is a world of electrons, chemical elements, and genes, among other things.*

Until the Enlightenment, the word "science" was used as a synonym of philosophy, and it referred to *systematic* knowledge. The English philosopher William Whewell (1794–1866) invented the word "scientist" in analogy with those who are dedicated to art, who are artists (art — artists, science — scientist) to separate what natural philosophers did from what the rest of philosophers did. Perhaps this was not a good move. Indeed, in German, science is *Wissenschaft*, (roughly the knowledge (*wissen*) ship (*shaft* — form old German). This term is much broader than our "science" and includes any type of systematic study. German also has *Naturwissenshaft* and *Geisteswissenshaft* (using the anachronistic "Geist" or soul), the first related to knowledge about nature, and the second to knowledge about almost everything else.

What is considered necessary is the systematic and scholarly investigation, be it related to matter, life, or historical events. The chasm that we invented between the humanities and science (the "two cultures") is harmful. As noted by distinguished scientist and author Lewis Thomas[15]: *We would be better off if we had never invented the terms "science" and "humanities" and then set them up as if they represented two different kinds of intellectual enterprise.*

In 1933 famous Polish and naturalized-French physicist and chemist, Marie Skłodowska Curie[16] (1867–1934) participated in a meeting of the Committee of Letters and Art of the League of Nations in Madrid on the "Future of Culture," in which some participants, including well-known scholars Paul Valéry, Gregorio Marañón, Salvador de Madariaga and Miguel de Unamuno, blamed science for the "crisis of culture." In her presentation, she explained: *I am among those who think that science has great beauty. A scientist in his laboratory is not a mere technician: he is also a child confronting natural phenomena that impress him as though they were fairy tales. We should not allow it to be believed that all scientific progress can be reduced to mechanisms, machines, gearings, even though such machinery also has its beauty. Neither do I believe that the spirit of adventure runs any risk of disappearing in our world. If I see anything vital around me, it is precisely that spirit of adventure, which seems indestructible and is akin to curiosity.*

Instead of discussing the *content* of science (this is not what is essential here), I will concentrate on what everyone should cultivate: *the scientific temper.* Understanding what is scientific and how science works will hopefully make you less prone to believe certain things that are not supported by scientific studies and others that are bullshit, and therefore should be handled with care.

"Science" is frequently used in the wrong way, often referring to something different, like technology (see below) and sometimes referring to nothing: a mere word to add authority to rubbish, as is the case for the many times we hear "scientifically proven" when it is not.

In a world necessarily and increasingly dependent on science and its technological applications, democracy can only work with people and governments that respect science, heed scientific advice, and accept that it is the best way to know.

Without this understanding, it is impossible to consider the value of sending a human expedition to Mars (for me a waste of resources), the urgent need to reduce carbon dioxide emissions, the use of nuclear energy, the effectiveness of vaccines, or to evaluate genetically modified organisms (GMO). The situation is complicated because many things are said for or against a specific issue, and the public sits in between, unable to understand. This is not due to any personal flaws but because many problems are pretty complicated to understand even by experts, and most people do not have the time or means to dig deeper and decide for themselves if a specific "scientific" claim is valid or at least plausible. Related to this is that some critical issues we face are unprecedented, work in progress, without a clear scientific consensus (yet). But once consensus is established, it is foolish not to listen. *We regard as "scientific" a method based on deep analysis of facts, theories, and views, presupposing unprejudiced, unfearingly open discussion and conclusions. The complexity and diversity of all the phenomena of modern life, the great possibilities and dangers linked with the scientific-technical revolution and with a number of social tendencies demand precisely such an approach ....*

With these words began a memorable article[17] written in 1968 by the Soviet physicist (designer of the Soviet thermonuclear bomb), political activist, and Nobel Peace laureate for 1975, Andréi Dmítrievich Sakharov (1921–1989), later human rights activist (together with his wife Yelena Bonner (1923–2011)), persecuted by the Soviet government. Sakharov proposed the following:

- *The division of mankind threatens it with destruction. Civilization is imperiled by a universal thermonuclear war, catastrophic hunger for most of mankind, stupefaction from the narcotic of "mass culture," and bureaucratized dogmatism, a spreading of mass myths that put entire peoples and continents under the power of cruel and treacherous demagogues, and destruction or degeneration from the unforeseeable consequences of swift changes in the conditions of life on our planet.*
- *The second basic thesis is that intellectual freedom is essential to human society — freedom to obtain and distribute information, freedom for open-minded and unfearingly debate, and freedom from pressure by officialdom and prejudices. Such a trinity of freedom of thought is the only guarantee against an infection of people by mass myths, which, in the hands of treacherous hypocrites and demagogues, can be transformed into bloody dictatorship. Freedom of thought is the only guarantee of the feasibility of a scientific democratic approach to politics, economy, and culture.*

Science is not, then, just a method to describe nature or a way to extend our senses. Instead, it is about formulating a *coherent explanation* of what we observe and discovering the *hidden nature of reality*, laying to rest along the way all myths that do more harm than good.

Science provides a gradually more accurate and encompassing description of a reality whose nature is independent of the theories (just models) used to describe it. It offers us world models, a map of reality, which we test through empirical confrontation. Indeed, empirical confrontation of our ideas is the only way to assure that we are not fooling ourselves, something we are prone to do. The repeatable experiment or observation is the test necessary to convince the most stubborn or demonstrate that an error or even fraud was committed.

This is because we perceive our world through our senses and in modern times through our instruments. But we should be aware that our interpretations and theories are not independent of our cognitive processes and emotions, which might bias them. Thus, it is good to subject scientific results to peer review and experimental replication by independent researchers. The following diagram illustrates the path from the "real world" to our knowledge of it (Figure 6.3). Acknowledging this process, a scientist must take care not to fall prey to these often hidden or unconscious biases.

Here is an example, Shepard's Table[18] (Stanford psychologist Roger N. Shepard Born 1929), to convince you that not everything is as you perceive it (Figure 6.4). Look at the *tabletops* and decide if they are the same size or not. Now use a piece of paper and trace one and compare it to the other. Surprised?

Our intellectual tradition, already part of Greek thought, includes at its core the idea of theory — a logical, systematic scheme — that allows us to describe and explain a large part of reality. To this was added much later by Galileo (without underestimating Archimedes and his contemporaries) the idea of examining the correspondence between what the theory predicted and results of experiments and observations. "question" nature to provisionally accept or reject our theories. A theory contains natural laws and precise definitions summarizing what we have learned about the natural world thanks to many during the last centuries.

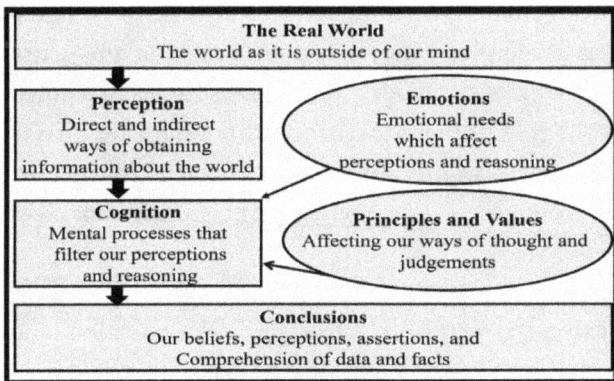

Figure 6.3. From the real world to our knowledge of the world.

Figure 6.4.   Shepard's Table.

"Theory" is used in science with a different meaning from its everyday use. When someone says: *my theory is that what happened was ...*, it refers to a hypothesis or a conjecture, often an opinion based on little evidence. Sometimes (particularly with the theory of natural selection), some say that "it is *merely* a theory." But the theory of evolution, the theory of relativity, or quantum theory, are much more than "*merely* a theory." They constitute the best map or model of reality that we have.

Our intellectual tradition, with its Greek roots, rests on a few fundamental principles. They have evolved over the centuries, and the canons of what we consider legitimate sources of knowledge have changed — the role of sacred writings, the weight we give to mystical visions, and the sense of the "supernatural" have changed (in decreasing importance) — but the essence of the principles has not changed. I list them following Searle.[19]

- Reality exists independently of human representation. Although we have mental and linguistic representations of the world in the form of beliefs, experiences, and theories, a world "out there" exists independent of these representations, even though it may be impossible to separate reality from its representation. There are social constructions (money, marriage, race, property, etc.) that are real but lose their reality if human beings disappear, but not so for the elliptical orbit of a planet, which will continue to be so with or without us. This principle can be summed up as one of *metaphysical modesty.*

- One of the functions of language is to communicate meaning (semantics) that allow sharing ideas that refer to the state of things and objects that exist independently of language. Language (including mathematics) allows us to represent the world and elaborate theories to describe and explain reality coherently. Despite its independence, we can formulate correct theories of the observable world (whether we have achieved it is a different matter), which we can summarize as *epistemic arrogance* that clashes with metaphysical modesty.

- The truth is a question of the fidelity of the representation. A proposition will be true or false depending on whether the things of the world really are as it is expressed. In other words, the proposition: "the Sun is at 150 million km from Earth" will be true if the Sun actually is 150 million km from Earth. (Although we know that it is an approximation). This is the idea of *truth as correspondence,* where we define "truth" in the following way: a proposition is true if and only if it corresponds to the facts.

- Knowledge is objective. By this, we mean that since what we know is true, by definition, it does not depend on the knower or on time or place. Moreover, since the truth is the fidelity of representation of an independent reality, it cannot be contingent on the knower's character, motivation, gender, and circumstances.

- Logic and rationality are formal, meaning that what matters is the form of the arguments and not their content. Rationality implemented by logic provides procedures, norms, and methods that evaluate competing affirmations, but it does not propose anything about the world. It does not indicate, by itself, what to believe or do. Logic is necessary but not sufficient to understand the world since the world consists of concrete and material things, not concepts. As rationality is formal, it cannot be refuted since it does not affirm anything refutable.

To argue that some reasoning is relative, we must do so from a firm platform, and the formal rules of reason that operate on this platform are general and universal, at least for human beings, who are the ones that matter. Moreover, without a starting point of universal objective principles on which to base our understanding of the world and the justifications of our actions, we would not be able to understand anything.

The difficulty lies, as philosopher Thomas Nagel[20] points out, in that: *To be rational we have to take responsibility for our thoughts while denying that they are just expressions of our point of view. The difficulty is to form a conception of ourselves that makes sense of this claim.*

With time we have adjusted or replaced our models to better fit new empirical data. As a result, we have made predictions that have been later confirmed (such as the recent discovery of the Higgs boson or more recent detection of gravitational waves produced by the interaction of distant black holes). In this sense, we think that we approach the truth of the world, which will always be tentative, but it is the best we have. This is because science, although conservative, is inherently self-correcting, and that is crucial for its advance. Quite different from that which is accepted by dogma, such as the idea that humans were created by a special divine act, "creationism" or the better sounding "intelligent design," which are nevertheless the same idea shared, according to surveys, by about half of the US population. This is alarming since it shows the level of scientific understanding in the country with some of the best universities and the largest military budget on the planet.

Unfortunately, in some areas, especially those related to human studies and due to a variety of factors, including the statistical analysis of data, some results which have been published turn out not to reproduce, and this only adds to widespread distrust of science.

This is why many ideas, especially some that occur in the so-called field of "alternative medicine," can be rejected since they are not based on an empirical test that can determine their effectiveness. Without this test, they are useless. In some cases, such as homeopathy or acupuncture, they are not effective even in principle, no matter how many people supply testimony. However, we also understand why despite this, it "works" for some.

We know things about the world that start with measurements or observations (data) interpreted according to what we wish to study. There are different kinds of data, more or less precise, that must be checked against other measurements or by various independent researchers (often in different countries) to make sure that there were no mistakes made and that the datum indicates that which it claims. For example, it would be foolish to claim that the planet is warming because a thermometer in your garden has shown an increase in average temperature (whichever way you take that average) over a period. On the other hand, if you wish to know if your baby has a fever, one measurement is enough (assuming a good thermometer, and if you are unsure, you can check with a second thermometer). Experimental scientists have been preparing for years to measure things the correct way. It is annoying when unprepared persons criticize scientific results and outright dangerous when politicians, decision-makers, and "influencers" do it.

The practice of science (formulating hypotheses, deducing consequences, contrasting with empirical data, etc.) is not very different from what we use during a criminal investigation or in finding fault with a car's engine. The difference lies in the rigor observed, the precision of language and the measurements, and the excellence of the test, combined with a coherent theory capable of explaining the results. Figure 6.5 illustrates the general relationship between our empirical findings and our theories or hypothesis. The formulation of a hypothesis is the essential creative step to explain the data, and this is known as an "abduction" (often an intelligent guess) or an "inference to the best explanation." But once a hypothesis or model is formulated, it must stick out its neck, predicting *new phenomena* (otherwise, it is no good), which will allow for a confrontation of new data with the predictions.

Without an agreement, the theory or hypothesis can be rejected (falsified) or needs modification. If there is an agreement, the theory is provisionally accepted. Julian Huxley (1887–1975)[21] wrote about the often misunderstood concept of "Natural Law": "*A law of Nature is not (and I wonder how often this fallacy has been exploded, only to reappear next day) — a law of Nature is not something revealed, not something absolute,*

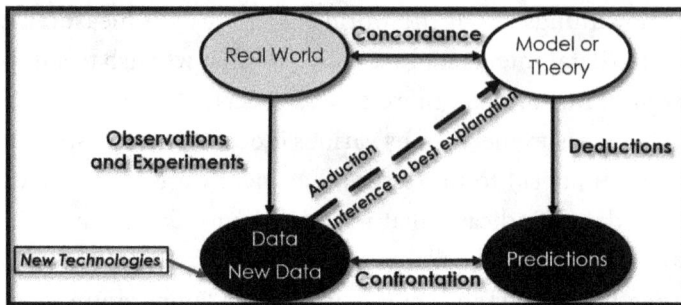

Figure 6.5. The empirical cycle.

not something imposed on phenomena from without or from above; it is no more and no less than a summing-up, in generalized form, of our own observations of phenomena; it is an epitome of fact from which we can draw several conclusions."

I will provide you with an example that will be useful later. In 1900 the eminent German physicist Max Planck (1858–1947), 1918 Nobel laureate for Physics, discovered a law that holds for a body at a specific temperature. Because it emits (thermal) energy (electromagnetic radiation), the intensity follows a very well-defined range of wavelengths (or frequencies) described by a precise mathematical formula. It is at the foundation of quantum mechanics and resolved a problem with classical theory. Figure 6.6 illustrates this.

It is called "black body" radiation, but we need not get into semantics. Suffice it to say that the Sun, for example, emits approximately as a black body at a temperature of about 5,500°C. Therefore, most of its energy is emitted in a narrow range of wavelengths that we call visible light (for obvious reasons), but with some invisible ultraviolet and infrared radiation. On the other hand, a much colder object like the Earth, at an average temperature of about 15°C (59°F) emits primarily infrared radiation. At a normal temperature of 37°C (98.6°F), your body also radiates in the infrared (but at shorter wavelengths than the Earth), which is why you can see people in the dark with special infrared goggles, especially since at night the earth is colder. This is also why your electric stovetop gets "red hot" as the temperature increases.

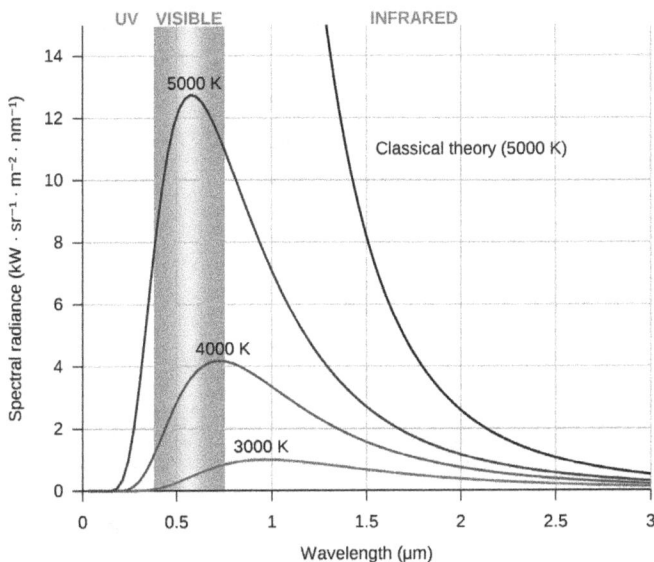

Figure 6.6. The spectrum of black body radiation. Intensity on the vertical axis against wavelength.

Figure 6.7 shows the black body distribution of energy for the Earth and the Sun (note that the Sun has been scaled vertically by a factor of one million to fit). Visible light is in the wavelength range of about 0.4 (violet) to 0.7 (red) μm (millionths of a meter), and the maximum is emitted at 0.5. The maximum energy radiated by Earth is at about 10 μm.

The question about what is and what is not, possible or impossible, is not limited to science. Searching for truth is cardinal in all aspects of human endeavor. In his book, *On Tyranny*, Timothy Snyder writes: *To abandon facts is to abandon freedom. If nothing is true, then no one can criticize power because there is no basis upon which to do so. If nothing is true, then all is a spectacle. The biggest wallet pays for the most blinding lights.*

The scientific enterprise is a social activity structured to be subjected to critical review by the scientific community (peer review). The results, having passed this process, are incorporated into "public knowledge." Finally, the consensus is established by agreement between the *experts*.

We can think of the scientific activity as the construction of a community puzzle. All scientists have access to the parts of the puzzle that

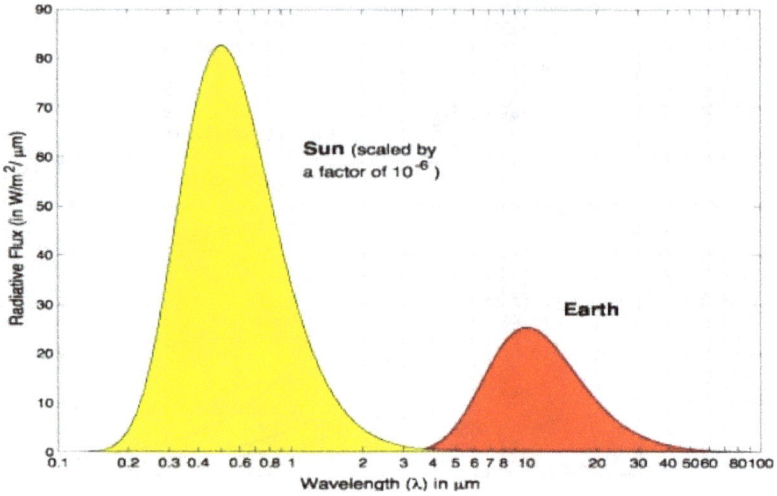

Figure 6.7. Emission from Sun and Earth.

are already assembled, it is public, and they can surmise the big picture and contribute new pieces if they find them. They know what others have contributed and where a lot of activity takes place, perhaps caused by a unique piece that serves to clarify the picture. They are willing to test whether a piece fits perfectly in a particular part of the puzzle. Those scientists' experts in that puzzle area examine the new piece in detail to determine whether it is acceptable. Each one works independently or in research teams, and the puzzle acts as a coordinator of their activities. Consensus is an essential ingredient of the scientific enterprise but does not *guarantee* results, and past consensus has proven wrong, but it is the best option we have.

In her book, Princeton professor of the history of science Naomi Oreskes writes[22]: *a crucial aspect of this process is revision; most peer-review papers are revised many times prior to publication. both informally as preliminary results are presented at conferences and workshops and drafts are sent to colleagues for comment, and then formally through editorial peer review. Papers are then revised in response to reviewers' suggestions of clarification and corrections. If errors are detected after publication journals may issue errata or retractions.*

This applies mainly to scientific research pursued in an academic setting where researchers strive to publish their results since their academic careers and prestige depend on it ("publish or perish"). But there is also a more somber reality where results are not published, at least not ASAP, and for everyone to see because they respond to military or industrial interests and are kept secret. Yes, the results must be validated, but in these cases, it is done in secrecy until someone blows the whistle or a spy gets to them. A famous case relates to the research that led to nuclear weapons, and as I will tell you later, research done by the oil industry about global warming.

Be that as it may, we should recognize that if the experts agree, it is reasonable to accept their judgment, even if there are some dissidents and occasionally the experts may be wrong. Science is inherently an international and democratic activity. But there is a final authority: nature. This implies that if there is no evidence for something, then science cannot study it.

And if there is no clear consensus, as is the case with some frontier science issues (which are *not* biological evolution or climate science), then the most reasonable attitude is to suspend judgment until better evidence is obtained. This, of course, assumes that the experts are objective and independent, which is not always the case. Nevertheless, an expert does know, and the non-expert should listen to her advice (making sure that the expert truly is an expert), something that is happening less as people think that what they see on YouTube makes them just as expert. This consensus process is exceptional, independent of national borders, political ideologies, and religious affiliations. However, in the real world, this becomes sometimes corrupted for the simple reason that scientists are human, after all. But after a while (sometimes a long one), truth prevails.

A long time ago, the US philosopher, logician, and scientist Charles Sanders Peirce (1839–1914) had this to say about the truth[27]: *Different minds may set out with the most antagonistic views, but the progress of investigation carries them by a force outside of themselves to one and the same conclusion. This activity of thought by which we are carried, not where we wish, but to a foreordained goal, is like the operation of destiny. No modification of the point of view taken, no selection of other facts for study, no natural bent of mind even, can enable a man to escape the predestinate opinion. This great hope is embodied in the conception of truth and reality. The opinion which is fated to be ultimately agreed to by all who investigate is what we mean by the truth, and the object represented in this opinion is the real.* (The caveat here is "by all who investigate".)

Ultimately, it is about determining which intellectual habits deserve respect and which do not, which are the paths that lead to knowledge or to intellectual ruin.

Lee McIntyre, in his excellent book, writes[23]: *No matter the biases beliefs or petty agendas that may be put forward by individual scientists, science is more objective than the sum of its individual practitioners.* Lewis Thomas writes[24]: *We have to be sure learned enough to know better than to say some things, about letters and about science but we are too reticent about our ignorance. Most things in the world are unsettling and bewildering, and it is a mistake to try to explain them away; they are there for marveling at and wondering at, and we should be doing more of this.*

This situation differs from other disciplines, where different "schools" of thought often arise, and even more in politics and religion, where the lack of consensus has often led to bloody murder. Science can claim precedence in terms of knowledge about the world we live in (epistemological privilege), what the facts are and put them to the test. It is more challenging or even impossible to judge the correctness of answers related to social and ethical questions and how we behave because the answers go beyond the concept of truth. Scientific epistemology is one thing, and social epistemology is another. The second one has become more complex with the advent of the ever-increasing availability of random or selected bits of information, and I shall look into this towards the end.

Science has transformed our world view (or could since most people are unaware, uninterested, or even opposed), and technology has transformed the world. The unquestionable success of science arises because it accepts only rational explanations with a high dose of empirical therapy. It offers us the possibility of discarding the yoke of superstition, fantasy, and illusion, even if fantasy and illusion are welcome and even necessary in other areas of human endeavor.

Bertrand Russell tells us[25]: *It is undesirable to believe a proposition when there is no ground whatever for supposing it true.*

Accepting invalid arguments, false premises, and erroneous data has consequences ranging from innocuous to disastrous. So, if you think that putting on a quartz necklace protects you from "bad vibes," this is relatively innocuous, and if it makes you feel better, if it helps your self-esteem, then there is no severe problem. On the other hand, if you accept the testimonies of others and believe that going to Medjugorje (or wherever) is going to miraculously cure you of a disease and therefore do not seek medical help in time, the consequences could be deadly. For obvious reasons, those who prayed for a cure and were not cured do not offer testimony.

In some circles, the entire scientific enterprise has been questioned, asserting that science is a social construction, that scientific knowledge is relative to the environment in which it arises, and that there is no single norm for the justification of our beliefs. This has caused in certain circles "science wars," with discussions that, although often got out of hand, also served for a serious reexamining of the scientific enterprise.

Part of a philosophical discussion has been misunderstood and used for all kinds of arguments to justify pseudoscience and reject science. Suffice it to quote from a book's introduction: *Supernature* to see how this argument is used as a license for nonsense[26]: *Science no longer holds any absolute truths. Even the discipline of physics, whose laws once went unchallenged, has had to submit to the indignity of an Uncertainty Principle. In this climate of disbelief, we have begun to doubt even fundamental propositions and the old distinction between natural and supernatural has become meaningless.*

Strangely, I agree in terms of the natural and supernatural but just in the opposite sense to what the author wishes to imply. The first part is nonsense, science never held absolute truths (although some are well established), and the uncertainty principle has nothing to do with dignity. In general, the mention of quantum mechanics and the uncertainty principle is common in this type of text. The authors demonstrate their ignorance of quantum mechanics and assume (correctly) that the reading public will not notice.

Scientific arguments are open to discussion and criticism, and some may be (and have been) proven wrong (by scientists themselves) with arguments based on evidence. A further and important aspect of scientific discourse, really of any discourse, is the requirement that the terms used be clearly defined. Otherwise, it is impossible to formulate a precise idea. Many arguments lead someone to say: *Well, that depends on what you mean by* .... Both sides will outline arguments for or against some issue, but they pass by without meeting, like ships in the night. For example, I have witnessed many discussions regarding the existence of God, with neither party trying to define the term (not easy, by the way), generating a lot of heat and little light. Some say something like: *God is love*, disarming any argument about existence since even atheists fall in love.

Let us be clear that *Homo sapiens* came first and then came the word. That we have a word for something does not necessarily mean that it exists as something concrete, and many words refer to powerful fictions that we have invented and that define and dominate culture. When the daily use of a term gives what it relates to an unjustified reality, we commit the fallacy of misplaced concreteness, or "reification." This fallacy also occurs as personification, attributing human qualities and intentions to things that cannot have these attributes. For example, *The Universe will not allow the human being to disappear* attributes to the Universe the capacity of an intention, and *US. has decided not to participate in the Kyoto agreement* attributes to a country the ability to decide (when those who make decisions are people — generally few). As a rhetorical device, it can work well: *After two hours of political platitudes, everyone grew bored. The delegates were bored; the guests were bored; the speaker himself was bored. Even the chairs were bored* (more about fallacies below).

It is common in pseudoscience and pseudo-medicine to misuse words that have been carefully defined by science. Thus, we get into "technobabble" with terms such as "electromagnetic field," "vibrations," "fifth dimension," "energy," or "quantum effect," used without respecting their definition.

One of the most eminent physicists of all times: James Clerk Maxwell (1831–1879) had already commented on this in 1871 in his inaugural lecture at King's College: *Such indeed is the respect paid to science that the most absurd opinions may become current, provided they are expressed in language the sound of which recalls some well-known scientific phrase.*

Science does not know everything, but it knows more today than it did in the past and will know even more in the future if we get there. But, also, when a scientific result or even a whole theory (as it has happened historically) is discarded, this in no way implies that alternative theories (if any) become true. As some argue, the veracity of creationism (in its various manifestations) does not depend on natural selection (Darwin's theory of evolution) being wrong. Evolution is a fact (no alternative), and the best explanation of this fact is natural selection, as we shall see.

Scientists work under very competitive conditions, and there are political and economic interests that influence scientific enterprise. Sometimes scientists cheat or get paid handsomely to deny results which is inconvenient for some. But from this, it cannot be concluded that facts and scientific theories are social constructions independent of empirical reality, as some sociologists of science have proposed.

On a planet where we were condemned (in myth and reality) not to understand each other: *And the Lord came down to see the city and the tower, which the children of men build. And the Lord said, Behold, the people is one, and they have all one language; and this they begin to do: and now nothing will be restrained from them, which they have imagined doing. Go to, let us go down, and there confound their language, that they may not understand one another's speech.*[28] (nice huh,). Science offers the only universal language that allows people of very different cultural backgrounds to understand each other, thus overcoming the tragedy of Babel.

Unfortunately, the representation of scientists in politics is almost nil, and a high fraction of those who govern have legal backgrounds. (German chancellor Angela Merkel holds a Ph.D. in Physics.) As a result, they are ill-prepared to deal with the severe and complex problems we face since these are not related to human laws but natural laws. And the record shows their ignorance.[29] We must ensure that those who act in politics adhere to the highest intellectual and ethical standards, respect science, and do not politicize it, as is happening. Delegitimizing science when we increasingly need it is alarming, witness all the bullshit we have heard related to COVID-19 and the heavy toll this has taken. The filter of critical judgment and reason must be the first one through which we decant our ideas and analyze our beliefs, and then follow with a distillation considering ethics, justice, and the lessons of history.

Science discovers the profound and extraordinary beauty of nature. Along with other cultural expressions that exalt our existence, literature, music, and other expressive arts, science is part of our culture and gives us a unique way of knowing. Arguably, this is its most significant contribution to humanity and not its material consequences that we so much praise. Science has given us a new vision of the world in which we live, one that has liberated the human spirit, or at least has the potential to do so, despite all the mobile phones.

Science has discovered the immensity of the universe, the unimaginable extension of time, the weird behavior of the microscopic, the fundamental structures that make us alive, and the brotherhood of all living beings on the planet. This has a very high value and goes beyond practical considerations, meeting a need that arises from the fact that we are thinking creatures.

I summarize three descriptive characteristics of science:

- **Empiricism:** The source of knowledge (not exclusive to science) is experience. At least some of the terms and propositions of a scientific theory refer to objects, events, and facts that are obtained from experience. Scientific experience contrasts with everyday experience because experiments and observations are carried out under carefully designed, controlled, and specific conditions so that the experience can be replicated by others. How to go from experience to knowledge is the fundamental problem of epistemology.
- **Mathematization:** Science is characterized by using, wherever possible, mathematical tools to investigate and analyze empirical data and to formulate and develop its theories in a rigorous way. But the use of mathematics does not define science, nor is it necessary to do science. However, it contributes to its precision and rigor, theoretical development in some highly mathematized areas, and the correct handling of empirical data (through statistics). In fact, the most important theory in biology — natural selection — was formulated in its early days without

mathematics, although many subsequent mathematical developments have been of great importance.

- **Social Structure:** Science is a social activity structured in a network of cooperation, competition, and trust and critique, leading to results accepted by consensus. Scientific activity is cumulative despite revolutionary events. The rules of thought are carefully followed so as to make inferences that produce the best explanation of what it is desired to understand. Originality and novelty are encouraged, and although there is a healthy resistance to change, orthodoxy is not dogmatically upheld.

Although scientific studies have many fundamental aspects in common, starting with critical thinking, they also have their differences depending on the subject matter of study and on the canons accepted by each discipline. Thus, for example, Astronomy is not an experimental science (in the strict sense of what an experiment is) since it is impossible to control the observed events' boundary conditions. Still, the theoretical ideas of Astrophysics are empirically testable by observations. Some scientific disciplines lend themselves to reductionist analysis, while others are integrative and still others are historical. There is a comprehensive and mathematical theoretical framework for some fields (Physics and Chemistry — which is applied physics) that allows theoretical developments and the formulation of universal laws that lead to testable predictions. In others (Biology), the theoretical base is less broad, it is not clear what its universal laws are (since biological events are contingent), and mathematization is only partial. These differences mean that we find differences in how scientific research is carried out in each area. Some are more empirical while others are more theoretical and sometimes the same discipline is divided into applied and theoretical. There is no one scientific method characteristic of science in general — a widespread belief — but each scientific discipline has developed its own canons of research and validation. Above all, however, there is a critical coherence because no scientific discipline assumes something that contradicts what is known in another. For example, an explanation of a biological process that violates the laws of thermodynamics would not be accepted.

As already said by Francis Bacon[30]: *Cientia potentia est,* perhaps inspired by: *A wise man is strong; yea, a man of knowledge increaseth strength* of the Bible.[31] Scientific knowledge provides us, through technology, power, which has allowed us to reach the Moon, communicate instantly with the whole world (increasingly to say less), generate nuclear energy (either for civil or military uses), fight diseases, and modify plants for purposes that can be debated.

Science has a dual purpose: to explain and to predict. Predicting responds to the question: How? Explaining responds to the more difficult question: Why? We want to know how a projectile flies to predict where it will land, but this is not enough. We also want to know the reasons why it flies that way and not another. If it were only about predicting, then any formulation that did it properly (within acceptable errors) would be admissible, a position known as "instrumentalism." The history of science teaches us that instrumentalism is unsatisfactory. The historical struggle between the Copernican and the Ptolemaic system is a good example. From an instrumental point of view, both systems (at that time) produced similar predictions, but the dispute went beyond instrumentalism, seeking the "true" reason for the phenomena.

Many value science, above all for its intellectual content — knowledge for its own sake, but there has always been a pragmatic purpose. Plato himself valued knowledge for its transformative action on the knower. In the Middle Ages, the Church valued astronomy (despite its brief conflict with Copernicanism) because it needed to determine the exact date of Easter. (The Sunday following the first full moon of spring in the northern hemisphere.) Undoubtedly in modern times, an increasingly important motivator of scientific research is the technology that it can foster.

I present you with an astonishing result that illustrates the empirical cycle. I discussed above how Max Planck in 1900 derived a mathematical formula to precisely describe black body radiation. In 1989 the COsmic Background Explorer (COBE) satellite was launched by NASA to study the remnant radiation from the big bang. Figure 6.8 shows the measurements superimposed on a theoretical Planck curve for a temperature of 2.725 K. Theory and data agree precisely! It is a striking example of the great power of physics and mathematics to describe nature. It is possible to understand what happens in distant places of the universe, with the laws of nature discovered in our terrestrial laboratories.

I close by quoting the great science communicator, Carl Sagan, who 25 years ago wrote as if he were a psychic[32] (no, there aren't any): *Science is more than a body of knowledge; it is a way of thinking. I have a foreboding of an America in my children's or grandchildren's time — when the United States is a service and information economy; when nearly all the key manufacturing industries have slipped away to other countries; when awesome technological powers are in the hands of a very few, and no one representing the public interest can even grasp the issues; when the people have lost the ability to set their own agendas or knowledgeably question those in authority; when, clutching our crystals and nervously consulting our horoscopes, our critical faculties in decline, unable to distinguish between what feels good and what's true, we slide, almost without noticing, back into superstition and darkness. The dumbing down of America is most evident in the slow decay of substantive content in the enormously influential media, the 30 second sound bites (now down to 10 seconds or less), lowest common denominator programming, credulous presentations*

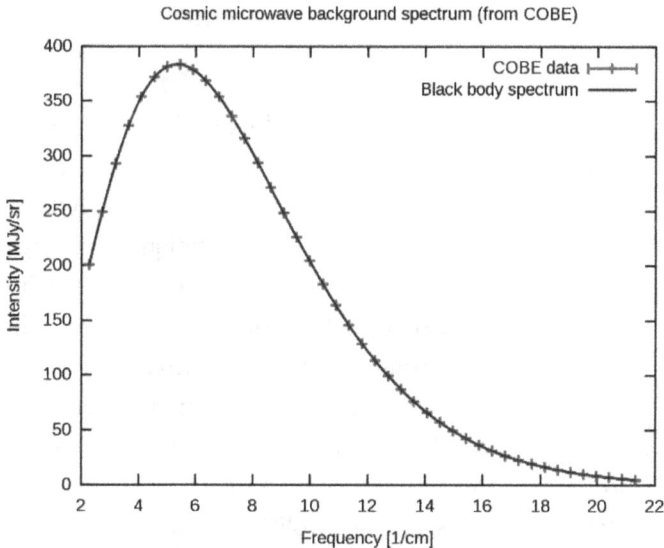

Source: NASA/COBE Science Team.

Figure 6.8. COBE black body measurements and theoretical curve.

on pseudoscience and superstition, but especially a kind of celebration of ignorance. I cannot say it any better.

6.3 Technology and the Golem

Technology always has unforeseen consequences, and it is not always clear, initially, who or what will win, or what will lose. ...Gutenberg thought his invention would advance the cause of the Holy Roman See, whereas, in fact, it turned out to bring a revolution which destroyed the monopoly of the Church.

— Neil Postman[33]

They blame technology for what they consider the failings of industrial civilization, instead of blaming a society which while responsible for the development of technology is often unwilling to control its application and unable to foresee the consequences of its massive deployment.

— Julius Lukasiewicz[34]

Pure science does not remain pure indefinitely. Sooner or later, it is apt to turn into applied science and finally into technology. Theory modulates into industrial practice, knowledge becomes power, formulas and laboratory experiments undergo a metamorphosis, and emerge as the H-bomb.

— Aldous Huxley[35]

Aristotle already differentiated between the knowledge obtained by reason he called *episteme,* which had no useful purpose, and practical knowledge acquired from experience, which he called *techne* (which gives rise to our term "technology"). Before technology, we had "technique," applying practical knowledge obtained by trial and error to create useful artifacts (such as the ax, the water wheel, or the steam engine). Technology refers to the application of scientific knowledge to build artifacts. (A horse-drawn carriage can be created by technique, but it requires technology to build a modern car.)

As the search for knowledge, science is of neutral ethical value. This, with the restraint that the motivation, planning, and financial support of science are determined by social and political forces that are not neutral. (In the US, to a large extent co-opted by military and industrial interests.) But the knowledge obtained is objective (its truth value does not depend on who proposes it) and neutral. However, technology is not neutral since somebody *decides* how to use knowledge, leading to morally acceptable or objectionable purposes. Scientific knowledge belongs to humanity, but technology belongs to national or industrial interests and can be patented.

Albert Einstein discovered a straightforward relationship, a consequence of his theory of relativity, which expresses the equivalence between mass and energy, the famous $E = mc^2$. This relationship is fulfilled independently of borders, cultures, or political systems and does not possess a moral value. The field of nuclear energy that leads either to the design and manufacture of terrible weapons (see below) or to its peaceful use, is a consequence of this discovery. The decisions regarding the use of nuclear energy are of tremendous ethical import and are crucial for the future of humanity in the face of the energy crisis that is looming because of the damage our current energy generation system inflicts on the atmosphere and biosphere.

The boundary between science and technology is somewhat blurred, and science itself is nourished by technology in a productive feedback loop. For example, consider the revolutionary discovery that nature has discontinuous fundamental characteristics (quantum mechanics). The energy levels of atoms and molecules and their interactions with electromagnetic waves occur with discontinuous (quantized) energy values. This is at the foundation of spectroscopy, a tool of great importance contributing to many new scientific discoveries, including the recent discovery of extrasolar planets.

The discoveries about the atomic structure of matter and its interactions with light also led to the transistor, LEDs, and the laser, significant inventions for modern technology. These new technologies, including processing and storing information embodied in computers and communication networks, facilitated new scientific discoveries. Indeed, without modern instruments and computers, it would be difficult, if not impossible, to discover and less understand the phenomena of ozone depletion or climate change. But on the other hand, these same technologies allow for increased control of societies and endanger democracy.

The result of the technological development of the last centuries can be put on the balance, especially the accelerated growth of the previous decades. It is reasonable to question if it contributed more to violence than to peace and if, in the end, it did more harm than good, and I do not doubt that both sides will have their arguments, just because technology is a *double-edged sword*. But the edge we use depends on us (primarily non-scientists) and entails an ethical decision. We discovered electricity and used it to light the darkness or darken a soul; you choose. We learned to genetically modify living things to create modified rice which helps the body to produce vitamin A (golden rice) therefore saving the young from blindness.

Vaccines resulting from genetically modified viruses (or the recent mRNA vaccines) or glyphosate (roundup) ready plants cause discord. You choose, but each must be evaluated on its own merits, risks, and benefits. The problem: hindsight is 20–20, whereas foresight is almost blind.

John Boyd Dunlop (1840–1921) invented the pneumatic (inflatable) tire (initially for his son's tricycle in 1887) and could not have imagined that the quest for rubber would lead to genocides in the Congo (see below). Henry Ford could not have foreseen (no one could) what the massive industrial production of automobiles would lead to; he did not imagine all the adverse effects of this individualized form of transportation. It is unnecessary to make a long list; suffice it to say that it is the tenth cause of death worldwide (the first: coronary heart disease) and the first one not related to a disease. (And I am not even thinking of atmospheric pollution.) But we are already on this path. There seems to be no easy alternative to the automobile and all the related activities (gas stations, mechanics shops, battery, tire sales, etc.) that form a complex, interrelated system. The car (and the rest of transport vehicles: trucks, airplanes, and ships) contribute significantly to the CO_2 that accumulates in the atmosphere. More recently, in 1989, the inventor of the World Wide Web (Sir Tim Berners-Lee, a British computer scientist) could not have imagined the social and cultural transformation this invention is causing. Renowned academic Neil Postman[36] (1931–2003) wrote about the surrender of culture to technology.

An industrial society based on technology imposes an implicit ideology that subordinates humans to industrial needs; human beings are dehumanized, becoming "human capital," exploiting the world to cover its requirements moved by its own dynamics.

An example of "own dynamics," in this case minor, is the use of the "QWERTY" keyboard, which, compared to other configurations (Dvorak), is inefficient.[37] It has been estimated that a person's fingers travel a significantly greater distance on the Qwerty keyboard than on a Dvorak keyboard. But this technology is already implemented on such a scale that it is difficult to change. (Qwerty was designed to be slow to avoid jamming the mechanism of early mechanical typewriters.) Something similar happened with a switch to a metric system, talked about for years in the US, but it will not happen. So we are slaves to what was already done in economics, technology, bureaucracies, or evolution.

The character of the scientific enterprise has changed in important aspects beginning in the second half of the last century with the Manhattan Project for the construction of a nuclear bomb. (A consequence of the discovery of nuclear fission in pre-war Germany by Lise Meitner and Otto Hahn in 1938 — Lise had to flee because she was Jewish.)

By the way, in this case, it was a blessing that Hitler was so ignorant of science as to believe that there was a difference between Jewish science and German science. He replied to Max Planck (1858–1947, Nobel Prize in Physics 1928) as he pleaded with him not to expel Jewish scientists with[38]: *If science cannot do without Jews, then, we'll have to do without science for a few years.* Several of those who had to flee were important for the Manhattan Project. (After the war, the US and the Soviet Union competed to locate German scientists and technicians, regardless of their war activities,[39] to work for them. Wernher von Braun being one of thousands — so much for ethics.)

The cost of modern particle accelerators, gigantic telescopes, large brain activity trackers, and multinational consortiums organized to decipher the human genome or molecular structures is very high and out of reach for individual researchers. The era in which a monk in an isolated monastery (Gregor Mendel (1822–1884)) could make an essential contribution to science has passed (this does not apply to theoretical work). Inevitably, competition for state funds has changed the character of science, and the relationship between science and other academic disciplines — the humanities and the social sciences (where science is often not present) — has also changed.

The priority of scientific research is not independent of the political, social, and economic context and generally responds to the interests of those in power. In the case of research funded by the private sector, short-term practical results are sought to benefit investors through patents and new products. Among the most prominent investors in research and development, we find corporations such as Apple, Google, Microsoft, Amazon, and Facebook (the big five of the US).

In the case of support by government agencies, the direction of research is increasingly determined by what is considered national interests and military needs (with some interventions by religious groups). Instead of pursuing research to understand and solve the many problems we face and thus improve everyone's lives, it is increasingly directed at finding better ways to kill, in what US historian of science Clifford Conner (born 1941) calls[40] "The Tragedy of American Science." The NSF, an independent agency of the United States government that supports fundamental research and education in all the non-medical fields of science, has a paltry $8.3 billion budget (2020). NASA's budget is $22.6 billion, and that of the National Institutes of Health (NIH), supporting medical research, is $42 billion. Conner writes: *In 2017, 48.5% of federal R&D spending went to the Department of Defense alone, and that doesn't include the vast expenditures on nuclear research in the department of energy's budget.[…] Imagine, by contrast, what could be accomplished if all of that money and all of that scientific talent were instead directed toward finding solutions to the crucial problems facing the human race today — poverty, hunger, disease, and environmental devastation. But they are not, and if that isn't a tragedy, the word holds no meaning.*

Scientists no longer follow their curiosity exclusively but pursue the funds. There is a moral decision that must be made (if at all) not different from that made by the one who accepts a job in an armament factory. The ideal of disinterested science and scientific knowledge as the heritage of humanity is rapidly fading.

Some years ago, University of Texas professor Lloyd Dumas wrote[41]: *Our brilliant technological accomplishments have made us too complacent, too arrogant about our ability to control even the most dangerous technologies we create and permanently avoid disaster. But we will not because we cannot. We humans are fallible. We are not perfect and never will be. Understanding the many-sided nature of our unavoidable fallibility and how it interacts with the most dangerous technologies we create is the key to the fundamental change that can lead us away from disaster.* Nobody has heeded his warning if they have read it.

The figure of the Golem,[42] a being of medieval Jewish mythology, a mud humanoid who, through invocations, became alive and followed the orders of its master protecting him, serves as an analogy to technology. On his forehead was engraved the word "emet," which means truth, but he was ignorant. Every day, Golem's strength increased, and without strict control, it could become dangerous. To kill Golem, one had to remove the first letter from his forehead (not easy) to read "met," which means dead. It is a kind of proto-Frankenstein, a precursor of Hulk. It occurs to me that nuclear weapons are very real Golems.

But lest you think that I am arguing against technology, let me say the following: Science and technology have led to the partial conquest of many fatal diseases a 100 years ago that are now cured in a couple of weeks if the medicine is available. In developed societies (to label them somehow), children's death rate dropped significantly, and life expectancy increased with medical and public health advances in the last two centuries. Not long ago, if you cut yourself when shaving in the morning, you could die the next day because of an infection. Existence was precarious in a world in which it was not known what caused an infection. If you were infected with *Streptococcus*, *Staphylococcus*, or *Pneumococcus* (the "coccus" comes from their spherical shape) and dozens of other diseases caused by bacteria, the probability of being cured was low. In the Middle Ages, the European population was decimated by typhus outbreaks, and syphilis and gonorrhea were causing havoc. Plague pandemics constantly threatened, causing millions of deaths in the fourteenth century, in some parts killing over half the population (caused by the bacterium *yersinia pestis*). At that time, antibiotics and vaccines were unknown, and people searched for reasons and scapegoats, finding them in foreigners and minorities (Jews, of course, were a primary target).

Bacteria played an important role in many historical events. In 1812, Napoleon suffered his great defeat in Moscow, succumbing to the cold, hunger, and typhus transmitted by lice. His great army of 650,000 men became a miserable caravan of 90,000 weak and sick soldiers who, in their retreat, infected the inhabitants of the places through which they passed or died. The typhus epidemic killed hundreds of thousands of civilians. More French soldiers died of typhus than were killed by Russian soldiers.

Together with the superior armament of the conquerors, bacteria caused the catastrophe faced by the indigenous populations of America (the continent). A century after the arrival of Christopher Columbus, the populations succumbed to various epidemics caused by pathogens unknown until then in America and against which indigenous people lacked immunity. In Mexico, it is estimated that the local population descended from some 20 million (a very uncertain number) when Hernán Cortés arrived on its shores in 1519 to 1.5 million 100 years later. The "propagators of the faith" were rather propagators of disease that contributed to the extermination of indigenous peoples.[43]

All this changed when in 1928, the Scottish bacteriologist Alexander Fleming observed that the fungus *penicillium notatum* killed a culture of *Staphylococcus aureus*. He had discovered penicillin, a toxin that attacks bacteria, and started a new era of scientific medicine. With the discovery of antibiotics, today, almost nobody dies after being cut while shaving. (Unless the barber cuts your throat in an attack of irrationality.)

But, slowly, especially in hospitals where penicillin was used, strains of *staphylococcus aureus* resistant to penicillin began to appear. What happened was that by exposing a population of bacteria to an antibiotic, those vulnerable died. Due to the variability in natural populations, some resisted the antibiotic, and when they found themselves without their natural competitors, eliminated by the antibiotic, they multiplied. An excellent example of evolution at work. Today, strains of *staphylococcus aureus* resistant to multiple antibiotics populate hospitals, sites in which, by necessity, many antibiotics are used. Without an effective antibiotic, the mortality caused by these infections will increase. It is a significant problem and difficult to control. Something similar is happening with new strains of COVID-19.

In 1954 two million pounds of antibiotics were produced in the US. Today it is tens of millions. They are miraculous drugs, a scientific triumph of great magnitude, a treasure to save lives that we are regrettably squandering. Unfortunately, more than half are used in animal husbandry (pigs, chickens, cows) in a prophylactic and non-therapeutic manner, creating a problem that endangers our health. The use of antibiotics in animal husbandry is abused to promote their growth and produce cheap meat in industrial quantities and compensate for the poor hygienic conditions. It would be much better in the long term to reform animal farms, which still function as in the Middle Ages, with animals, water, food, and excrement mixed in a fruitful and dangerous breeding ground. Superbugs resistant to all antibiotics form a dark cloud on the horizon.

We exalt the technological progress of the last century. We fly, some went to the Moon, we drive cars (soon they will not need us), we almost venerate the internet, and medical advances have been quite astonishing. Much suffering has been alleviated by technology and another fraction induced by it. But we do not realize the price we will have paid for all of this. Technology is the grandchild of the industrial revolution, a sort of slow-motion accident fed by energy mainly obtained from the burning of fossil fuels. And now we have received the bill, payable in increasing yearly installments. The total price: The collapse of civilization, perhaps the end of humanity.

Globally, science and technology are dominated by countries that have the resources to engage in it, and its development has been in response to their needs and priorities. Many of their basic problems have been solved (for the short term), and scientific and technological development is moving towards issues that could be described as "luxury." New high-cost biomedical technologies, new computer and space exploration technologies, and, unfortunately, new weapons of little use to the mass of humanity.

But in countries where their basic problems have not been solved (most of them), research and development should have different goals. They need to resolve local issues related to public health and sustenance and support the development of home-grown technologies that are of little interest in developed countries. It does make little sense to send a spacecraft to the Moon if you cannot find remedies for local problems. Thus, it becomes necessary to promote local scientific institutions, however costly this may seem; otherwise, they will always depend on science and technology from industrialized countries, which could be called "technological imperialism."

6.4 Causality

Our terror and our darkness's of mind
Must be dispelled, not by the sunshine's ray,
No by those shining arrows of light,
But by Insight into nature, and a scheme
Of systematic contemplation. So
Our starting point shall be this principle:
Nothing at all is ever born from nothing
By the god's will. Ah, but men's minds are frightened
Because they see, on earth and in the heaven,
Many events whose causes are to them
Impossible to fix; so, they suppose
The god's will be the reason. As for us,
Once we have seen that Nothing comes from nothing,
We shall perceive with greater clarity
What we are looking for, whence each thing comes,
How things are caused, and no "god's will" about it.
— Lucretius[44]

We have a strong tendency to perceive patterns and assume a cause rather than a coincidence. When we perceive a pattern, we explain it by saying that it is someone's face or that it means something (from divine messages to human conspiracies), and we do not think that it can be merely a random pattern. It is not a matter of taste; our brain is a generator of stories — call it theories — that serve to understand why things happen — their causes — and thus control our environment. Some of these stories are good — scientific theories — others are fiction. Perceiving patterns is, in most cases, something beneficial, allowing us in a second to conclude something that if we had to analyze, it would take too long.

In both magic and the paranormal or pseudoscience (the distinction is that the magician does not pretend to do more than illusions), our tendency to seek causes is used to accept the unacceptable. The magician uses what he calls "closing all doors" to leave the viewer without an explanation. He acts differently and explicitly shows the viewer that "it is not here" or "see that there is nothing here." When the spectator formulates an explanation of how he does it, for example, thinking that he "put the coin in his mouth," the magician opens his mouth to show that it is empty. If he later thinks "he has it in his left hand," The magician refutes it by opening his left hand. So, in the end, there is no choice but to accept magic. I have heard people ask a magician: But is it just a trick?

Similarly, when seeing lights in the night sky, it is argued that it cannot be an airplane, helicopter, or balloon (they do not move that way), etc. In the end, the incredible is accepted without more options: they are visitors from Orion, since we have an aversion to "I don't know."

In an important book, Quine y Ullian state[45]: *In general, we tend to believe not only that explanations exist, but that ones that would enlighten us exist. [...] it often happens that when we look for an explanation, we reasonably believe that it will be found within certain narrow limits. We believe that one of some small number of conceivable explanations must be right. In this situation, the elimination of some of the possibilities increases the plausibility of those that remain. Sometimes even an explanation that was initially held to be implausible is accepted because it explains something that can be explained in no other conceivable way. People have been hanged for want of plausible alternatives. [...] Sometimes an explanation has no evidence at all to support it apart from the fact that it would, if true, explain something we want to be explained, and it can draw high credibility from this source alone.* This last alternative leads us to think of supernatural, paranormal, and conspiracy theories as explanations.

When a clear answer is not available, and because we are averse to uncertainty, a response is invented with no evidence, perhaps citing some obscure, not well done, or even non-existent study. This might make the snake oil merchant rich, but you might pay with your life. We seem wary to admit not knowing something, and some prefer to believe anything, even if it is not true; think of Jimmy Kimmel's "Lie Witness News."

An explanation is based on something previous that, in turn, needs an explanation. At some point, this recursion ends with specific facts that are accepted without cause, analogous to mathematical axioms. For example, we can explain that the planets move according to Einstein's gravitational theory because the mass of the Sun and that of the planets curve space to determine their movement. But ultimately, we cannot explain why space deforms with the presence of a mass. They are simply basic premises of the theory — this is how the world is.

Although we think that nothing arises without antecedents or disappears without consequences, something already proposed by Lucretius (read the epigraph if you have not already), random events are not caused. The notion of cause fades in quantum mechanics. There is no discernible cause for why a dice comes up with two (however much you concentrate) and less a cause for a radioactive atom to decay at a specific moment. We can also think of an event without an antecedent, something which always was. (Nothing says the universe is not eternal.)

Before looking for a cause, it is necessary to establish an effect that needs to be explained. Most people who believe in some God do not ask for its origin or cause; it leads nowhere. There are two different conditions between cause and effect:

- Sufficient condition: An event A is sufficient for another event B to occur, if and only if when A occurs, B occurs. When sufficient event A occurs, event B must occur. If we know a sufficient condition, we can infer the effect from the cause. Taking a glass of arsenic is sufficient to die, but it is not necessary (you can die from other causes)
- Necessary condition: An event A is necessary for another event B to occur, if and only if when B occurs, A occurred. When the necessary event A does not occur, event B cannot occur. If we know a necessary condition, we can infer the cause from the effect. Oxygen is necessary for combustion, but luckily it is not enough, not sufficient. When there is oxygen, not always there is combustion. A bacterium is necessary for an infection, and by eliminating the necessary cause, we eliminate the infection. Transmission

is necessary for an epidemic; if we eliminate transmission, we stop the epidemic (so, wear a mask).

If we are interested in *producing* an effect, then it is reasonable to look for sufficient causes. If, on the other hand, we are interested in avoiding an effect, then we must look for the necessary causes, since if we manage to eliminate a necessary cause, then the effect will not happen.

There are cases for which the cause is *necessary and sufficient*. It is a necessary and sufficient condition to apply a force to cause an object to change its speed. The spirochete *treponema pallidum* is a necessary and sufficient condition to develop syphilis.

A cause is not generally sufficient to generate its consequence since several other factors influence a cause-effect relationship. For example, although exposure to a pathogen causes a disease, it does not always occur because of several additional conditions (the person's immunity, general state of health, the extent of exposure, the availability of drugs, etc.). Therefore, for a cause to generate its effect, additional conditions must be met that become sufficient for the effect to occur. The determination of additional and relevant conditions is obtained in controlled experiments where it is possible to eliminate factors that are not relevant to the cause-effect relationship (say, the phases of the Moon).

Not all causal inferences are similar. Some are immediate and obvious, and even a child quickly infers that the cause of its burns is the hot object it touched. The causal relationship between sex and having children was perhaps not evident to our ancestors (necessary but not sufficient). Not always when she had sex, she became pregnant, and in the months that separated the cause and the awareness of pregnancy, countless other things occurred that could have been taken as a cause. Other cases are more tenuous, and the cause is difficult to establish, as in the case of clinical trials. In medical research, controls are established to determine the cause with high probability. The results will be reliable if *double-blind randomized studies with placebo control* are conducted and adequately analyzed. The pharmaceutical industry usually contracts so-called contract research organizations to do this. We have seen this at work recently as trials demonstrated the effectiveness of COVID-19 vaccines. Still, sometimes things can go wrong when corrupt CEOs care more about the bottom line than public health.

Although we have developed clear methods to establish cause in science, this is not so easy for human affairs. No historian can establish the cause of war (in part because of a conjunction of causes). Many wars have started by an accumulation of bungling and missteps by those in power (this is the main reason why bungling idiots in power cause alarm). We do not fully understand the causes of human behavior beyond some general ideas about nature and nurture; we need more *science* in social science, often thwarted by outdated beliefs. Cognitive psychology has made some exciting discoveries about our mostly irrational choices,[46] but we are far from understanding why we often act like perfect idiots. And perhaps we are afraid of the answer. Lee McIntyre makes a strong case for a *science* of human behavior.[47]

We talk about the cause when we understand that something has been the *immediate* factor of an effect. For example, when there is a fire, it is not a pertinent answer to say that there was oxygen; even if without oxygen, there is no fire (it is a necessary condition but not sufficient). In this case, we look for the *proximate cause*: a short circuit, lightning, or a pyromaniac.

Establishing that A causes B does not mean that we know *how* it happens; that is a different question. In other words, we can establish that smoking causes an increase in the probability of developing cancer without knowing what the mechanism is. On the other hand, a causal relationship will be more acceptable and controllable if we understand the mechanism. Global warming is caused by greenhouse gases, and we know and understand the mechanism. (As I will explain in detail.)

The first step in determining a causal relationship is to formulate relevant hypotheses for the cases studied, and that one of these hypotheses is the cause.

Early in the 19th century, childbirth was the single most common cause of female death because of "puerperal fever" (today, we know it was caused by an infection with *streptococcus pyogenes*). Not one woman survived childbirth during a whole year in a small Italian hospital. Dr. Ignaz Semmelweis (1818–1865), a Hungarian physician, discovered in 1847 that the occurrence of this fever was drastically reduced if doctors and nurses delivering babies washed their hands in an antiseptic solution[48] as they went from the morgue to the delivery room. Many years passed before the mechanism, infections by invisible germs, was understood by German physician and microbiologist Robert Koch (1843–1910) and French biologist and chemist Louis Pasteur (1822–1895).

Today, scientists have found ways to study and understand the functioning of invisible pathogens, and we can stop their attack if we know how they propagate and what might stop them. But research takes time and resources, and results arrive too late for many.

6.5 Ockham's Razor and Newton's Laser Sword

That which can be asserted without evidence can be dismissed without evidence.

— Christopher Hitchens[49]

Given the variety of phenomena we observe, science seeks the *best explanation*, which is frugal and coherent with what we already know, following Ockham's principle[a] *not to suppose the existence of more things than necessary.* Whenever several hypotheses claim to explain a phenomenon, choosing the most frugal that explains it entirely with a minimum number of premises, causes, and variables is preferable. This process of inferring the best explanation is not a *deduction* nor an *induction* (two ways of proceeding) but an *abduction*, as we have seen.

If you cannot find your car at the mall's parking lot, you can hypothesize that it was snatched by an alien spacecraft or that you forgot where you left it, the second option being more frugal. Coherency with what we already know and conceptual integration is part of frugality. An explanation that is not coherent with what we already know and uses discordant concepts is not simple. There is an essential coherence between different scientific disciplines because none supposes something that contradicts what is known in another. Pseudoscience is incoherent with what we know about the world, with well-established natural laws, appealing to concepts such as "life force" or "telepathy" that are discordant. Astrology is not consistent

---

[a] Formulated by William of Ockham, an English Franciscan friar and scholastic philosopher and theologian (c. 1285–1349).

with astronomy or physics, homeopathy is not coherent with what we know about biology and chemistry, reincarnation does not accord with what we know about life, and visions of the nonexistent future, nor do ghosts agree with logic and physics.

For the case of the always popular "paranormal" phenomena, we can propose explanations that go against everything we know, or we can think that a mistake was made or that we are facing fraud or a case of hallucination. The paranormal is normal. If we find a dead and desiccated animal in the field, it is simpler and more coherent to assume that it died of some disease than a mysterious being, "the *chupacabra*" (goatsucker) was the cause. A psychic will get a lot of information about you by looking at you and chatting with you (Cold reading). If they could actually perceive the future or talk to the dead, it would be pretty challenging to fit this into what we know about the future (nothing) and the dead (who do not respond).

Related to the above, we have "Ockham's broom," invented by biologist Sidney Brenner. It refers to the process by which inconvenient facts are swept "under the carpet." The broom is commonly used by pseudoscientists, creationists, and adherents to conspiracy theories. Occasionally, even a scientist has used it, and it is a valuable tool for politicians.

Mathematician Mike Alder[50] offers a modern and sharper version that he dubbed Newton's laser sword: "We must not dispute propositions unless it can be demonstrated by precise and mathematical logic that they have observable empirical consequences." Ask one who talks about Noah's Ark as if it were history, where all the water came from, or how the penguins came to board it somewhere in Turkey to hear what they tell you.

6.6 The Responsibility of Scientists and Citizens

To insist on acting as a responsible individual in a society that reduces the individual to impotence may be foolish, reckless, and ineffectual, or it may be wise, prudent, and effective. But whichever it is, only thus is there a chance of changing our present tragic destiny

— Dwight Macdonald[51]

I believe that despite the enormous odds which exist, unflinching, unswerving, fierce intellectual determination, as citizens, to define the real truth of our lives and our societies is a crucial obligation that devolves upon us all. It is, in fact, mandatory. If such a determination is not embodied in our political vision, we have no hope of restoring what is so nearly lost to us — the dignity of man.

— Harold Pinter[52]

We have become, by the power of a glorious evolutionary accident called intelligence, the stewards of life's continuity on earth. We did not ask for this role, but we cannot abjure it. We may not be suited to it, but here we are.

— Stephen Jay Gould[53]

Physicist J. Robert Oppenheimer (1904–1967) was the scientific director of the Manhattan Project to develop the "atomic bomb." (I write it this way since the so-called atomic bombs are *nuclear bombs*, that is, reactions occur between nuclei and not between atoms — the latter is chemistry). On July 16, 1945, at 5:29:45 am, at a desert site in New Mexico aptly named "Jornada del Muerto" (journey of the dead) 40 miles south of Los Alamos, the first nuclear bomb ("the gadget," plutonium-based) was detonated to see if the contraption worked, (codenamed Trinity). In an interview in 1965, recalling that day, Oppenheimer said: *I remembered the line from the Hindu scripture, the Bhagavad-Gita; Vishnu is trying to persuade the Prince that he should do his duty and, to impress him, takes on his multi-armed form and says, "Now I am become Death, the destroyer of worlds." I suppose we all thought that one way or another.*[54](There are many different translations from the Sanskrit). He *remembered* it. His brother, Frank Oppenheimer (1912–1985), also a renowned physicist, who was with him, recalled that all he *said* was "It worked."[55] We know from many studies in cognitive science that our memories are unreliable.

A famous letter addressed to President Truman[b]: — A PETITION TO THE PRESIDENT OF THE UNITED STATES — was written the day after, signed by Hungarian American physicist Leo Szilard (1898–1964) and 69

---

[b] U.S. National Archives, Record Group 77, Records of the Chief of Engineers, Manhattan Engineer District, Harrison-Bundy File, folder #76.

co-signers who were involved with the Manhattan Project. It said in part: *The development of atomic power will provide the nations with new means of destruction. The atomic bombs at our disposal represent only the first step in this direction, and there is almost no limit to the destructive power which will become available in the course of their future development. Thus a nation which sets the precedent of using these newly liberated forces of nature for purposes of destruction may have to bear the responsibility of opening the door to an era of devastation on an unimaginable scale* and ended with: *in view of the foregoing, we, the undersigned, respectfully petition: first, that you exercise your power as Commander-in-Chief, to rule that the United States shall not resort to the use of atomic bombs in this war unless the terms which will be imposed upon Japan have been made public in detail and Japan knowing these terms has refused to surrender; second, that in such an event the question whether or not to use atomic bombs be decided by you in light of the considerations presented in this petition as well as all the other moral responsibilities which are involved.*

Eight days after that, on July 25, 1945, a memo from the War Department Office by the Chief of Staff to General Carl Spaats, the Commanding General of the United States Strategic Air Forces, reads[56]: *The 509 Composite Group, 20th Air Force will deliver its first special bomb as soon as weather permits visual bombing after about 3 August 1945 on one of the targets: Hiroshima, Kokura, Niigata, and Nagasaki.*

It was at the time not understood what the "special" bomb would produce beyond the huge explosion. But, in the words of writer and philosopher Dwight Macdonald (1906–1982)[57]: *In any case, it was undoubtedly the most magnificent scientific experiment in history, with cities as the laboratories and people as the guinea-pigs.*

In the fall of 1945, shortly after giving up the directorship of the Los Alamos bomb project, Oppenheimer commented that[58]: *atomic weapons were actually made by scientists, even ... by scientists normally committed to the exploration of fairly recondite things. The speed of the development, the active and essential participation of men of science in the development, have no doubt contributed greatly to our awareness of the crisis that faces us, even to our sense of responsibility for its resolution.*

Shortly before he died in 1955, Albert Einstein said to Nobel laureate chemist and peace activist Linus Pauling (1901–1994), about a letter he had written to President Franklin D. Roosevelt on October 11, 1939: *I made one great mistake in my life — when I signed the letter to President Roosevelt recommending that atom bombs be made.* The letter had been written in fear that the Germans might achieve a nuclear weapon. It said in part: *In the course of the last four months it has been made probable — through the work of Joliot in France as well as Fermi and Szilárd in America — that it may become possible to set up a nuclear chain reaction in a large mass of uranium, by which vast amounts of power and large quantities of new radium-like elements would be generated. Now it appears almost certain that this could be achieved in the immediate future.*

This new phenomenon would also lead to the construction of bombs, and it is conceivable — though much less certain — that extremely powerful bombs of a new type may thus be constructed. A single bomb of this type, carried by boat and exploded in a port, might very well destroy the whole port together with some of the surrounding territory. However, such bombs might very well prove to be too heavy for transportation by air.[59] (In this case, his brilliant imagination fell short.)

The warnings about nuclear weapons were ignored, and today, they are a real and present danger, sitting there until one goes off. There are no fail-safe mechanisms for the simple reason that humans are not fail-safe. Accidents *do* happen.

The Fourth National Climate Assessment to congress and the President was issued on November 23, 2018 — double black Friday indeed — (perhaps so that no one would pay attention), endorsed by NASA, NOAA, the Department of Defense, and 10 other federal scientific agencies. It begins with[60]: *Earth's climate is now changing faster than at any point in the history of modern civilization, primarily as a result of human activities.* A bit further down: *The assumption that current and future climate conditions will resemble the recent past is no longer valid.*

Given the fondness for Twitter by former President Trump and his aversion to reading, one of the authors tweeted the following summary:

It's real, it's us, and the impacts are becoming increasingly serious.

— Katharine Hayhoe (@KHayhoe)

The 45th POTUS responded: *I don't believe it*, just as he did not believe that the orders to kill journalist Jamal Khashoggi came from the Saudi regime, as established by US intelligence. (I invite you to read the Summary Findings in the reference.[61])

This is what you get with a president and "commander in chief" who suffers from the *Dunning-Kruger* bias. It is alarming. Some of his friends resorted to conspiracy theories (see below) to discredit the report, claiming that people are getting rich with this, expressing things like *The reality is that a lot of these scientists are driven by the money that they receive*[62] (US Senator Rick Santorum). If anything, it is the other way around; measures to alleviate the looming climate catastrophe imply great losses for the fossil-fuel industry. Responsibility is talking truth to power; keeping quiet is irresponsible and worsens when you participate due to ignorance or greed.

The Asilomar Conference on Recombinant DNA held in 1975 was organized by Paul Berg (born 1926) (Nobel Prize in chemistry for 1980). It was convened to discuss guidelines to ensure the safety of new technology: recombinant DNA or, more generally, genetic engineering. The worry was that biohazardous materials could spread from laboratories to the environment without proper controls, with unforeseen and damaging consequences for human health and Earth's ecosystems. These worries have resurfaced with the grim pandemic caused by the SARS-CoV-2 virus.

In a compelling article in the *Bulletin of the Atomic Scientists*,[63] Nicholas Wade contends that the virus *could have escaped* from the Wuhan laboratory (research funded by the NIH). If this is proven to be true (and we shall eventually know), it will be a reminder about Paul Berg's concerns so many years ago and must alert us to future mishaps. Some modern research looks at "gain of function," basically modifying pathogens to increase their pathogenesis to better understand potential propagation and develop remedies — entering ethical quicksand.[64] That some deranged mind might think about bioweapons is not to be taken lightly. The use of genetic engineering now dominates research in biology. But fear (sometimes well-founded) has remained in the public mind, to such a point that many foods ("Frankenfoods") are labeled "Contains no GMOs" as if this were equal to "Contains no arsenic." No hazard due to this technology (applied to crops) has been documented (not so for certain agricultural practices where crops are modified to resist herbicides and insecticides). There is also opposition by some religious groups who liken these technologies to "playing God," disregarding the fact that we have modified the genetic properties of plants and animals (by different means such as radiation) for quite a while. For eons, humans used sharp stone tools to cut things, today, we have scalpels, and no one suggests that scalpels are bad because they cut better. Genetic engineering "cuts better."

Public fear of new technologies is nothing new, and there will always appear someone to fan the flames appealing to what often is a distortion of facts and theories, which today can become "viral" in a moment. Presently we no longer worry about "Edison electric light," now there is 5G (Figure 6.9).

Although scientists often have speculated about their discoveries' effects on our societies and even on our planet, it was mostly an academic exercise. This has changed with the increasing awareness that new technologies could seriously impact our civilization and survival.

In this context, a new revolutionary player has arrived, which needs to be looked at carefully. I am referring to the discovery of CRISPR[65] (clustered, regularly interspaced short palindromic repeats), enabling us to intervene with high precision in an organism's genome. This can either prevent some illnesses with a genetic component or modify a future human being, leading to the possibility of "designer babies." Do we want

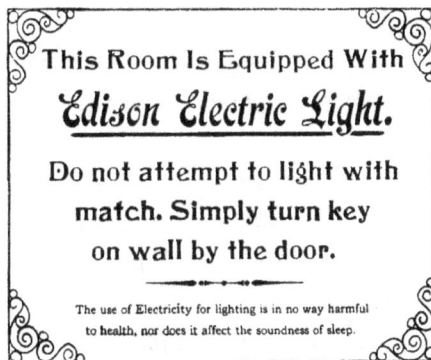

Figure 6.9. Public fear of new technologies.

this? This ethical question has been put on the table by the discoverers of this technology, US biochemist Jennifer Doudna (born 1964), French biochemist Emmanuelle Charpentier, and Lithuanian biochemist Virginijus Šikšnys (born 1956). (I would not be surprised if they win a Nobel Prize.) They just did! (2020) (without Virginijus Šikšnys).

A different example[66] is the invention of the blue LED, considered so important that the Japanese researchers (Isamu Akasaki, Hiroshi Amano, and Shuji Nakamura) who accomplished this were awarded the 2014 Nobel Prize in Physics. The committee stated that this feat *enabled bright and energy-saving white light sources.* Red and green LEDs were already in use, but with this addition, you now had RGB, and so you could control the colors of LED sources, which today are ubiquitous and save much energy. However, biologists have discovered the adverse effects on human health of blue light, particularly our sleeping patterns, but physicists do not know biology. At any rate, this would not have stopped them from their research. (You can and should use your device's blue light settings.)

The gap between scientific knowledge, which feeds new technologies, and public understanding of science, is broadening. When high government officials ignore the scientific community's concerns or even urgent warnings about pursuing some public policies ultimately against the common good, scientists must engage the public and politicians, even if they do not wish to hear, and scientists dislike the idea. These days, some scientists are bullied and threatened for speaking about controversial (according to the public) science. Dr. Anthony Fauci needs bodyguards!

Bertrand Russell had this to say in 1955[67]: *I come at last to a question which is causing considerable concern and perplexity to many men of science, namely:* **what is their social duty towards this new world that they have been creating?** *I do not think this question is easy or simple. The pure man of science, as such, is concerned with the advancement of knowledge. And in his professional moments, he takes it for granted that the advancement of knowledge is desirable. But inevitably, he finds himself casting his pearls before swine. Men who do not understand his scientific work can utilize the knowledge that he provides. The new techniques to which it gives rise often have totally unexpected effects. The men who decide what use shall be made* *of the new techniques are not necessarily possessed of any exceptional degree of wisdom. They are mainly politicians whose professional skill consists in knowing how to play upon the emotions of masses of men. The emotions which easily sway masses are very seldom the best of which the individuals composing the masses are capable. And so, the scientist finds that he has unintentionally placed new powers in the hands of reckless men.*

Richard Feynman (1918–1988), considered one of the most brilliant physicists in recent times and Nobel Prize winner in 1965, had this to say[68]: *Our responsibility is to do what we can, learn what we can, improve the solutions, and pass them on. It is our responsibility to leave the people of the future a free hand. In the impetuous youth of humanity, we can make grave errors that can stunt our growth for a long time. This we will do if we say we have the answers now, so young and ignorant as we are. If we suppress all discussion, all criticism, proclaiming "This is the answer, my friends; man is saved!" we will doom humanity for a long time to the chains of authority, confined to the limits of our present imagination. It has been done so many times before. It is our responsibility as scientists, knowing the great progress which comes from a satisfactory philosophy of ignorance, the great progress, which is the fruit of freedom of thought, to proclaim the value of this freedom; to teach how doubt is not to be feared but welcomed and discussed, and to demand this freedom as our duty to all coming generations.*

Yes, knowledge is power, as is often repeated. But the present conundrum is that those who know rarely have power, and those who have power rarely have enough knowledge to make correct decisions. The saying that "the pen is mightier than the sword" is obsolete; it no longer applies. Recently the 45th president of the US stated in an interview with *The Washington Post*[69] (concerning a Federal Reserve decision) said that: *They're making a mistake because I have a gut, and my gut tells me more sometimes than anybody else's brain can ever tell me.* Well, my grasp of biology tells me that the brain usually generates knowledge, and the gut generates shit.

The tireless Noam Chomsky wrote in a memorable essay in 1967[70]: *Intellectuals are in a position to expose the lies of governments, to analyze actions according to their causes and motives and often hidden intentions.*

In the Western world, at least, they have the power that comes from political liberty, from access to information and freedom of expression. For a privileged minority, Western democracy provides the leisure, the facilities, and the training to seek the truth lying hidden behind the veil of distortion and misrepresentation, ideology and class interest, through which the events of current history are presented to us.

Distinguished philosopher of science Karl Popper (1902–1994) wrote in 1968[71]: *It could be questioned whether there is such a thing as a responsibility of the scientist that differs from that of any other citizen or any other human being. I think the answer is that everybody has a special responsibility in the field in which he has either special power or special knowledge. Thus, in the main, only scientists can gauge the implications of their discoveries. The layman, and thus the politician, do not know enough. This holds for such things as new chemicals for increasing the output of farming products as much as for new armaments. Just as, in former times, noblesse oblige, so now, as Professor Mercier[c] has put it, sagesse obligé: it is the potential access to knowledge that creates the obligation. Only scientists can foresee the dangers, for example, of population increase, or of the increases in the consumption of oil products, or the dangers inherent in atomic waste, and thus even in atoms for peace. But do they know enough about it? Are they conscious of their responsibilities? Some of them are, but it seems to me that often they are not. Some, perhaps, are too busy. Others, perhaps, are too thoughtless. Somehow or other, the unintended repercussions of our heedless general technological advance seem to be nobody's business.*

As I have mentioned before, knowledge has no ethical value. A scientist (and intellectual) pursues her profession seeking new knowledge, ideally for the sake of knowledge, but increasingly with some practical goal that will not be neutral and might be controversial.

A scientist or intellectual possesses knowledge that the general public and politicians lack, knowledge that can be important when it comes to establishing public policy (be it nuclear power, vaccinations, climate, pandemics, and an increasing number of concerns). Therefore, her responsibility as a human being is to tell the citizen how and what science

---

[c] He refers to the physicist: André, M, author of *Science and Responsibility*, Torino, 1969.

understands and help avoid going down the wrong path. This, even if the FIRM complex pushes in the wrong direction and at risk of losing her job or support to continue her research. As Popper said: *Sagesse oblige*.

Of course, there is a practical problem. People who need to understand scientific issues, such as those discussed in this book, often do not read books, not even those who have government jobs (be they elected or appointed). There exists an extensive collection of books, written at all levels and read mostly by those who do not need to be persuaded. The WWW is not very useful either since, as discussed later, people will navigate inside mutually excluding bubbles that only strengthen their wrong ideas and nurture polarization.

The scientist and author C. P. Snow (1905–1980) had already expressed this idea in a famous lecture in 1959, referring to what he perceived as two cultures that did not communicate, an issue he described as one of "life or death."[72]: *In our society (that is, advanced western society), we have lost even the pretense of a common culture. Persons educated with the greatest intensity we know can no longer communicate with each other on the plane of their major intellectual concern. This is serious for our creative, intellectual, and, above all, normal life. It is leading us to interpret the past wrongly, to misjudge the "present, and to deny our hopes of the future". It is making it difficult or impossible for us to take good action. [...]. It is dangerous to have two cultures that can't or don't communicate. In a time when science is determining much of our destiny, that is, whether we live or die, it is dangerous in the most practical terms.*

What is worrying is that his words are still valid 60 years later (with due clarifications) even more in our globalized world in which there are many cultures, and *advanced western society* can be questioned.

To confront the growing inability to accept scientific consensus and act accordingly, scientific institutions and universities need to do more than just write press releases, no matter how many Nobel Prize winners sign them. It is time to get out and vigorously defend science, and scholarship in general, for the benefit of the people and cease being tolerant of ignorance. Mainstream news outlets should also do their part, and some do. A good example is CNN's Dr. Sanjay Gupta, a neurosurgeon, acting as a chief medical correspondent, clarifying for the audience issues of great importance. But then we had the likes of Rush Limbaugh (1951–2021), a conservative radio host listened to by millions, who said the following: *The four corners of deceit: government, academia, science, and media. Those institutions are now corrupt and exist by virtue of deceit. That's how they promulgate themselves; it is how they prosper.* (a bit circular since he was a prominent *media* figure).

Many institutions worldwide are desperately trying to hold back this avalanche of senseless discourse, but inflammatory rhetoric seems to have the upper hand.

The anti-scientific breeze which blows through many societies, and surprisingly also in the US, warns us of a possible storm and possible stormtroopers. Only enlightened education will protect us. Regardless of their course of studies, every citizen must get the opportunity to learn about the eternal questions and today's answers, very different from the familiar ones, devoid of obsolete ideas that confuse the mind.

If I said "surprisingly" above, it is because the US is, without a doubt (up to now) and however you wish to measure it, the world's leader in scientific research (on the current path, it won't be long before others become number one).

But paradoxically, a large sector of its population does not believe important scientific facts because they do not like them or are misinformed. Recently there was a "march for science," something unthinkable just a decade ago. Otto[73] wrote: *With every step away from reason and into ideology, the country moves towards a state of tyranny in which public policy come to be based not on knowledge but on the most loudly based opinion.* We can see it happening.

Behind this suicidal anti-science mood of a large segment of the population is an ever-increasing distrust of authority, and for good reasons. The tragedy arises once scientific authority based on facts and consensus is confused with political authority, which is all the contrary. Presently it is anti-vaxxers and climate change deniers who place us in danger.

The tenacious Noam Chomsky writes, reviewing his ideas regarding the responsibility of intellectuals[74]: *As for the responsibility of intellectuals, there does not seem to me to be much to say beyond some simple truths:*

intellectuals are typically privileged; privilege yields opportunities and opportunity confers responsibilities. An individual then has options.

The National Academy of Sciences (NAS) was established by an Act of Congress signed by President Abraham Lincoln in 1863. It provides independent, objective advice to the nation on matters related to science and technology. It publishes many relevant reports, mostly read by those interested in scientific issues, which is not enough. The NAS and other scientific societies must dedicate resources to present the scientific consensus and results of studies to a much broader audience. This can only be achieved through the media and the participation of scientists in public forums. It is the responsibility of scientists and scholars to insert themselves into the public discourse, like it or not, and not just in the large urban areas. It might require a lot of resources (money and people), but the cost of not attempting this will be much higher. The greatest difficulty resides in that people who reject or distrust science might not even be interested. I have experienced this discouraging attitude myself when speaking to the public about a subject of great public interest: Alien visitations. I try to be as open as possible, relate what we know, talk about the history of this idea and the science behind it. It amazes me how invariably there are people (and not just a few) who tell me in so many words that I do not know the truth, and that they have seen enough on YouTube, or have had personal experiences to know better than I. (I do not argue).

Pay enough, and I would suppose that even Fox would allow a presentation about the looming climate crisis or the danger of the anti-vaccine movement (now — April 9, 2019 — I just read the NYT reporting the largest measles outbreak in 100 years in Brooklyn). Or do we need to rely on Stephen Colbert, to point out and ridicule the 45th POTUS for saying to the Associated Press, concerning climate that: *I have a natural instinct for science, and I will say that you have scientists on both sides of the picture.* All evidence shows that he had no such instinct (whatever that means). It is irrelevant that his uncle was a distinguished scientist at MIT or that his grandfather's fortune came from running a brothel in Bennett, British Columbia[75]; you are what *you* are.

The democratic notion of giving "both sides" of a story equal time is a good one, but only if there *are two sides*, which in the case of natural selection, vaccination, or climate change (and many other issues) is not the case. "Gut feelings" and instincts are our primitive reactions to a world that is difficult to understand, and that is why we need science. The NAS webpage recently published[76]: *Natural changes in the Sun and Earth cannot explain today's global warming. Human activities are causing Earth to heat up in ways that are different from warm periods in the past.*, but who goes there? It should take out a full-page ad in all major newspapers, place it there, and repeat it every month. Too much to ask?

But citizens also carry a burden of public ignorance in a democracy, which should compel them to seek knowledge and accept expert advice instead of watching a screen full of bullshit for hours on end. It leads to the end of democracy. This worry is not new, but it is becoming more urgent. Spanish philosopher Ortega y Gasset wrote long ago in 1932[77]: *The mass crushes beneath it everything different, everything that is excellent, individual, qualified and select. Anybody who is not like everybody, who does not think like everybody, runs the risk of being eliminated. And it is clear, of course, that this "everybody" is not "everybody." "Everybody" was normally the complex unity of the mass and the divergent, specialized minorities. Nowadays, "everybody" is the mass alone.*

Without knowledge, citizens will continue to be gaslighted by all sorts of publicists and demagogues and be unable to choose intelligently among several options. Citizens will respond and act according to exaggerated fears and alluring stories instead of being moved by the idea that a better world is possible if they choose with knowledge. For this, they should listen and understand what those with expertise are saying. *I don't believe it* is plain idiocy. Understandably, the public is skeptical of authority and experts, but this cannot become a norm across the board.

Schools should be at the forefront of this effort to educate, to teach *how to think independently*, and the media would do well to have science experts at their service. Journalists should demand answers to fundamental questions from politicians rather than gossip. As is evident by the results, they have mostly failed, but it is unclear how to change this, given the increasing threat to journalists by states who do not like what they publish.

We should not take this lightly — it had happened before when civilization gave way to the Middle Ages, and after more than a millennium and a lot of spilled blood emerged to catch a bit of light. But unfortunately, no law enforces adherence to scientific knowledge, pursuing the truth, and it is easier to return to myths, magical thinking, or not thinking at all. Otto writes[78]: *The one thing we do know about science — the one thing that is predictable about it — is that if we don't value it, or if we become inhospitable to the tolerance, freedom, and open exchange of ideas that stimulates it, if we wall it off and call it a separate culture instead of something we all should do, if we seize funding it, if we try to be overly directive of it, if we insist on certainty, if we elevate ideology in our public policies, if we attack scientist or call entire fields a hoax, if we stomp on it and call it weird, in short if we become authoritarian, we will stifle creativity and science will suffer or disappear. We won't get the big breakthroughs or cures. We won't get the economic booms. We won't get the national defense advantages. We won't get a clean environment or healthy children.*

6.7 Fallacies

There are infinite possibilities of error, and more cranks take up fashionable untruths than unfashionable truths.

— Bertrand Russell[79]

"Are there dragons?" she asked. I said that there were not. "Have there ever been?" I said all the evidence was to the contrary. "But if there is a word dragon," she said, "then once there must have been dragons.

— Penelope Lively[80]

Essential to the scientific temper is the use of careful and impeccable reasoning and procedures. Sadly, there are situations, particularly outside of science, when this is not accomplished, in part because it is difficult to do. Let me start with one of the most insidious and important "flaws" that we all share in our way of reasoning. Much research[81] on cognitive biases shows that people unconsciously resist changing their beliefs in the light of evidence that refutes them and, in general, tend to pay attention and look for evidence that confirms them — known as *confirmation bias.*

If you are convinced that global warming is a hoax, you will resist all evidence clearly showing the contrary. If you firmly believe that an election was rigged, you will not accept the results, no matter the lack of evidence. Likewise, although it was evident that the COVID pandemic in the US began to fade after vaccinations started, anti-vaxxers will explain it away as a result of the weather, cosmic rays, or whatever.

Social psychologist Leon Festinger (1919–1989)[82] wrote: *A man with a conviction is a hard man to change. Tell him you disagree, and he turns away. Show him facts or figures, and he questions your sources. Appeal to logic, and he fails to see your point.* Political commentator David Frum writes about a debate he had with notorious Steven Bannon[83]: *People do not think; they feel. They do not believe what is true; they regard it as true that which they wish to believe. A lie that affirms us will gain more credence than a truth that challenges us.* This single cognitive bias (among many) explains a lot about our strange behavior. It is how we are. Different from biases that affect our decisions systematically is a less-known problem related to noise, that is, random variations affecting decision making. An example will illustrate the difference. Interventions by police are systematically higher for non-white people due to a built-in bias (racism) well entrenched in US society. On the other hand, sentences handed out by judges for very similar violations of the law among people of the same ethnicity vary widely; a similar crime is punished by 30 days in jail or 10 years. That is noise,[84] contributing to injustice.

We also prefer stories and anecdotes and are emotional. At the same time, statistics, evidence, and clear reasoning are less palatable. All this allows others to gaslight us with dubious arguments: fallacies (from the Latin *fallatia,* meaning deception).

By way of example, I will consider just a few of the most insidious cases of muddled misleading arguments.[85] They are reasoning traps, which have the ability and, in many cases, the intention to persuade the unwary, traps into which we all fall from time to time, especially if we are not aware of them.

Formal fallacies represent patterns of erroneous *reasoning* that arise from the invalidity of the *form* of a deductive argument. For example, the following argument is *not* valid.

> All living things need water
> Roses need water
> *Therefore*, Roses are living things.

I assume you understand why it is invalid. But for *informal arguments* (which are the most common), the concept of validity is not generally applicable, and some fallacies do not even imply an argument, such as the *ad hominem fallacy*, as I will discuss.

The error of an informal fallacy arises from the language, the content, the rhetorical and psychological resource used, and the truth or falsity of the initial premises or data. Not all fallacies necessarily lead to a wrong conclusion, and not this characterizes them, but procedural error. For example, in the case of the *ad hominem* or *ad verecundiam* it is possible, depending on the context, that the assertion has merit, but it makes no sense to speak of its validity.

Although it is possible to systematize the valid ways of formal reasoning, it is difficult to do so with fallacies because we can be wrong in innumerable ways. Moreover, different authors classify these fallacies differently, and there is no consensus about their taxonomy. But what really matters is not the classification but the analysis of erroneous patterns to understand what fallacious reasoning consists of, which increases our sensitivity to the incursion of fallacies in all types of discourse.

Some fallacies are so common that they are known by their Latin names since they have been known for a long time. Aristotle was the first to categorize fallacies in his *Sophistic Refutations*. In the words of Jamie Whyte[86]: *The modern world is a noxious environment for those of us bothered by logical error. People may have become no worse at reasoning, but they now have so many more opportunities to show off how bad they are. If anyone cared about our suffering, talk radio and op-ed pages would be censored. Even Congress is now broadcast, as if no torment were too great.*

Philip Ward writes[87] in response to the question as to why we should pay attention to errors in judgment that are so common: *Why indeed! As if it were not provocation enough to read newspapers and magazines still containing horoscopes in the 1970s, to see shelf upon shelf of fashionable occult "literature" in otherwise reputable bookshops, fanatic religious sects springing up to make claims of miracle-working and Messianism, extremist political groups seeking converts among the badly educated and the confused, and pseudo-sciences making untestable and incredible claims.*

Hasty Generalization

We all do it. It is part of the scientific temper in the form of induction. It is impossible to study all instances of a specific event, say the color of swans or the mass of a proton. If, after a while, in different places, we see white swans, we generalize that "all swans are white," although we have only seen a sample of them. If we measure a proton's mass as 1.673×10^{-27} kg., we generalize assuming that all protons are the same and say "the proton's mass is 1.673×10^{-27}". Should we find a proton of a different mass (but we probably would call it something else) or a black swan, then our induction fails. An induction proceeds from the particular to the general. We incur a *hasty generalization* if our starting sample is not representative, not random, or too small. Often we make generalizations about people of the form all A's are B's (where often B is offending), starting from a few A's.

Aided and abetted by confirmation bias, hasty generalizations feed our prejudices. Every time someone tells you all A's are B's, just ask: do you know them all?

Cum hoc ergo propter hoc and Post hoc ergo propter hoc

If a person a) is poorly, b) receives treatment intended to make him better, and c) gets better, no power of reasoning known to medical science can convince him that it may not have been the treatment that restored his health.

— Peter Medawar[88]

These are ubiquitous and sometimes dangerous. We are susceptible to the fallacy of attributing false causes. In the variety *cum hoc ergo propter hoc* (together with this, then because of this), we argue that if one thing is observed together with another (we say that there is a correlation), they are causally related. There may be a correlation between the Moon's phases and some terrestrial events, but that does not mean that the Moon is the cause of the events. This has often been claimed, but it is not true. But our brain is predisposed to think this way. This is usually encapsulated in the dictum: *Correlation does not necessarily imply causation*. It is a *necessary* condition for causation (but not *sufficient*).

The form of this fallacy is:

A varies together with B
Therefore, A causes B (or perhaps B causes A)

A more common variant is the *post hoc ergo propter hoc* fallacy (after this, then because of this).

B happens after A
Therefore, A causes B.

Although an effect always does follow its cause, it does not imply a causal connection. A pernicious example is a belief that vaccines cause autism, based on testimonies by people whose children are diagnosed with this condition *after* vaccination. But the MMR (measles-mumps-rubella) vaccine is usually administered to babies at 12–15 months, but autism is difficult to diagnose before 24 months, so you see the problem. Moreover, the community of experts has shown through many studies that there is no link, but once the idea got wings and "anti-vaxxers" formed their bubbles, the damage was done.

This fallacy reigns in matters of health and so-called alternative medicines and therapies. Someone feels sick and testifies to feel better after a treatment with quartz crystals or magnets. Conclusion: the treatment is effective. However, there is no way to determine if the improvement results from the therapy without considering placebo effects. For the case of quartz or magnets, a plausible mechanism is missing. The coincidence of an improvement after treatment, together with confirmation bias, fuels the belief in the effectiveness of the treatment. Given individual variability, *the efficacy of a therapy can only be verified by carefully conducted double-blind randomized studies with placebo control.*

There may be several other explanations for the perceived correlation: There may be a third factor (a hidden variable) that is the cause of the correlation between A and B, or the correlation might be accidental.

The natural history of most diseases is self-limited and cyclical, and many cases of spontaneous remission are known (a miracle?). The *post hoc* fallacy is frequently incurred: the patient turns to a quack (often because his condition worsened despite being subjected to a traditional treatment) and then improves, concluding that the improvement was caused by the snake oil. When the patient does not improve, rationalization is used, claiming that at least he did not get worse, and if he dies, it can be argued that he did not start snake oil treatment on time.

While considering vaccination, let me show you the data. In the US, vaccinations ramped up beginning in January 2021, reaching a peak in April 2021 (Figure 6.10) (the gray line is 7-day average reaching a peak of about 2 million fully vaccinated by April 2021).

At the same time, you can see daily cases starting to decrease sharply (dark line 7-day average) after the start of vaccination. Unfortunately, the downturn in vaccinations, encouraged by certain politicians who could care less and some Christian fundamentalists, does not allow

Daily Trends in Number of COVID-19 Cases and 7-day Moving Average of the Number of People Fully Vaccinated in The United States Reported to CDC

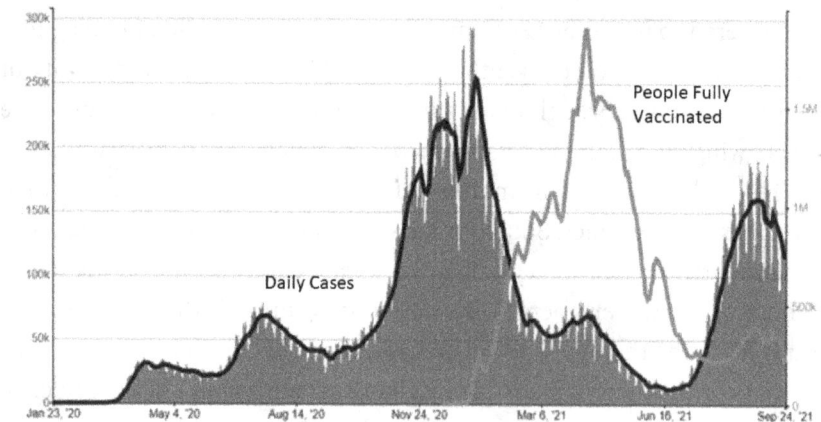

Source: CDC.

Figure 6.10.   Daily new COVID-19 cases (left) and vaccinations (right) in the US.

for optimism. Other variables are indeed in play, but as vaccination went down, cases resumed in September. If you are still skeptical, know that in mid-2021, over 95% of all COVID hospital admissions were unvaccinated.

Transfer of the burden of proof (ad ignorantiam)

A key logical consideration here is that if someone makes a claim, especially if it is an important one, then the burden of proof is upon him or her to show why it ought to be accepted as true. One cannot simply revert to a statement of faith. The fact that someone strongly believes that something is the case and dogmatically asserts it is only a description of his or her own psychological state of mind; it tells us nothing about the world.

— Paul Kurtz[89]

There is an asymmetry established by the fact that although it is possible to demonstrate the existence of something, it is impossible to demonstrate the non-existence of something that is logically possible. In other words, no one has demonstrated that aliens, telepathy, or gods do not exist, which is not proof that they exist. It is because of this asymmetry that what matters is the positive evidence. A proposition's truth is established by evidence in its favor and not by lack (ignorance) of evidence against it. Lack of evidence is not evidence. This asymmetry is the reason why the argument that says, *Since you cannot prove that what I say is not true and since everything is possible (which is not true), then it is true*, constitutes a fallacy.

In the case of pseudoscience, it is impossible to prove that everything alleged daily is not valid, and it would also be a waste of time. But it is unnecessary; the burden of proof — the *onus probandi* — falls on the person who claims a positive fact, not on the person who denies or questions that claim.

The general form is:

A is affirmed.
There are no ways to refute A. (we ignore)
Therefore: A is true.

This is also why in a court of law in the US, the prosecutor carries the burden of proof and must show beyond reasonable doubt that the defendant is guilty. It is not the responsibility of the defendant to *prove* his innocence. (Although it does not hurt to try, and there might be a good alibi.)

There are cases in which ignorance can form a valid argument for a conclusion or action. Thus, we can say: "A flood as described in the Bible would require the presence of an enormous volume of water on the Earth. However, the Earth does not have even a tenth of that water, even if we add that which is frozen at the poles (slowly melting). Therefore, such a flood could not occur" In this case, we argue that we are ignorant or unaware of any process that could have made nine-tenths of the flood water disappear.

We also argue that if we ignore the possible future effects of some action, it is prudent to avoid this action. A crucial precautionary principle in matters relevant to the environment. But it can also be harmful if we reject a vaccine because we ignore possible future adverse effects or claim that a few had allergic reactions (still better than the disease). Vaccines (discovered by Edward Jenner (1749–1823) to prevent smallpox) have worked in the past and saved millions of lives (in the 18th century in Europe, 400,000 people died annually of smallpox). If you do not know any cases of Polio, it is because of a vaccine. (The term "Vaccination" coined by Jenner derives from the Latin words for cow "Vacca" and cowpox "vaccinia".)

Ad hominem (against the person)

Arguments against a person do not refute an assertion by this person (an asshole can say something true). For example, the rebuttal: "That's what Hitler said!", Does not refute anything since not everything this criminal said was false. This is about converting approval or disapproval of a person into approval or disapproval of what the person expresses, a tactic on a psychological level. It is of the form:

A affirms X.
There is something questionable about A.
Therefore: X is false.

We can distinguish various types of *ad hominem*:

The ***abusive ad hominem*** directly attacks the person with insults or questioning of his character (character assassination) that are irrelevant in terms of the argument (much used by someone who played at being president): *You cannot believe Peter as a witness, he does not even have a job*, or *You say that this person is innocent, but you cannot be believed since you are a convicted thief.*

Although ad hominem attacks are irrelevant from a logical point of view, we must not lose sight that a person's political, economic, social, and religious ideas can question his impartiality if not invalidate his argument. In this sense, to give an example, an argument against an increase in corporate taxes by a member of a corporate entity or with ties to it, may be correct. Still, its origin justifies a high level of distrust since, presumably, we would not hear an argument in favor from this source. In other words, the *ad hominem* attack: "this person is a member of..." or equivalent, may be relevant when examining the argument, but it is not a reason to invalidate it. Likewise, an attack on a person because he is a sexual predator is not an argument against his mathematical abilities. Still, it is very relevant to employ him as a teacher in a school.

The tu quoque variant ("you too") tries to refute an argument by showing that the person who formulates it has acted or has expressed himself in a discordant way with the argument, which generally causes the person to become defensive. Thus, one person criticizes another for having lied, and the other responds with "you too" (you have lied on occasion) which is irrelevant. Even a compulsive liar might occasionally tell the truth. In general, someone is accused of doing or defending the same thing that he condemns or of not doing what he advises others. *I don't stop smoking, as my doctor recommended because he also smokes.*

A criticizes B for doing (or thinking) X.
But A has done (or thought) X.
Therefore: her criticism is invalid.

It is also common to attack an inconsistency:

"Now so and so is presenting himself against X, but a year ago, he was in favor of X."

But people can change their minds (even if it is difficult).

In pseudoscience, *tu quoque* is resorted to when there is nothing left. To explain quack therapy, when there is not a shred of evidence to support it, it is argued that there have been many questionable scientific cases in the past (you too) and that there are many scientific unknowns. You can also add several examples of scientific errors, such as the serious error of Thalidomide, a drug used between 1957 and 1961 to alleviate pregnancy symptoms but caused deformities in newborns. A modality is also used in which science is criticized because "scientists are arrogant" or "they think they know everything."

Although generally *ad hominem* is understood as an attack, there is also a positive variety. An argument is not true just because the proposer has many virtues: "Donald has never told a lie, therefore what he says must be true." (There is always a first time). In this case, the form is:

A affirms X.
There is something very positive about A.
Therefore: X is true.

The fallacy of inappropriate authority (ad verecundiam)

The AIDS crisis cannot be overcome through the distribution of condoms, which even aggravates the problems.

— Pope Benedict XVI (in a visit to Cameroon, March 2009)

Appealing to the authority of an expert is very common and a good thing. However, this does not *guarantee* the truth of a proposition since the expert can be wrong (or may not be an expert in the subject of concern). It would be much better to appeal to the facts rather than to some authority. In other words, an argument that says that Cobras (snakes) are poisonous because in an experiment, they bit 40 mice, and all died has more weight than saying that Cobras are poisonous because an expert on butterflies said so. If we believe that the universe has billions and billions of stars because Carl Sagan said so, this tells us why we believe so (and this may be acceptable since Carl Sagan was a recognized expert), but it is not the reason that this is true or false. But for many issues, we can only rely on experts, and the challenge is distinguishing real experts from those who are not.

In advertising, someone often provides "testimony" about the merits of what is to be sold, often some well-known person who is no expert.

The claim that high doses of vitamin C cured all kinds of ailments, from colds to cancer, was vigorously promoted by Linus Pauling (1901–1994). Pauling won the Nobel Prize twice (in Chemistry for his work on chemical bonds and the Peace Prize in 1962 for his campaign against nuclear tests). However, there is no proof of the therapeutic efficacy of high doses of vitamin C, although a large public sector believes so. In general, we can say:

A affirms X.
A is a prestigious and recognized expert (but A is perhaps not expert in X).
Therefore: X is true.
This fallacy can also be viewed as a particular form of *ad hominem.*

Argumentum ad Populum

As far as science is concerned, the authority of a thousand is not worth a scintilla of the reasoning of one man.

— Galileo Galilei[90]

In these cases, popularity is used to justify a proposition or action, widely used in commercial advertisements: "90% of respondents use product X." Therefore, you should also use product X. It is a variety of *ad verecundiam*, where authority is the most popular opinion. Many think: "It is not possible for so many to be wrong," but it is possible; just think of the 2016 US election.

It is of the following general form:

A affirms X.
Most people accept X.
Therefore: X is true.

The varieties of appeal to common practice (everyone does it), appeal to tradition (it has always been done that way), and appeal to belief (almost everyone believes this) are cases of this fallacy. They may be the reasons why you think or act in a certain way, but they do not justify it. That 85% of Americans believe in God does not make it true. Bertrand Russell tells us[91]: *The fact that an opinion has been widely held is no evidence whatever that it is not utterly absurd; indeed, in view of the silliness of the majority of mankind, a widely spread belief is more likely to be foolish than sensible.*

Naturalist Fallacy (Argumentum ad Naturam)

The claim of "naturalness," common in alternative medicine and foodstuff ("all-natural ingredients," "no GMO," etc.), is perceived by many as better. However, it is perplexing to enter a natural health store and see shelves filled with pills packed in plastic bottles (derived from fossil fuels).

It originates from the ambiguity of the word "natural" and, at the same time, resorts to a hasty generalization. Natural (and organic) is confused with good. But earthquakes are natural, snake poison is natural, and viruses are also natural (as far as we know). No one will argue that they are good, just as antibiotics, vaccines, and other drugs are artificial, and most will not argue that they are bad (if used correctly). Tobacco is a natural product, and although for many years the tobacco industry claimed that it caused no harm, in the end, they had no choice but to accept the facts. The box of Wendy's potato chips says: "Natural-cut Fries with Sea Salt." Potatoes with an artificial cut and with salt that does not come from the sea are surely worse. But sea salt is the same as artificial salt; both are Sodium Chloride.

The general form is:

This product (or process) is natural.
Natural things are good.
Therefore: This product (or process) is good.

As Whyte points out,[92] this fallacy also creeps into some moral or legal discussions. For example, some argue that homosexuality should not be accepted because it is not natural. Those who wish to refute this argument often try to show that it *is* natural, thereby admitting that it should be rejected if it were not natural, which is fallacious. Christians use some quotes in the Bible to argue against this but ignore that the Church is full of homosexuals and pedophiles.

Simpson's Paradox[93]

There are three kinds of lies: lies, damned lies, and statistics.

Here is an example of why you should be suspicious of statistics and aware of possible hidden variables. It is common to find comparisons of the value of certain statistics for different categories. Some categories are common — "sex" (I mean gender, not if you have it), "nationality," "occupation" — and others are defined for some purpose — "family income," or "years of study." If you are not careful when analyzing statistical results by comparing categories, you can fall into errors caused by "hidden variables," common in medicine and the social sciences.

Consider the following example: The results of diagnostic tests taken by students to evaluate a school's educational work were compared with the average results of all the students in the district. In addition, the students were grouped into two categories — group B — defined by those who receive financial assistance (students from families whose income was less than a certain amount) and — group A — those who did not. Table 6.1 summarizes the results.

The evaluators concluded that the school was doing a good job, especially with students receiving financial aid. However, another school did not group students and simply reported the following as in Table 6.2.

The evaluators decide to submit a negative report on the effectiveness of this second school, the results being below the district average. But beware! The two schools are the same. What happened? It is a case of Simpson's paradox, in which a result that is true for several categories is reversed when these categories are merged. The effect is the result of a hidden variable. In this case, students from low-income families perform worse on tests. If for the county the percentage of low-income students is 10%, but for the school in question, it is 50%, the previous result is

Table 6.1. Simpson's paradox. Evaluations by categories.

	District average	School results
Group A	80	85
Group B	50	60

Table 6.2. Simpson's paradox. Evaluations without categories.

	District average	School results
All students	77	72.5

Table 6.3. Simpson's paradox. Hidden variable.

	District total number of students	District test result	Students in school	Expected result	Test result
Total	10,000	77	100	77	72.5
A	9,000	80	50	80	85
B	1,000 (10% of total)	50	50	50	60

obtained. Observe how this happens in Table 6.3 in which the result is calculated if we apply the district averages for students at the school. The hidden variable is school *demographics*.

Gambler's Fallacy

It arises from the idea that for random events, if a particular result has not been obtained for a while, then the probability that it will happen increases. For example, some people study the lottery or roulette results, gamble on one that has not come out for a long time, and avoid those that have come out recently or repeatedly. But the probability of winning the lottery is the same if you play the same number each time or change it. If 24 has not come up on the (well balanced) roulette wheel for a long time, people think that "it is time for it to come up," and if five heads in a row come up when tossing a coin, they believe that it "must" come out tails on the next toss. But these random events are independent of each other, meaning that every time a (balanced) coin is tossed, the probability of heads is the same (1/2) no matter what happened before since the coin has no "memory" of what happened before. It is true that in the *long run*, a balanced coin will come up heads and tails half of the time, but it is also true that in the long run, you are dead.

The gambler's fallacy operates when events are independent and equiprobable. If they are not independent, then past results will affect future probabilities. If they are not equally likely (say a biased die), then some outcomes will be favored over others.

The previously mentioned fallacies are just a sample for you to get the general idea of the many ways our thinking can go wrong. There are books dedicated to this, and I would recommend Edward Damer's[94] and Jamie White's for a complete treatment of faulty reasoning which is becoming a growing problem when we need to think clearly.

I close with the words of physicist and journalist 1 David Grimes who in his book *Good Thinking*[95] writes: *The real challenge in this era of instantaneous information is to distinguish between the reputable and dubious, to reflect rather than react, and to question vigorously what we're told. This has never been more urgent nor more difficult.*

Chapter 7

Three Metaphysical Revolutions

He who desires to philosophize must first of all doubt all things. He must not assume a position in a debate before he has listened to the various opinions and considered and compared the reasons for and against. He must never judge or take up a position on the evidence of what he has heard, on the opinion of the majority, the age, merits, or prestige of the speaker concerned, but he must proceed according to the persuasion of an organic doctrine which adheres to real things, and to a truth that can be understood by the light of reason.

— Giordano Bruno (1548–1600 burned at the stake)[1]

I beseech you, in the bowels of Christ, think it possible you may be mistaken.

— Oliver Cromwell[a]

In a future that is as unavoidable as it is unwelcome, survival and sanity may depend upon our ability to cherish rather than to disparage the concept of human dignity.

— William R. Catton[2]

The history of science and scholarship is one of accelerating understanding of the world and our relation to it. Every generation of researchers builds on what the previous one has achieved. Still, sometimes revolutionary changes occur, some of them with significant impact within science (such as the transition from classical physics to relativity and quantum mechanics) and others that imply a significant change in our worldview. But whereas

---

[a] Oliver Cromwell in a letter to the General Assembly of the Kirk of Scotland; or, in case of their not sitting. To the Commissioners of the Kirk of Scotland: Musselburgh, August 3, 1650. http://www.cyberussr.com/hcunn/q-cromwell-beseech.html.

scientists understand and welcome these changes, the public resists or is unaware of the new views that in time have demoted humans from the lofty peak they thought (and think) they occupied.

It is difficult to accept that we are not God's gift to the Universe, but merely an animal that perchance is conscious and can think (not very well). But the change in metaphysical views cannot just remain inside university walls, while the great majority still adhere to the metaphysics of the Middle Ages. If we allow this, we shall return to them, or worse. I will review three essential cases.

7.1 Galileo Galilei's Case

Then spake Joshua to the Lord in the day when the Lord delivered up the Amorites before the children of Israel, and he said in the sight of Israel, Sun, stand thou still upon Gibeon; and thou, Moon, in the valley of Ajalon. And the sun stood still, and the moon stayed, until the people had avenged themselves upon their enemies. Is not this written in the book of Jasher? So, the sun stood still in the midst of heaven, and hasted not to go down about a whole day.

— Joshua 10:12-13

This being granted, I think that in discussions of physical problems, we ought to begin not from the authority of scriptural passages but from sense experiences and necessary demonstrations; for the Holy Bible and the phenomena of nature proceed alike from the divine Word the former as the dictate of the Holy Ghost and the latter as the observant executrix of God's commands. It is necessary for the Bible, in order to be accommodated to the understanding of every man, to speak many things which appear to differ from the absolute truth so far as the bare meaning of the words is concerned. But Nature, on the other hand, is inexorable and immutable; she never transgresses the laws imposed upon her or cares a whit whether her abstruse reasons and methods of operation are understandable to men. For that reason, it appears that nothing physical which sense-experience sets before our eyes, or which necessary demonstrations prove to us, ought to be called in question (much less condemned) upon the testimony of biblical passages which may have some different meaning beneath their words.

— Galileo Galilei[3]

If anyone thinks heavier bodies fall
More swiftly in their downward plunge, and thus
Fall on the lighter ones, and by this impact
Cause generation, he is very wrong.
To be sure, whatever falls through air or water
Goes faster in proportion to its weight,
For air's a frailer element than water.
Neither imposes quite the same delay
On all things passing through them, though both yield
More quickly to the heavier. But void
Can never hold up anything at all,
Never, its very essence is to yield.
So, all things, though their weights may differ, drive
Through unresisting void at the same rate,
With the same speed. No heavier ones can catch
The lighter from above, nor downward strike.

— Lucretius[4]

The first thing that comes to mind when Galileo Galilei (1564–1642) is mentioned is his conflict with the Catholic Church and his trial by the Inquisition in 1633 held at the convent of Santa Maria Sopra Minerva in Rome, under the papacy of Urban VIII. I will briefly review the metaphysical legacy this man left us, beyond his sad story. It is no longer an issue whether the Earth or the Sun is at the center — although a substantial part of the US population still does not know the correct answer.

Many consider it a case of conflict between religion and science. Still, it was more about political aspects related to the bloody 30-year war (1618–1648) between Protestants and Catholics, which threatened the authority of the Vatican. In his preface to his book about Galileo's case, renowned Italian-American philosopher and historian of science Giorgio de Santillana (1902–1974) says something which can be generalized to contemporary events[5]: *The real story affords a fascinating insight into the way in which such decisions actually take place and in which the ponderous state machinery is set into motion by what seem to be Reasons of State and perhaps later become so, but originate really as a constellation of accidents and personal motives. An objective account ought to be more relevant to a decent understanding than all the innuendoes, diversions, and stage sets invented around it on both sides. By pinpointing the culpability of a few individuals, it tends to absolve a far greater number who had stood hitherto under the darkest suspicions, and among them the Commissary-General of the Inquisition himself, who conducted the trial. Once recognized, the facts should lead us forward to the problems of present reality and disperse this perennial battle with windmills.*

In his time, no objective evidence for the motion of the Earth was available. Galileo, after the instruments of torture, were mentioned to him (some say he was shown them), decided to abjure and recited on his knees the text that had been prepared for him: *I, Galileo, son of the late Vincenzio Galilei of Florence, seventy years of age, arraigned personally for judgment, kneeling before you Most Eminent and Most Reverend Cardinals Inquisitors-General against heretical depravity in all of Christendom, having before my eyes and touching with my hands the Holy Gospels, swear that I have always believed, I believe now, and with God's help I will believe in the future all that the Holy Catholic and Apostolic Church holds, preaches, and teaches. However, whereas, after having been judicially instructed with an injunction by the Holy Office to abandon completely the false opinion that the sun is the center of the world and does not move and the earth is not the center of the world and moves, and not to hold defend, or teach this false doctrine in any way whatever, orally or in writing; and after having been notified that this doctrine is contrary to Holy Scripture; I wrote and published a book in which I treat of this already condemned doctrine and adduce very effective reasons in its favor, without refuting them in any way; therefore, I have been judged vehemently suspected of heresy, namely of having held and believed that the sun is the center of the world and motionless and the earth is not the center and moves.*

Therefore, desiring to remove from the minds of Your Eminences and every faithful Christian this vehement suspicion, rightly conceived against me, with a sincere heart and unfeigned faith I abjure, curse, and detest the above-mentioned errors and heresies, and in general each and every other error, heresy, and sect contrary to the Holy Church; and I swear that in the future I will never again say or assert, orally or in writing, anything which might cause a similar suspicion about me; on the contrary, if I should come to know any heretic or anyone suspected of heresy, I will denounce him to this Holy Office, or to the Inquisitor or Ordinary of the place where I happen to be.

Furthermore, I swear and promise to comply with and observe completely all the penances which have been or will be imposed upon me by this Holy Office; and should I fail to keep any of these promises and oaths, which God forbid, I submit myself to all the penalties and punishments imposed and promulgated by the sacred canons and other particular and general laws against similar delinquents. So help me God and these Holy Gospels of His, which I touch with my hands.

I, the above-mentioned Galileo Galilei, have abjured, sworn, promised, and obliged myself as above; and in witness of the truth, I have signed with my own hand the present document of abjuration and have recited it word for word in Rome, at the convent of the Minerva, this twenty-second day of June 1633.

He was condemned by the tribunal: *We condemn thee to the prison of this Holy Office during Our will and pleasure; and as a salutary penance, we enjoin on thee that for the space of three years thou shalt recite once a week the Seven Penitential Psalms.* This was commuted promptly to house arrest since no one wanted blood in their hands in the case of Galileo. It was also accepted that his daughter Virginia would recite the psalms of penance (Sister Maria Celeste of the Franciscan convent of San Mateo, in Florence).[6] By the way, there is no evidence that Galileo said: Eppur si muove (and yet she moves), although I can imagine that it went through his mind. (Everything indicates that Galileo was arrogant but not foolish.)

Thus, ended a fascinating and miserable story with the character of a Greek tragedy. The protagonists inevitably confront each other and are led to a fatal outcome as if guided by an invisible force, against which the main character rebels with audacious pride. How is it that the life of one so illustrious ended so poorly, recognized by many who held important positions, both civil and ecclesiastical? One of them, Ferdinand II of Médici, Grand Duke of Tuscany, described him as *the true light of our times.* Pope Urban VIII, when he was still Maffeo Cardinal Barberini (1568–1644), had sent a letter to Galileo: *I pray the Lord God to preserve you because men of great value like you deserve to live a long time to the benefit of the public.*

In 1609 Galileo pointed a small telescope of his making at the sky. Although it was much less potent than modern telescopes, he opened a new window to the universe, and he published what he saw in 1610 in a book

Sidereus Nuncius. It began with: *Great indeed are the things which in this brief treatise I propose for observation and consideration by all students of nature. I say great, because of the excellence of the subject itself, the entirely unexpected and novel character of these things, and finally because of the instrument by means of which they have been revealed to our senses.*

Galileo speaks of lunar mountains, where until then, it had been thought that the Moon was a perfect sphere made of "quintessence" (a fifth element). He reported on countless stars in the Milky Way and, more importantly, announced the discovery of Jupiter's moons (which Galileo astutely called "Medicean stars"). It implied that not everything revolved around the Earth, as was supposed. Galileo also observed the phases of Venus and its apparent size change, which falsified the Ptolemaic model. Everything indicated that the Aristotelian separation between a sublunary and a superlunary world was false.

His observations caused great interest, and he was invited to Rome by Barberini to present his results, to the astronomers of the Roman College, by then the most prestigious Jesuit university (founded by Ignacio de Loyola). Galileo became a celebrity. But the most intransigent Aristotelians refused to accept what Galileo had observed, some arguing that these were defects in the telescope's lenses, others refusing to look through it.

One such was Cesare Cremonini (1550–1631), a highly regarded professor of Aristotelian philosophy at the University of Padua. Galileo writes to Johannes Kepler: *My dear Kepler, I wish that we might laugh at the remarkable stupidity of the common herd. What do you have to say about the principal philosophers of this academy who are filled with the stubbornness of an ass and do not want to look at either the planets, the moon, or the telescope, even though I have freely and deliberately offered them the opportunity a thousand times? Truly, just as the ass stops its ears, so do these philosophers shut their eyes to the light of truth.*

Santillana writes[7]: *Official astronomy, represented by the great Tycho Brahe, had declared against it, [the Copernican system] and Tycho had advanced instead a compromise system of his own in which the Earth remained at the center of things. The philosophers of the schools dismissed Copernicus because his theory could not be reconciled with their physics. The Protestants were against him because they felt he cast doubt upon the literal truth of Scripture. As for the Catholic hierarchies, they held Copernicus in respect as a churchman and scholar, but they considered his system as one more of those ingenious mathematical devices which could lay no claim to physical reality. Mathematics was rated at the time as a thing for technicians and virtuosi, as they were called, with no claim to philosophical relevance; and the mystical and metaphysical speculations of some adventurous minds who searched for the "divine secret" in proportion and number were not such as to compel the assent of responsible scholars. Added to this, the Churchmen derived some good reasons for their reserve from the book of Copernicus himself, which had come to them provided with a spurious preface written really by Osiander, a Protestant clergyman, which disclaimed any pretension of physical validity for the theory.*

Galileo defends himself from the scholastic academics and astronomers of the Roman College who accused him of denigrating the Holy Scripture for lack of good arguments. Instead, they started a war of intrigue that entangled some Church officials. Thus, a battle was fought in writings, gossip, secret denunciations, and behind-the-scenes machinations. The Jesuit astronomer Father Christoph Grienberger (1561–1636) expressed after Galileo's trial[8]: *If Galileo had been able to retain the affection of the Fathers of this College, he would be living in worldly glory; none of his misfortunes would have happened, and he could have written at will on any subject, I say even of the earth's motions, etc.*

Galileo, as a good Catholic, tried to warn the Church of the problems to which it was exposed if it attacked a theory of the universe that was later to be proven true (as had done St. Augustine before-see below). In his letter to the Grand Duchess Christina of Tuscany (written in 1615 and published in 1636), he says about this situation[9]: *The reason produced for condemning the opinion that the earth moves and the sun stands still is that in many places in the Bible one may read that the sun moves, and the earth stands still. Since the Bible cannot err; it follows as a necessary consequence that anyone takes an erroneous and heretical position who maintains that the sun is inherently motionless and the earth movable.*

Regarding this argument, I think in the first place that it is very pious to say and prudent to affirm that the holy Bible can never speak untruth whenever its true meaning is understood. But I believe nobody will deny that it is often very abstruse and may say things which are quite different from what its bare words signify. Hence in expounding the Bible if one were always to confine oneself to the unadorned grammatical meaning, one might fall into error. Not only contradictions and propositions far from true might thus be made to appear in the Bible, but even grave heresies and follies. Thus, it would be necessary to assign to God feet, hands, and eyes, as well as corporeal and human affections, such as anger, repentance, hatred, and sometimes even the forgetting of things past and ignorance of those to come.... (Even in present times many have not understood this.)

In a letter of 1615 addressed to Father Antonio Foscarini, the influential Jesuit theologian Roberto Cardinal Bellarmino (1542–1621) (one of those who condemned Giordano Bruno) expresses the position of the Church[10]: *First, I say that it seems to me that your Paternity and Mr. Galileo are proceeding prudently by limiting yourselves to speaking suppositionally and not absolutely, as I have always believed that Copernicus spoke. For there is no danger in saying that, by assuming the Earth moves and the sun stands still, one saves all the appearances better than by postulating eccentrics and epicycles; and that is sufficient for the mathematician. However, it is different to want to affirm that in reality the sun is at the center of the world and only turns on itself, without moving from east to west, and the earth is in the third heaven and revolves with great speed around the sun; this is a very dangerous thing, likely not only to irritate all scholastic philosophers and theologians, but also to harm the Holy Faith by rendering Holy Scripture false.*

The resistance to Earth's movement was not only based on biblical texts, although it was understood that if God ordered the Sun to stop, that meant that the Sun was moving. But there were other questions: If Earth moves, how do we not feel it? If Earth is not the center of the universe, why do things fall to the center of the Earth? There were no clear answers for all this since it was necessary to change Physics before obtaining them. That is what happened with the change that we know as the "scientific revolution."

The "Copernican revolution" supported by Galileo was not empirically based. The same Galileo points out in his Dialogue, in the words of Salviati[11]: *No, Sagredo, my surprise is very different from yours. You wonder that there are so few followers of the Pythagorean opinion* (that the Sun is at the center), *whereas I am astonished that there have been any up to this day who have embraced and followed it. Nor can I ever sufficiently admire the outstanding acumen of those who have taken hold of this opinion and accepted it as true; they have through sheer force of intellect done such violence to their own senses as to prefer what reason told them over that which sensible experience plainly showed them to the contrary. For the arguments against the whirling of the earth which we have already examined are very plausible, as we have seen; and the fact that the Ptolemaics and Aristotelians and all their disciples took them to be conclusive is indeed a strong argument of their effectiveness. But the experiences which overtly contradict the annual movement are indeed so much greater in their apparent force that, I repeat, there is no limit to my astonishment when I reflect that Aristarchus and Copernicus were able to make reason so conquer sense that, in defiance of the latter, the former became mistress of their belief.* It is an interesting thought about the relation between theory and evidence.

Returning to history; The Church under the papacy of Paul V, after much consultation, decided in 1616 that Copernicanism was heretical and forbade the publication of Copernicus' book until specific corrections were made to clearly indicate that Copernicus was merely proposing a hypothesis.

At the request of Pope Paul V, Bellarmino informed Galileo in February of 1616 in a confusing incident (some think that documents were falsified) that the congregation of the Holy Office had declared that Copernicanism was opposed to the Holy Scriptures. It indicated that the Inquisition had prohibited maintaining, defending, and teaching the Copernican hypothesis. However, in a meeting between Galileo and the Pope, he noted that he understood the situation and that Galileo did not need to worry.

In 1625 Galileo began to work on his: *Dialogo Sopra i Due Massimi Ssistemi del Mondo, Tolemaico e Copernicano.* When it was published in 1632, it caused a storm despite having been authorized by ecclesiastical authorities. Urban VIII, named Pope in 1623, felt betrayed by Galileo because his treaty clearly defended the Copernican system, ignoring the instructions of 1616. Furthermore, in the figure of Simplicio, he seemed to mock the Pope. This led to the trial in Rome.

During the last years of his life, under house arrest, he wrote his most important treatise: *Discorsi e Dimostrazione Matematiche Intorno a Due Nuove Scienze*. This work, secretly taken out of Italy and published in Leiden in 1638, established the foundations of mechanics. Together with the discoveries of Johannes Kepler (1571–1630), they inspired Isaac Newton (1642–1726). His monumental work published in 1687: *Philosophiae Naturalis Principia Mathematica* achieved the grand synthesis, uniting the heavens and the Earth under the same fundamental laws. Galileo, born in 1564, died on January 8, 1642, at 78, shortly before Newton's birth. He had lost the battle, but in the long run, he won the war. He was a better theologian than the theologians were scientists.

The "scientific revolution" inspired by the publication of *De Revolutionibus Orbium Coelestium* by Nicolaus Copernicus (1473–1543) is not primarily a methodological revolution characterized by using mathematics and experimentation. Many cases are known in the Middle Ages in which experiments and calculations were used (within the technical limitations of the time) to corroborate or refute specific ideas. An interesting case is that of John Philoponus of Alexandria (~490 to ~570), who a 1,000 years before Galileo, wrote that heavy bodies did not fall faster than light ones (as Aristotle thought but never went to the trouble of finding out) throwing stones from some tower much older than the one at Pisa (from which Galileo did not officially throw anything). Also, the Spanish Dominican Father Domingo de Soto (1494–1560), in a book on the physics of Aristotle published in 1551, presents an accurate description of how bodies fall, and read what Lucretius wrote in the epigraph.[12]

The revolution occurred instead in a metaphysical and cosmological arena that changed the foundations of natural philosophy. The rupture was with the Aristotelian metaphysical conception of an organic teleological world (with a final cause). The imperfect sublunary world, with the Earth at its center, was different from the perfect rest. In its place arises a uniform and possibly infinite world of mathematical law. The Earth is nothing more than one of many planets, a mechanical world of inert matter in perpetual motion.

Galileo's case is a forced conflict between science and religion that arose mainly from the petty personal struggles of the protagonists. It is also a case of authoritarian repression of a minority idea. Galileo intended (for the good of the Church) to separate matters of faith from science matters. Galileo quotes Caesar Cardinal Baronius (1538–1607): *That the intention of the Holy Ghost is to teach us how one goes to heaven, not how heaven goes.* [13]

Tommaso Campanella[14] (1568–1639), a contemporary defender of Galileo, Dominican friar and philosopher (imprisoned for 27 years and tortured for his views), wrote: ...*any attempt to forbid Christians to study the book of nature is a crime against Christianity itself. For if the Christian religion is true, then it not only has no fear of other truths, but also should welcome any further knowledge of the natural world as additional insights into the wisdom and goodness of God. In short, the Church damages itself if it cuts off any access to God which may be found in the book of nature.* That echoes the fundamental lesson of Galileo: study the book of Nature. Still valid.

Galileo's friend, the painter Ludovico Cigoli (1559–1613) (who is the first painter to portray the Madonna standing upon a pockmarked Moon), wrote to him: ... *I have an idea for an emblem those pedants could put on their shingle. A fireplace with a stuffed flue, and the smoke curling back to fill the house, where people have gathered to whom darkness comes before dusk, (gente a cui si fa notte innanzi sera)* (we still have a lot of those).

In the end, the irony of the case is that neither the Earth nor the Sun turned out to be immobile. The whole battle was for an illusion (as are most of them). Several "Copernican revolutions" followed him, each one changing our perspective in the sense of making us less significant in the scheme of things.

7.2 Darwin's Case

> Another fallacy comes creeping in
> Whose errors you should be meticulous
> In trying to avoid — don't think our eyes
> Our bright and shining eyes, were made for us
> To look ahead with; don't suppose our thig-bones
> Fitted our shinbones, and our shins our ankles,
> So that we might take steps; don't think that arms
> Dangled from shoulders and branched out in hands

With fingers at their ends, both right and left
For us to do whatever need required
For our survival. All such argument,
All such interpretation, is perverse,
Fallacious, puts the cart before the horse.
No bodily thing was born for us to use
Nature had no such aim, but what was born
Creates the use. There could be no such thing
As sight before the eyes were formed, no speech
Before the tongue was made, but tongues began
Long before speech was uttered, and the ears
Were fashioned before long before a sound was heard,
And all the organs, I feel sure, were there
Before their use developed; they could not
Evolve for the sake of use, be so designed.

— Lucretius[15]

Today we find people bothered by the ideas this man had over a century and a half ago. The name of Charles Robert Darwin (1809–1882), along with that of his colleague and friend Alfred Russell Wallace (1823–1913), evokes "evolution," but the ideas of the evolution of life precede Charles. His grandfather, Erasmus Darwin (1731–1802), poet and naturalist, wrote in his 1794 book Zoonomia[16]: *From thus meditating on the great similarity of the structure of the warm-blooded animals, and at the same time of the great changes they undergo both before and after their nativity; and by considering in how minute a proportion of time many of the changes of animals above described have been produced; would it be too bold to imagine, that in the great length of time, since the earth began to exist, perhaps millions of years ... that all warm-blooded animals have arisen from one living filament, which THE GREAT FIRST CAUSE endued with animality ... and thus possessing the faculty of continuing to improve by its own inherent activity, and of delivering down those improvements by generation to its posterity, world without end?*

What is revolutionary is the *theory* proposed by Darwin for the process: *natural selection*, as explained in his famous book[17]: *On the Origin of Species by Means of Natural Selection: Or, The Preservation of Favoured Races in the Struggle for Life* which sits alongside Newton's *Principia* as two great scientific works.

According to Darwin, no divine intervention was necessary for the process of the creation of species. Darwin's long argument went against the orthodox idea of a divine creator, an "intelligent designer," in his time proposed by the natural theology of William Paley (1743–1805) (see below) that Darwin had studied. In his autobiography, Darwin expresses[18]: *Although I did not think much about the existence of a personal God until a considerably later period of my life, I will here give the vague conclusions to which I have been driven. The old argument of design in nature, as given by Paley, which formerly seemed to me so conclusive, fails, now that the law of natural selection has been discovered. We can no longer argue that, for instance, the beautiful hinge of a bivalve shell must have been made by an intelligent being, like the hinge of a door by man. There seems to be no more design in the variability of organic beings and in the action of natural selection than in the course which the wind blows.*

Natural selection, an algorithmic process (as I will explain) without a guide or plan, generated by random changes, was contrary to certain religious beliefs (and still is). It offers a metaphysics that postulates a universe without purpose nor creator. It is enough to see how Darwin changed the last paragraph of his great work, to realize that the conflict was latent in his mind.

At the end of the first edition, we read: *There is grandeur in this view of life, with its several powers, having been originally breathed into a few forms or into one; and that, whilst this planet has gone cycling on according to the fixed law of gravity, from so simple a beginning endless forms most beautiful and most wonderful have been, and are being, evolved.*

In the second edition, we can read: *There is grandeur in this view of life, with its several powers, having been originally breathed **by the Creator** into a few forms or into one; and that, whilst this planet has gone cycling on according to the fixed law of gravity, from so simple a beginning endless forms most beautiful and most wonderful have been, and are being, evolved* (my emphasis).

This was added to appease some people, particularly his wife Emma and her family. In passing note that this last paragraph is the only one of the entire corpulent work in which the term "evolution" is used (in "evolved").

In a letter[19] written in 1863 to his friend, the botanist Joseph Hooker, Darwin writes: *It will be some time before we see "slime, snot or protoplasm" generating a new animal. But I have long regretted that I truckled to public opinion & used Pentateuchal term of creation, by which I really meant "appeared" by some wholly unknown process. It is mere rubbish thinking, at present, of origin of life; one might as well think of the origin of matter.*

The main issue is the random component of evolution, which is a consequence of the game between chance and necessity.[20] It just does not fit with the idea that God made man "in His image" (it is the other way around). It is the reason for the bitter opposition to Darwinian evolution by certain religious groups, who have clearly understood what natural selection implies.

The Catholic Church had no choice but to accept that which could no longer be doubted, resorting to a somewhat elaborate solution to the problem. With Darwin, what is at stake is nothing less than the divine creation of human beings, the existence of the soul, the purpose of life, and ultimately immortality. Why go to church otherwise? The conflict between Darwin and religious ideas is inescapable. It is not resolved as easily as Galileo's case, which was also more personal than theological, as we saw. Moreover, it is offensive to many people who react: Are you telling me that there is nothing else?

In a speech to the Pontifical Academy of Sciences on October 22, 1996,[21] Pope John Paul II, although partly admitting biological evolution, also postulated that there was divine intervention in terms of human consciousness. (There *is* something else.)

In the speech, he refers to the encyclical *Humani generis*, issued by the pathetic Pope Pius XII in 1950, about false opinions against the foundations of Catholic doctrine. In *Humani generis*, the compatibility between Catholic religious beliefs and scientific research is affirmed, but, at the same time, the immutability of the fundamental postulates of the Christian religion is reiterated[b]: *Some imprudently and indiscreetly hold that evolution, which has not been fully proved even in the domain of natural*

---

[b] Encyclical *Humani Generis* of the Holy Father Pius XII. http://w2.vatican.va/content/pius-xii/en/encyclicals/documents/hf_p-xii_enc_12081950_humani-generis.html.

sciences, explains the origin of all things, and audaciously supports the monistic and pantheistic opinion that the world is in continual evolution. […], Whatever new truth the sincere human mind is able to find, certainly cannot be opposed to truth already acquired, since God, the highest Truth, has created and guides the human intellect, not that it may daily oppose new truths to rightly established ones, but rather that, having eliminated errors which may have crept in, it may build truth upon truth in the same order and structure that exist in reality, the source of truth. As I mentioned before, nothing is "fully proven" in science. A good example of dogmatism.

In his speech mentioned above, John Paul II said: *Today, more than a half-century after the appearance of that encyclical, (Humani Generis) some new findings lead us toward the recognition of evolution as more than a hypothesis.*

So far, so good, but then further down: *Pius XII underlined the essential point: if the origin of the human body comes through living matter which existed previously, the spiritual soul is created directly by God ("Animas enim a Deo immediate creari catholica fides non retimere iubet"). (Humani Generis)*

As a result, the theories of evolution which, because of the philosophies which inspire them, regard the spirit either as emerging from the forces of living matter or as a simple epiphenomenon of that matter, are incompatible with the truth about man. They are therefore unable to serve as the basis for the dignity of the human person.

That is to say that biological evolution is accepted in terms of the human body, but consciousness (spirit), that property so difficult to explain, requires divine intervention. One can understand his position: if there is "nothing else" there is no religion.

In his fascinating book neuropsychologist, Paul Broks[22] writes: *The brute fact is there is nothing but material substance: flesh and blood and bone and brain. I know, I've seen. You look down into an open head, watching the brain pulsate, watching the surgeon tug and probe, and you understand with absolute conviction that there is nothing more to it. There's no one there. it's a kind of liberation.*

But as difficult as it might be to explain a phenomenon, it does not follow that it is divine. *Not explicated does not mean inexplicable.* It would constitute the God of our ignorance (referred to as "the god of the gaps"), but then it would be temporary since, over time, our ignorance diminishes. Thus, the movement of the planets, for a long time inexplicable, is not caused by angels pushing, as was believed. Dante writes in Convivio Book 2.2[23]: *Here, certain Intelligences, or Angels, as we are more accustomed to call them, which preside over the revolution of the heaven of Venus as its movers, are invited to listen to what I intend to say.*

But, furthermore, admitting evolution, to what humans did the Pope refer to? *Homo erectus*? *Homo habilis*? *Homo neanderthalensis*? Recent studies of Neanderthal DNA indicate that it was not a separate species. There was genetic mixing between Neanderthals who migrated from Africa about 400,000 years ago and Sapiens who emigrated some 80,000 years ago without a visa. Through hybridization, we all have a bit of Neanderthal in us.[24] So how did the Almighty decide when consciousness and humanity began? *Homo habilis* of a million years ago, now extinct, was not a divine creation? And if it was, was it a bad design without a soul?

Evolution by natural selection is evidenced by research in areas as diverse as taxonomy, geology, and molecular biology, and only those who ignore all this work can doubt this fundamental fact. Indeed, if it were not so, we would not be now suffering from the "delta variant," and let us hope we do not get to omega. Research demonstrates the evolutionary process and integrates different facets of the biological world under a theoretical mantle that allows us to understand the history of life on this planet. The eminent Ukrainian-American biologist Theodosius Dobzhansky (1900–1975) summed it up in the following way[25]: *The business of proving evolution has reached a stage when it is futile for biologists to work merely to discover more and more evidence of evolution. Those who choose to believe that God created every biological species separately in the state we observe them but made them in a way calculated to lead us to the conclusion that they are the products of evolutionary development are obviously not open to argument. All that can be said is that their belief is an implicit blasphemy, for it imputes to God appalling deviousness.*

7.2.1 *Not very intelligent design*

> Creationists make it sound as though a "theory" is something you dreamt up after being drunk all night.
>
> — Isaac Asimov[26]

> To put it bluntly but fairly. anyone today who doubts that the variety of life on this planet was produced by a process of evolution is simply ignorant — inexcusably ignorant. in a world where three out of four people have learned to read and write.
>
> — Daniel Dennett[27]

The influential English theologian and philosopher William Paley (1743–1805) argued in his *"Natural Theology"* of 1802 subtitled: *Or, Evidence of the Existence and Attributes of the Deity, Collected from the Appearances of Nature*, that when we come across a clock, whose clear purpose is to tell time, we would conclude that such a complicated object was the result of a design. His premise was that there can be no design without a designer[28]: *In crossing a heath, suppose I pitched my foot against a stone and were asked how the stone came to be there, I might possibly answer that for anything I knew to the contrary it had lain there forever; nor would it, perhaps, be very easy to show the absurdity of this answer. But suppose I found a watch upon the ground, and it should be inquired how the watch happened to be in that place, I should hardly think of the answer which I had given, that for anything I knew the watch might have always been there. Yet why should not this answer serve for the watch as well as for the stone; why is it not admissible in that second case as in the first? For this reason, and for no other, namely, that when we come to inspect the watch, we perceive — what we could not discover in the stone — that its several parts are framed and put together for a purpose, e.g., that they are so formed and adjusted as to produce motion, and that motion so regulated as to point out the hour of the day; that if the different parts had been differently shaped from what they are, or placed in any other manner or in any other order than that in which they are placed, either no motion at all would have carried on in the machine, or none which would have answered the use that is now served by it.*

The argument by analogy can be summarized as follows:

1. We observe an ordered world. ("Ordered world" can mean the DNA molecule, a cell, or the human eye, among other things.)
2. All the ordered things we observe require an intelligent designer to produce them.
3. *Therefore*: That someone is God.

The argument fails in two aspects. First, if by "order" we mean biological order (cats beget cats, dogs breed dogs), then this order has a good explanation in the laws of genetics. The second premise is also false. The "order" of the planetary orbits or the structure of a crystal is the product of natural laws. Although many things (like Paley's clock) are ordered and designed, order does not require intelligence. The analogy fails. The great merit of Darwin's explanation for evolution is that it explains the observations without resorting to imaginary supernatural influences.

Although we cannot explain all the details and there are controversies regarding some (a natural thing in any scientific theory), errors have been made, and frauds committed, corrected, and pointed out by scientists themselves, they do not invalidate the theory. These controversies and difficulties are used by creationists to imply that the whole theory is wrong, whereas, for scientists, it is a motivation for further studies. This happens with all scientific theories, problems to be solved by future researchers as has occurred in the past. We do not rule out Physics because we do not understand the nature of dark matter, and we do not rule out Neuroscience because we cannot explain memory.

Furthermore, the conclusion is invalid because if someone designed the world, it is not *necessarily* the Christian God as the creationists claim; it could have been other gods. In any case, given the obvious defectiveness of the product, the best one could argue for is that it was the result of a committee of Gods.

The fallacy of *false dichotomy* used by those who subscribe to the idea of intelligent design as a scientific alternative to biological evolution (who with few exceptions are not scientists) has the following form:

1. Either intelligent design theory or natural selection theory is true.
2. The theory of natural selection is false (because it does not explain some detail).
3. *Therefore*: The theory of intelligent design is true.

In this case, both premises are false, ignoring the possibility of a third theory and the fact that natural selection has not been falsified, so the conclusion does not follow.

The idea of "intelligent design" has played an important role, particularly among Christian fundamentalist groups. Even the courts have had to determine whether this is scientific (something quite astonishing), and therefore should be taught in biology classes and have ruled that it is not science.

The most famous legal case in the US (after the historic 1925 "monkey" trial of John Scopes in Tennessee for teaching evolution, violating state law) is that of the Dover school district in Pennsylvania.[29] In 2004, the school board made up of a Christian fundamentalist majority (in the US, these are local elective positions), adopted the following resolution: *Students will be made aware of the gaps/problems in Darwin's theory, and other theories of evolution, including intelligent design in class. Note: Origins of Life is not taught.*

(It is not clear what the *other theories* are.) It was further resolved that ninth-grade students would be read in class a document that stated among other things: *Intelligent design is an explanation of the origin of life that differs from Darwin's opinion.* (But the theory of natural selection is as much an "opinion" as that there are atoms, or that light travels at 300,000 km/s).

The legal case (Tammy Kitzmiller *et al.* vs. Dover Area School District *et al.*) was heard in court before Judge John E. Jones III. After more than a month of argumentation, the judge presented his decision on December 20, 2005, in a 139-page document devastating for the proponents of intelligent design. In his conclusions, he wrote30: *Both Defendants and many of the leading proponents of ID make a bedrock assumption which is utterly false. Their presupposition is that evolutionary theory is antithetical to a belief in the existence of a supreme being and to religion in general.*

Repeatedly in this trial, Plaintiffs' scientific experts testified that the theory of evolution represents good science, is overwhelmingly accepted by the scientific community, and that it in no way conflicts with, nor does it deny, the existence of a divine creator.

To be sure, Darwin's theory of evolution is imperfect. However, the fact that a scientific theory cannot yet render an explanation on every point should not be used as a pretext to thrust an untestable alternative hypothesis grounded in religion into the science classroom or to misrepresent well-established scientific propositions. [...] It is ironic that several of these individuals, who so staunchly and proudly touted their religious convictions in public, would time and again lie to cover their tracks and disguise the real purpose behind the ID Policy. [...] The breathtaking inanity of the Board's decision is evident when considered against the factual backdrop which has now been fully revealed through this trial.

In its most modern form, intelligent design postulates that there are biological structures (macroscopic such as the eye, or microscopic such as the bacterial flagellum) or processes (such as blood coagulation) that are what is called "irreducibly complex." It is argued that its parts only make sense if they are coupled with the other parts of the structure or process to shape its function, which implies a design (and therefore a designer).

The question of why many natural systems *appear* to be designed is valid. The answer is found in the theory of evolution by natural selection that shows how the eye (for instance) can be a consequence of the adaptive development of an initial photo-sensitive cell, as Figure 7.1 shows. Half an eye is better than none; an eye with a better definition is better than one without definition in terms of survival. Natural selection takes care of the rest.

Darwin himself was aware of what those who subscribe to intelligent design propose: *If it could be demonstrated that any complex organ existed, which could not possibly have been formed by numerous, successive, slight modifications, my theory would absolutely break down. But I can find no such case.*

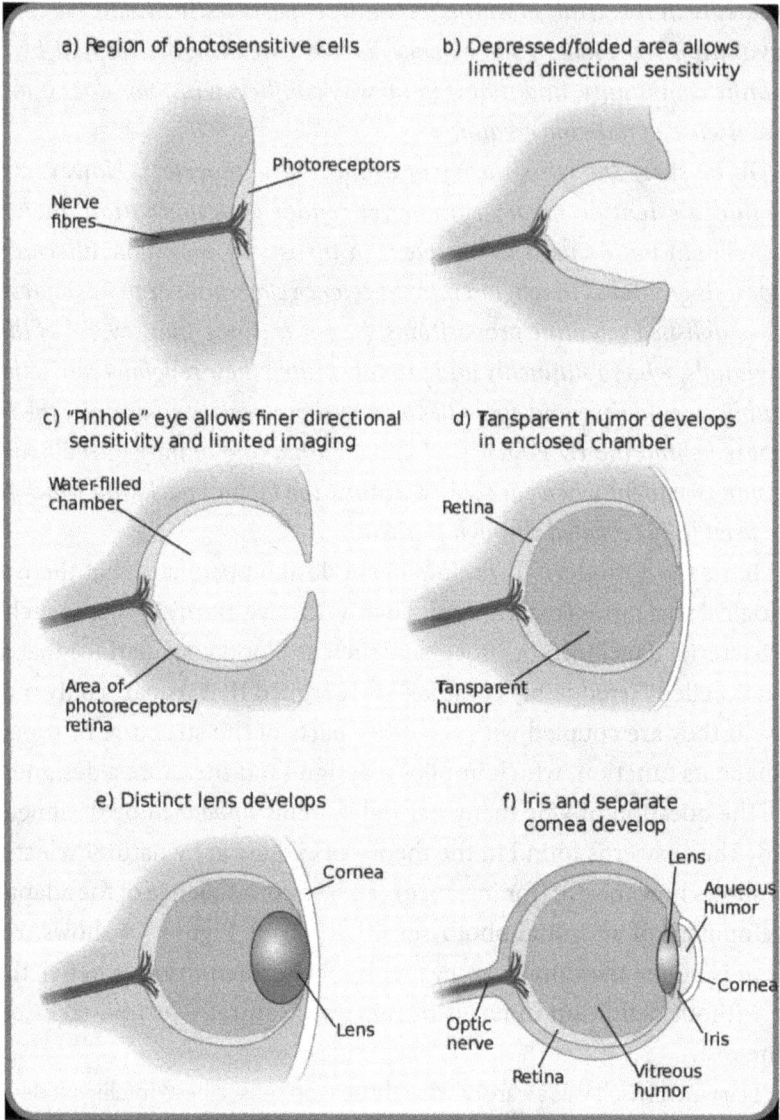

a) Region of photosensitive cells

b) Depressed/folded area allows limited directional sensitivity

Photoreceptors

Nerve fibres

c) "Pinhole" eye allows finer directional sensitivity and limited imaging

Water-filled chamber

Area of photoreceptors/ retina

d) Transparent humor develops in enclosed chamber

Retina

Transparent humor

e) Distinct lens develops

Cornea

Lens

f) Iris and separate cornea develop

Lens

Aqueous humor

Cornea

Optic nerve

Iris

Retina

Vitreous humor

Source: Matticus78 at the English language Wikipedia.

Figure 7.1.   Stages in the evolution of the eye.

The argument that "all this" could not arise at random is common, denoting a lack of understanding of the basic elements of biological evolution since this is not what the theory of natural selection says. It also is not about "survival of the fittest" if by fittest you understand the strongest.

This wrong idea is behind "Social Darwinism," used in the past to justify the supremacy of one group over another, to justify violence and war. It is hogwash, except that it still seems to be part of the social genes of not a few. Biological evolution is based on the following factors that together produce the *appearance* of design.

- *Variation* — all life forms are susceptible to genetic variability within a population (by mutations and genetic mixing) [random process].
- *Inheritance* — genetic properties are passed on to descendants [descent with modifications].
- *Natural selection* — organisms with characteristics valuable to their survival in the complex environment they inhabit are more likely to have offspring and spread these characteristics [deterministic process].
- *Time* — the process of evolution needs many generations, a lot of time. The age of the Earth is about 4 billion years, and not 6,000 as creationists claim against all the archaeological and geophysical evidence. *It is the availability of all the time in the world that allows almost impossible things to happen.*

Darwin wrote: *It is not the strongest of the species that survives nor the most intelligent but the one most responsive to change.* Unfortunately, the world is changing, and we are not responding.

Not understanding that the evolutionary process has two components, one that is random and another that is deterministic, produces the wrong idea of the random generation of the biosphere. The evolutionary process is not about assembling a molecule or an organism *in toto* from its parts but is a cumulative process of small changes throughout billions of years, caused by random genetic changes and guided by the process of natural selection.

One might also ask that if forms of life were designed by an omniscient and omnipotent intelligence; then why the great majority has failed. Were they flawed designs? Why create something that your omniscience tells you will not survive? Why not provide us with the necessary intelligence not to do what we are doing? Is it a bad joke?

In the end, an inevitable question remains. If a designer (God) is invoked to explain what seems inexplicable (although it is only unexplained), the question of who designed the designer is ignored. And if the answer were that He was there forever, then one could ask what He was doing before the creation so many billion years ago. But if He can be there forever, then you might as well think of the universe (or multiverse) existing eternally and get rid of the middleman.

Francis Collins, director of the Genome Project and a Christian, said[31]: *As someone who's had the privilege of leading the human genome project, I've had the opportunity to study our own DNA instruction book at a level of detail that was never really possible before. It's also now been possible to compare our DNA with that of many other species. The evidence supporting the idea that all living things are descended from a common ancestor is truly overwhelming. I would not necessarily wish that to be so, as a Bible-believing Christian. But it is so. It does not serve faith well to try to deny that!*

7.2.2 Darwin's insult

We have many people even here who hasten to condemn evolution without having the remotest conception of what it is that they are condemning, nor the slightest interest in an objective study of the evidence in the case which is all that 'the teaching of evolution' means, men whose decisions have been formed, as are all decisions in the jungle, by instinct, by impulse, by inherited loves and hates, instead of by reason. Such people may be amiable and lovable, just as is any house dog, but they are a menace to democracy and to civilization because ignorance and the designing men who fatten upon it control their votes and their influence.

— Robert Millikan[32]

The fundamental attributes of life that we expect will arise elsewhere in the universe are few and general. The first thing we observe is that an organism constitutes an open system (that is, it exchanges matter and energy with its environment) that is far from thermodynamic equilibrium (and so can decrease entropy locally). This is a physical characteristic and should not be confused because many humans seem unequilibrated psychologically. Living beings maintain a higher temperature and a concentration of molecules different from that of their environment. With death, equilibrium returns, the dead body cools and disintegrates into its component atoms, the distinction between inside and outside disappears, and entropy increases.

A constant exchange of material and energy — metabolism — is needed to maintain the living state. Thus, the laws that govern the behavior of the living are the same ones that govern the behavior of the dead, with no need for a "vital force" or "breath of life" or anything that resembles it, *nothing else*. Nor is the second law of thermodynamics violated (as argued by some who clearly do not understand it) because it applies to *closed systems*.

If the above were all, there could be worlds of organisms that are formed, live, and reproduce (or not) and die (or not) in which nothing interesting will happen. The ingredient that would transform this boring world into one of great interest is *error*. What Darwin called *descent with modification* is the key to the evolutionary process that allows, once life is established, for changes and *endless forms most beautiful* to emerge. Without hereditable modifications (mutations), there is no evolution.

We can also discard the widely held idea that evolution means progress and, even worse, that progress has a goal, and to top it off, the goal is us. Instead, suppose there was a purpose (a teleological cause). In that case, there is no doubt that it should aim for something much better than us.

The process of biological evolution is of great explanatory power because of its fundamental logical character. The algorithmic character to which I alluded above — Daniel Dennett[33] calls it Darwin's dangerous idea — gives it a universality that puts it on a par with some other universal laws such as those of thermodynamics.

An algorithmic process is characterized by three basic properties:
- It is independent of its specific implementation (either by humans who manipulate or by a computer), and its power comes from its logical structure.
- It can be broken down into a series of automatic and straightforward steps that do not require any kind of thought or skill, no matter how complex the total process is, and interesting its result.
- The process guarantees the outcome.

The biological algorithm is one of selection and construction. It begins with random changes of the genome (the genotype), which are translated into changes in the organism's characteristics (the phenotype). The process continues by selecting those phenotypes that are better at surviving and reproducing in the local environment.

Because of this universal character of the evolutionary process, it is reasonable to think that, given the appropriate conditions, biological evolution also operates on any other planet on which organisms that reproduce with inheritable variations arise. But by no means does this imply extraterrestrial intelligent and humanoid aliens, nor biochemistry identical to that of terrestrial organisms.

We observe that even though millions of species have arisen on Earth, intelligence emerged only once in the last 0.005% of the planet's history. From what we observe, it may not last beyond 0.006% of its history. As Steven Jay Gould points out[34]: *Wind back the tape of life to the early days of the Burgess Shale;* [500 million years ago] *let it play again from an identical starting point, and the chance becomes vanishingly small that anything like human intelligence would grace the replay.*

Daniel Dennett summarizes as follows: *Here. then, is Darwin's dangerous idea that best accounts for the speed of the antelope. the wing of the eagle. the shape of the orchid, the diversity of species. and all the other occasions for wonder in the world of nature. It is hard to believe that something as mindless and mechanical as an algorithm could produce such wonderful things. No matter how impressive the products of an algorithm, the underlying process always consists of nothing but a set of individually mindless steps succeeding each other without the help of any intelligent supervision; they are "automatic" by definition: the workings of an automaton. They feed on each other.*

Shortly after the publication of his work, Darwin, in a letter to his critic, the geologist Charles Lyell, expresses: *We must under present knowledge assume the creation of one or of a few forms, — in the same manner as philosophers assume the existence of a power of attraction, without any explanation. But I entirely reject as in my judgment quite unnecessary any subsequent addition "of new powers, & attributes & forces"; or of any "principle of improvement," except in so far as every character which is naturally selected or preserved is in some way an advantage or improvement, otherwise, it would not have been selected. If I were convinced that I required such additions to the theory of natural selection, I would reject it as rubbish. But I have firm faith in it, as I cannot believe that if false it would explain so many whole classes of facts, which if I am in my senses it seems to explain.*[35]

The biosphere is a delicate veneer of life sheltered by an atmosphere that nourishes and protects it. It is bounded to a large extent by the temperature at which water remains liquid. A few kilometers above the surface of the Earth, the temperature is so low that water freezes. Let's go in the opposite direction towards the interior of the Earth. The crust's temperature increases to reach the temperature at which the water boils at about four kilometers.

The earliest evidence of life (the initial microscopic life left no fossil remains) is provided by chemical residues of biological processes found in ancient rocks. Thus, for example, the Australian fossils of the Vendian (543–650 million years ago) are the oldest remains of multicellular organisms we know. Still, older fossils are found as stromatolites, the remains of bacterial colonies that populated the planet over a billion years ago.

Most organisms are not as versatile as we are and inhabit a limited ecological niche of the biosphere. Except for the cockroaches, of course, and the invisible bacteria, which always were, always will be, and are undoubtedly the dominant and most common form of life on the planet (and possibly the universe). They are the organisms that present the greatest biochemical diversity of all and the ability to live in spaces into which you would not dare to put your finger and will continue to live after the Anthropocene Transition.

We must distinguish (as Darwin already suggests) between the *evolution* of life once we have replicating organisms and the *origin* of life. This process passes from chemical compounds to replicating polymers and some bacterial precursor (proteobacteria), perhaps in a world dominated by autocatalytic RNA molecules.

Today we know a lot more than Darwin did, who could not talk about genes or karyotypes, never mind DNA. The differences between species and between individuals of a species (such as *Homo sapiens*) are determined by differences in the genome (DNA). Closely related species have a small but important fraction of the chemical base sequences that are different. For example, comparing the human and chimpanzee genome sequences shows that they differ by about 1.2%, so they are our closest cousins, followed by the gorillas. Whereas the genetic difference between individual humans is about 0.1%, sufficient to identify criminals in some cases. The difference between us and a banana is about 50%, so we are more closely related to a Chimp than to a banana, as is evident.

We have a variety of dogs, selectively bred for some characteristic we desire (a case of artificial selection as opposed to natural selection), size, aggressiveness, or whatever, and the difference encoded in their DNA causes the difference between a Chihuahua and a Great Dane. But they all are *dogs*, none superior to another except that a Great Dane cannot escape through a small hole whereas a Chihuahua can. So, you could say the Chihuahua is superior, but only in that restricted sense for that particular situation. A Chihuahua cannot confront a Rottweiler, whereas a Great Dane could, and so on, but there is no sense in which one variety of dog is superior to another.

Hitler did not understand this, and so, although he was of average height and dark-haired and dark blue-eyed, he subscribed to the concept of a mythical superior Aryan race that originated somewhere in Asia. The French aristocrat Count Joseph Arthur de Gobineau (1816–1882) distinguished between "white," "yellow," and "black" races, writing that[36]: *The white race originally possessed the monopoly of beauty, intelligence and strength and that whatever of the positive qualities the Asians and blacks possessed was due to subsequent miscegenation*, and so much garbage up to the present. For Hitler, Jews, and gypsies, although white, were not Aryan, were inferior and could be eliminated (which he did). He did not bother with black people because there weren't many in that part of Europe, although the few were discriminated against, abused, and murdered.

The point is this: the difference in skin color between people is skin deep, as unimportant as eye color. Skin color has to do with survival in the sunny climate of Africa, which is where we all came from, as established by a plethora of scientific studies, and whether you like it or not, our ancestors were dark-skinned. The whole idea of "race" is a fabrication, perhaps understandable from the perspective of a human that lived in northern latitudes a few 100 years ago and felt superior but deplorable today. Racism is one of those human afflictions that is the cause of a great deal of suffering worldwide. Racists are not even wrong. I've yet to hear of people of one "race" arguing that *theirs* is inferior! And although progress has been made, it is evident that racism is alive and well and is not eliminated by passing laws. It just gets repressed until someone encourages its resurgence or looks the other way. Not long ago, there were thousands of towns in the US that were known as "sundown towns,"[37] not because of their beautiful sunsets, but because crossing the city limits, you could find a sign which warned non-whites to *not let the sun go down on you while in this town*. The signs are gone, but not the sentiment, and today there are other ways to segregate a part of a city or a whole town. Just go for a drive, and you will see it.

I quote from the inspiring book[38] written by Benjamin Ferencz (born 1920), the last surviving prosecutor of the Nuremberg trials: *You are 99% genetically identical to every human being on the planet, whatever they look like, wherever they came from, wherever their great-great-great-grandparents came from, whatever language they speak, whatever believes they hold, whether they are criminals, evil or psychopaths. It cannot be one rule for them and another for you or for me — under any circumstances — but especially when it comes to justice, which should be the pinnacle of civilization and serves to protect us all.*

Nothing else is needed to understand how we came to be, and this, for many, is something offensive. However, they cannot explain what is offensive, merely complaining that humans do not "descend from monkeys." But that is not what evolution says; if it were, there would be no monkeys. It says that in the deep past, millions of years ago, there was a common ancestor of monkeys and humans, and in any case, if someone should be offended, it would be the monkeys. For those who ignore science, it seems miraculous. For those who know science, it is wonderful.

7.3 Ramón y Cajal's Case

> Since mind, in fact, is part of man, one part,
> Fixed in one definite place, like ears and eyes
> And other senses regulating life —
> Since hands or eyes, or nostrils, have no feeling
> Apart from us, and no existence either
> Except a rapid wasting, so the mind
> Cannot exist without a human body
> To serve as an urn, as vessel, for it. Make
> A better metaphor, if you can, to serve you!
>
> — Lucretius[39]

> Much stranger than the textbook picture, the brain is a cryptic kind of computational material, a living three-dimensional textile that shifts, reacts, and adjusts itself to maximize its efficiency. The elaborate pattern of connections in the brain — the circuitry — is full of life: connections between neurons ceaselessly blossom, die, and reconfigure. You are a different person than you were at this time last year because the gargantuan tapestry of your brain has woven itself into something new.
>
> — David Eagleman[40]

This is ongoing, a metaphysical transformation whose extent we still do not understand. Neuroscience of the brain has a direct, profound, and very personal impact on the conception of what we are and could strike a fatal blow to religions if everyone understood it.

Spanish neuroscientist and pathologist Santiago Ramón y Cajal (1852–1934; Nobel Prize in Physiology or Medicine for 1906) is considered the founder of neuroscience. (Ramón was his father's last name and Cajal that of the mother as customary.)

What began with him continues with fascinating new research. Today we have no choice but to reject something that is the product of thousands of years of previous thought: the idea that the mind is one thing and the brain another, something that many still believe. It generates a pseudo-problem since two ontologically separate entities *cannot* interact since they have nothing in common. Something is posited that simply cannot be, and then analyzed as to how it can be.

French philosopher René Descartes decided that the small pineal gland (which is responsible for producing melatonin) located at the center of the brain was the organ in which the interaction between the body and the soul occurred (or between the brain and the mind). However, he did not solve anything since the question was not *where* the interaction occurred but *how*. British philosopher Gilbert Ryle (1900–1976) called it "the dogma of the Ghost in the Machine."

The difficulty was already evident in his time. In a letter written to Descartes by the erudite Princess Elizabeth of Bohemia (1617–1680), the third of thirteen children of Frederick V, elector of the Rhenish Palatinate, and Elizabeth Stuart, in 1643 she asks[41]: *So I ask you please to tell me how the soul of a human being (it being only a thinking substance) can determine the bodily spirits, in order to bring about voluntary actions. For it seems that all determination of movement happens through the impulsion of the thing moved by the manner in which it is pushed by that which moves it, or else by the particular qualities and shape of the surface of the latter. Physical contact is required for the first two conditions, extension for the third. You entirely exclude the one [extension] from the notion you have of the soul, and the other [physical contact] appears to me incompatible with an immaterial thing.* She could have written it yesterday.

Modern studies of the brain's functioning clarify that brain activity and mental activity are the same, caused by neuronal patterns regulated by internal and external states through chemical signals. One of the many experimental indications that the brain and the mind are the same things was done by Roger Sperry (1913–1994, Nobel Prize in Physiology or Medicine 1981). He showed that individuals whose two hemispheres of the brain had been surgically separated (callosotomy — surgery that cut the corpus callosum to treat severe cases of epilepsy) acted as if they had two independent brains. The mind also had divided. In his words[42]: *The left and right hemispheres following their disconnection, function independently in most conscious mental activities. Each hemisphere, that is, has its own private sensations, perceptions, thoughts, and ideas all of which are cut off from the corresponding experiences in the opposite hemisphere. Each left and right hemisphere has its own private chain of memories and learning experiences that are inaccessible to recall by the other hemisphere. In many respects, each disconnected hemisphere appears to have a separate "mind of its own."*

In light of the evidence, despite the desire of religious authorities and their dogmas, we will have to accept the disturbing fact that we are simply a complex of atoms that form molecules that form cells, among which are the neurons. These are connected in a surprising way creating a "connectome" from which our mind emerges. If we could know the structure of the connectome in detail, it would define who we are (at that moment). From that arises a mind capable of thinking and trying to understand how it occurred. We are not an embodied spirit, but we are the activity of our brain. Francis Crick (1916–2004) (Nobel Prize winner in 1962 for the discovery with James Watson (born 1928) and Maurice Wilkins (1916–2004) of the structure of DNA) writes[43]: *You, your joys and your sorrows' your memories and your ambitions, your sense of personal identity and free will, are in fact no more than the behavior of a vast assembly of nerve cells and their associated molecules. As Lewis Carroll's Alice might have phrased it: "You're nothing but a pack of neurons".*

Transcendence matters to us so much that despite the total lack of evidence and logic, we insist on the idea that "there must be something else" and that hidden influences are out there.

But a disembodied mind that could survive the total destruction of the brain contradicts the experience that the mind does not survive anesthesia (that is, it remembers nothing of what happened). It is also affected by local brain lesions that cause agnosias (the inability to process certain sensory information).

The reason our brain thinks, behaves, and reacts to certain inputs, in one way and not another, has a strong component determined by natural selection (and another that is cultural). Evolutionary and cognitive psychology explains these characteristics, and those interested in manipulating the public mind are the first to study these findings that most individuals are unaware of. Why do you think most prices end in 95 or 99?

Psychologist Hank Davis writes[44]: *Evolutionary psychology is a relatively new field that offers a scientific, indeed a biological approach, to understand human behavior. Unlike other fields of psychology, an evolutionary approach attempts to understand humans as part of the biological world in which they evolved. Many of those puzzling, irrational behaviors may stem from adaptations made by our ancestors. If so, we are stuck with mental modules the weigh us down in both laughable and dangerous ways. The mental equipment we carry in our modern skulls is over 100,000 years out of date.*

For our ancestors who inhabited a dangerous world they could not control, the idea of *agency* and *intentionality* in the phenomena they could not explain had survival value, and as collateral damage paved the way for religious thinking.

It will cause headaches and emotional torment to many because, if they still fight about Darwin, the above implications are stunning, which in boxing would be judged as technical "knock out." A religion without something that lasts after death has little reason to be. But in the end, as I indicated, the world is as it is and not as we would like it to be. The torment is so intense that some societies will not support or even prohibit investigating this issue. And thus, return us to the Middle Ages (and it is happening also in the West), just as the past US government (I refer to Trump) was trying to curtail research into Climate Change.

The metaphysical problem that all this implies is summarized by Patricia Churchland, one of the founders of "neurophilosophy"[45]: *Here is the implication that may not be welcome: In dementing diseases and in normal aging, neurons die, brain structures degenerate. In death, brain cells quickly degenerate, with massive loss of information. Without the living neurons that embody information, memories perish, personalities change, skills vanish, motives dissipate. Is there anything left of me to exist in an afterlife? What would such a thing be? Something without memories or personality, without motives and feelings? That is no kind of me. And maybe that is really okay after all.*

Although we do not know how memory functions, we have learned that our memory is fallible. It is influenced by events that occur after memory is created and what is remembered is a reconstruction (and sometimes simply a construction) based on some remembered events. So, the one who says: "I clearly remember," is clearly wrong. Something true remains in the memory; otherwise, there would be no memory, but the fact is that the details are not remembered correctly; things that did not happen are introduced, and things that did happen are forgotten.

We do not even remember our grandparents clearly, and we do not remember (if "remember" is the right word) John Kennedy, Mahatma Gandhi, Giordano Bruno, or Jesus. We only remember what others have written about them who possibly did not know them either, but at least we have what they wrote and perhaps old photographs or TV recordings (not of the latter two), and in modern times what they left on Twitter or Facebook.

It becomes challenging to know the characters of antiquity, whose stories were written after many years of an unreliable oral tradition, memories plagued by errors, distortions, and motivated thinking. The gospels are a good example, written decades after the death of Jesus, reproduce an oral tradition that altered the facts, that multiplied the errors, that invented events that never happened, and cited phrases that were never said. But the authors did not write history but were motivated to convince the pagans, displacing the Jews and ingratiating themselves with the authorities. So, it makes no sense to interpret them literally.[46] Whenever I hear someone say, "Jesus said" or, "Muhammad said," I wonder how they know.

The scientific study of the mind enters an interesting dimension since it is the mind that studies itself. How consciousness arises from the constitutive materials of an organism is a mystery. It is not even a problem. Ramon y Cajal wrote: *as long as our brain is a mystery, the universe, the reflection of the structure of the brain, will also be a mystery.* Many mysteries of the past were reduced to solved problems once we forged the conceptual and mathematical tools to deal with them. It is a mistake to suppose that we have the methods and concepts necessary to transform the mysteries of the present into solvable problems at this point in history. The history of knowledge shows us that it is an unfounded position. We need new paradigms to understand the mind if it were possible to understand oneself. Some think that, although we can know in detail how the brain works, it might be impossible to understand what consciousness is, to know how it feels like to be a bat.[47] Paul Brooks, a neuropsychologist, puts it this way in his fascinating book (a la Oliver Sacks)[48]: *An ocean of incomprehension heaves beneath the textbook-confident surface of plain facts and technicalities that I present to my colleagues and patients. I have a clear picture of the material components of the brain, and I'm prepared to add lip at length about features of its functional architecture — the interlocking systems and subsystems of perception, memory, and action. But quite how our brains create that private sense of self-awareness we all float around in a mystery. I have no idea how the trick is achieved.*

We will see.

Paul Churchland points out[49]: *Recall our early attempts to understand the nature of Life, and the many dimensions of Health, prior to the many achievements of modern Biology, such as macroanatomy, cellular anatomy, metabolic and structural chemistry, physiology, immunology, protein synthesis, hematology, endocrinology, molecular genetics, oncology, and so forth. The medieval and premodern attempts, we can all agree, were downright pitiful, as were the medical practices that were based on them. But why should we expect our understanding of the nature of Cognition (cf. Life) and the many dimensions of Rationality (cf. Health) to be any less pitiful, prior to our making comparable achievements in penetrating the structure and the activities of the biological brain?*

Not long ago, it was considered that four humors flowed in the body (a concept invented by Hippocrates): blood, black bile or melancholy, phlegm, and yellow bile or cholera, in analogy with the four elements: air, earth, water, and fire. All the evils resulted from an imbalance between these humors, and the cures consisted of painful purges, bleeding, induction of vomiting, or sweating, to restore balance. Sometimes an improvement was achieved, since inducing vomiting, for example, relieved the effects of poisoning. Our language still remembers this when we speak of "bad humor" and melancholic choleric or phlegmatic tempers. In terms of the brain and the mind, we find ourselves in the Middle Ages.

There is also another possibility. We are starting to produce "intelligent" devices: Alexa this, Alexa that, quite useful if it is too difficult for you to get up and turn off the light, and maybe also quite useful to spy on your habits and preferences. We place listening devices in our homes and later complain about our privacy! Who knows if a non-biological intelligence created by us will take over in the future, the mother of all Golems, and perhaps it is so on some other planet a million years ahead of us newcomers? It is unclear if artificial intelligence will have consciousness, but it will not be limited by our environmental problems; it will be much sturdier and unethical. There is no reason to think that we are the pinnacle of possible intelligence; indeed, machines can surpass us in many respects. We are at the highpoint only because the next one down is a Chimp.

We might someday understand (if we are still around) and will become soulless. But we will not, therefore, become "heartless," we will continue to fall in love, we will continue to reject reprehensible behaviors (except for those doing it), and we will be moved by our good and bad experiences. But it will change our illusion of eternal life after this one, and it will free us from the eternal boredom in paradise or the everlasting torment in hell. On the contrary, it will invite us to live the only short life we have in a fuller and more dignified way. Who knows, it might even take away our urge to kill.

I end on a black humor note (I can't help it). After being captured in Argentina by Mossad agents, living under the false name of "Ricardo Klement," Adolf Eichmann, one of Hitler's henchmen who had escaped, was tried in Jerusalem and sentenced to death. His last words before being hanged in 1962 were: "Long live Germany, long live Argentina, long live Austria. *I shall never forget them*".[50]

Chapter 8

Beliefs and Knowledge

It is wrong always, everywhere, and for anyone, to believe anything upon insufficient evidence.

— William K. Clifford[1]

The influence of our wishes upon our beliefs is a matter of common knowledge and observation, yet the nature of this influence is very generally misconceived. It is customary to suppose that the bulk of our beliefs are derived from some rational ground, and that desire is only an occasional disturbing force. The exact opposite of this would be nearer the truth: the great mass of beliefs by which we are supported in our daily life is merely the bodying forth of desire, corrected here and there, at isolated points, by the rude shock of fact.

— Bertrand Russell[2]

I believe in God, the father almighty, creator of heaven and earth.
I believe in Jesus Christ, his only Son, our Lord.
He was conceived by the power of the Holy Spirit and *born of the Virgin Mary*.
He suffered under Pontius Pilate, was crucified, died, and was buried.
He descended to the dead.
On the third day he rose again.
He ascended into heaven and is seated at the right hand of the Father.
He will come again to judge the living and the dead.
I believe in the Holy Spirit,
the Holy Catholic Church,
the communion of saints,
the forgiveness of sins,
the resurrection of the body,
and the life everlasting. Amen.

— Credo of the Catholic Church

It is healthy to critically analyze our fabric of beliefs that constitutes the framework from which we look at our world. It is a spiny ivy rooted in our mind, which clings to our being by intercepting light, and which needs a gardener who knows where to cut. It has grown fed by the continuous process of formal and informal education, together with the media, and determines our way of thinking and acting.

Many say: "seeing is believing," although it would be more appropriate to say that "believing is seeing." What you see or smell is behind your eyes and nose, and what you hear is between your ears. There is more to seeing than meets the eye. Our brain processes our perceptions and often leads us astray (see Figure 6.4). Much has been written about why many believe unbelievable things, why nonsense is so prevalent, why stupidity became a virtue,[3] why people believe weird things.[4]

A great deal of what people think they know is the product of unreliable, misleading, or biased sources, increasingly influential and overwhelming, increasingly controlled by fewer individuals or organizations with lots of money, and the "free press" struggles to remain free. Kevin MacKay writes[5]: *The ability to spin and suppress news in their own interests provides economic and political elites with a powerful mechanism of control. Citizens in western democracies may have the right to vote, but if we have limited access to critical perspectives and unbiased news sources, we will almost certainly lack the ability to cast an informed vote. This, in turn, makes it difficult to understand and effectively counter the action of elites that are oppressive to working people and minorities, and destructive to the environment.*

The public is left defenseless, lacking the tools to distinguish truth from falsehood, the possible from the impossible, the scientific from that which is not. They are not prepared to face the imbroglios proposed by those who want their vote, money, or devotion. And if this has never been easy, today it is most difficult, exposed to lies and half-truths as we are most of the time. The scientific temper is the way out. In the end, *he who knows nothing must believe everything.* An increasing fraction of the public even thinks that science and opinion are the same things. Many cannot think properly and are fooled by fallacies. This is also the case for educated people because education increasingly emphasizes *what* to think and not *how* to think.

It is common to say that one has the right to believe whatever one wants, a right to one's opinions, just as one has the right to free expression (with some well-known but hard to understand exceptions). And many end up an argument expressing their right to an opinion, which was not what was argued about. The right to an opinion does not make the opinion true.

What concerns me is that our beliefs drive our actions and everyday decisions, often leading to disastrous results: personally, if you believe a swindler who practices some kind of "alternative medicine" or in the efficacy of homeopathic products, which could lead to a tragic end, and socially if you believe that by killing a bunch of innocent people you will gain access to some heaven or save the nation. History is filled with tragedies based on the mistaken belief by whites that people of a darker complexion are inferior or that people of another religion must be converted or killed. So, no, you do not have the right to believe anything you wish. I would go further: as a human being, you must examine your beliefs and justify them and have the courage to discard them if wrong. Unfortunately, there are people so gullible that they believe everything that crosses their ever more ubiquitous screens, and others, so skeptical that they do not believe even that which is obvious. Both positions are noxious.

To this, we can apply the following written by William Clifford many moons ago[6]: *If a man, holding a belief which he was taught in childhood or persuaded of afterward, keeps down and pushes away any doubts which arise about it in his mind, purposely avoids the reading of books and the company of men that call into question or discuss it, and regards as impious those questions which cannot easily be asked without disturbing it — the life of that man is one long sin against mankind.*

Our degree of confidence in something that we believe is variable; we believe some things with certainty (like the Sun will rise tomorrow) and other things we believe with some uncertainty (like coffee is good or bad for health).

There is a subtle psychological distinction between "believing *in*" and "believing *that*, which sometimes leads to futile arguments. Welsh philosopher, H. H. Price[7] (Henry Habberley, 1899–1984), proposed the difference. When we say that we believe *that* — the sky is blue, Ghana is in Africa, Russell was a great philosopher, etc. — we refer to our belief that a proposition is true subject to an empirical test. On the other hand, when we believe *in* — God, Santa Claus, ghosts, or the virgin birth of Jesus — this is not dependent on an empirical test. Beliefs *in*, maintained by intuition, and sentiments have a much higher emotional value than beliefs *that*. Intuition is a capacity to get immediate comprehension without reasons or observations. Intuition honed by eons of evolution is, in many cases, a good guide, but sometimes it fails catastrophically.

One thing is to believe *in* God and another to believe *that* God exists. The first one is not analyzable; with the first one, you cannot do much; with the second one, you can argue. The person who believes *in* the "vital force" (in whatever form) believes in this without further ado and does not care that there is no evidence for such a force. But when this person believes *that* it is possible to heal a person by manipulating their "vital force" (however this be done), he expresses a belief subject to empirical testing. A belief *in*, with nothing else, has little ontological value since it only indicates an internal psychological state so that it becomes the same to believe *in* God than to believe *in* the fairy godmother. When people repeat the credo of the Catholic Church (see the epigraph), they do not propose that they believe *that* all such propositions are true and subject to empirical verification (they know more than that — I hope); they simply believe *in*.

One difficulty is that if you believe in the virgin birth of Jesus and think a bit about that doctrine, you must conclude that Jesus was a female. And while on the subject, anthropological studies tell us that Jesus was dark-skinned and probably brown-eyed, very different from the whitewashed images you see in any church. Ah, she also was a Jew!

Religions provide a support community, moral guidance, a heartwarming tale, hope facing dangers, and a neat and easy story. Unfortunately, absolute answers to all our questions, especially the meaning of our lives that we all wish to understand, are mostly wrong (the answers, and sometimes the questions). Historian Teofilo Ruiz writes[8]: *Lack of belief in God and most certainly mistrust of religion, does not mean that as a historian or a human being I do not understand and cannot be sympathetic (within limits of course) to the power of religious belief in shaping human existence or in providing solace, escape, and meaning to untold millions throughout the world.*

Professor of religion at Princeton University Cornel West writes[9]: *The religious threats to democratic practices abroad are much easier to talk about than those at home. Just as demagogic and antidemocratic fundamentalisms have gained too much prominence in both Israel and the Islamic world, so too has a fundamentalist strain of Christianity gained far too much power in our political system, and in the hearts and minds of citizens. This Christian fundamentalism is exercising an undue influence over our government policies both in the Middle East crisis and in the domestic sphere and is violating fundamental principles enshrined in the Constitution; it is also providing support and "cover" for the imperialist aims of empire.*

That is the R in the FIRM complex.

What we hear coming from some pulpits is not Christian, not a message of love. Jesus would turn in his grave (well, let us not get into this). Islam is still in its own Middle Ages; after all, Muhammad died in 632 CE, so this religion is 600 years younger than Christianity. Think about what Christians were doing at that time, bludgeoning each other, burning witches and heretics, and engaging in brutal Crusades. Perhaps, given enough time, Islam will evolve in the future to a more open and less extreme religion, but "given enough time" is the problem.

On the other hand, science is messy and complicated, has no absolute answers but only the best ones to date. Ethical questions are a different matter altogether and must be answered by our humanity. Europe was paralyzed for a long time caused by religious dogma that branded any other thought as heretic (with dire consequences for those found guilty). The Enlightenment arose as an attempt to get rid of those shackles.[10]

We do not usually subject our beliefs to a critical evaluation; we simply believe and do not worry why unless something happens that requires this analysis. Therefore, the most important exercise that you can do to discover and analyze your beliefs is verbalization, that is, to examine through an explicit statement some belief that you hold but never questioned because you believed *in* and did not believe *that*. Many fear doing this exercise because they suspect that they will find unpleasant things that cause cognitive dissonance. Others do not care; they just believe.

The justification of a belief must respond with good arguments to all critics who think that the belief is not justified. A belief without justification, without evidence, a *belief in*, is maintained by faith, pretending, in the words of Boghossian, *that we know something we do not know.*[11]

One thing is the evaluation and justification of a belief (the belief *that*) employing critical reasoning and evidence, and quite another is understanding the origin of a belief. Beliefs form an interwoven system, a complex network, that determines our understanding of the world and our behavior under dynamic circumstances. Spanish philosopher Ortega y Gasset writes[12]: *Beliefs are the basis of our life, the terrain over which it happens. Because they present what for us is reality itself. All our behavior, even an intellectual one, depends on the system of our authentic beliefs. In them, "we live, we move, and we are." Therefore, we do not usually have an express awareness of them, we do not think of them, but they act latently as implications of what we expressly do or think. When we truly believe in a thing, we do not have the "idea" of that thing, but simply "we count on it".*

Some argue that everything ultimately rests on some basic belief that we hold without proof. Belief in reason is primordial (the belief *that* reason is the best way to understand and operate in the world), and some argue that it amounts to a religion that we have "faith" in reason. But it is not like that. The evidence to "believe" in reason consists of the inescapable fact that it is impossible to function outside of it (when this happens, you are sent to a clinic). Any argument that can be proposed against reason will be structurally contradictory and self-refuting as it pretends to use the reason it denies, producing a "performative contradiction" (as happens if someone claims, "I am dead").

Logic cannot be avoided or overcome. Logic supported by the fundamental principle of non-contradiction has an evolutionary origin. A mind that could not distinguish between A and not-A would not have survived. That is, we are descendants of those who had the capacity for logical thinking. Also, in the same sense, we are descendants of those with the ability to see patterns that connect events. This causes us to see channels and faces on Mars and virgins on grease patches and toasts (pareidolia) and that we often tend to perceive order in random arrangements and to invent a conspiracy when there is none (see below).

Many dedicate their time to disseminating information, repeating what others have written or said, without applying Clifford's dictum, as stated in the epigraph. They repeat, without criticism, what is said by many clumsy "experts," swindlers, or at least misinformed sources (whether knowingly or not). Repeating what the Bible, the Torah, or the Koran says is a ubiquitous example of this. Thus, it is an invalid circular argument that it is a sacred book or a divinely inspired book because the same book or an old legend says so.

The issue is not harmless, mainly because the mind that operates in this way is predisposed to believe other things without evidence if adequately presented or repeated enough times. Daniel Kahneman, the winner of the Nobel Prize in Economic Sciences in 2002, says[13]: *A reliable way to make people believe in falsehoods is frequent repetition, because familiarity is not easily distinguished from truth. Authoritarian institutions and marketers have always known this fact. But it was psychologists who discovered that you do not have to repeat the entire statement of a fact or idea to make it appear true.* It is the basis of propaganda and an integral part of what is called "public relations." Carl Sagan had this to say[14]: *One of the saddest lessons of history is this: If we've been bamboozled long enough, we tend to reject any evidence of the bamboozle. We're no longer interested in finding out the truth. The bamboozle has captured us. It's simply too painful to acknowledge, even to ourselves, that we've been taken. Once you give a charlatan power over you, you almost never get it back.*

The repetition over centuries that the Jews were the ones who killed Jesus, including early indoctrination in children's books, fed hatred. For example, consider the publication of *Der Giftpiltz*[15] (The poisonous mushroom), a children's book published in 1938. In it, you can read: *Just as it is often hard to tell a toadstool from an edible mushroom, so too it is often very hard to recognize the Jew as a swindler and criminal...* The caption of one illustration says: *When you see a cross, remember the gruesome murder*

„Wenn ihr ein Kreuz feht, dann denkt an den grauenhaften Mord
der Juden auf Golgatha..."

Figure 8.1.   When you see a cross…

by the Jews on Golgotha. That is why we worry about "hate speech," which collides head-on with "free speech" (Figure 8.1).

The adherence to demonic anti-Semitism was the primary motivation of the European Shoah.[16] A significant fraction of the German people participated (not just a small group of sadistic murderers like some think). I will talk about this later.[17]

Think of the witch-hunts that caused thousands of women to suffer between about 1450 and 1650 (Including Kepler's mother[18] — and about 20% of men) according to the best estimates[19] — 60,000 — cruelly killed (hanged, drowned, or burned) for supposed demon worship. Death only came after brutal torture, as illustrated by a manual (The Witch Hammer), where we can read[20]: *But if, after keeping the accused in a state of suspense, and continually postponing the day of examination, and frequently using verbal persuasions, the Judge should truly believe that the accused is denying the truth, let them question her lightly without shedding blood; knowing that such questioning is fallacious and often, as has been said, ineffective. And it should be begun in this way. While the officers are preparing for the questioning, let the accused be stripped; or if she is a woman, let her first be led to the penal cells and there stripped by honest women of good reputation. And the reason for this is that they should search for any instrument of witchcraft sewn into her clothes; for they often make such instruments, at the instruction of devils, out of the limbs of unbaptized children, the purpose being that those children should be deprived of the beatific vision. And when such instruments have been disposed of, the Judge shall use his own persuasions and those of other honest men zealous for the faith to induce her to confess the truth voluntarily; and if she will not, let him order the officers to bind her with cords, and apply her to some engine of torture; and then let them obey at once but not joyfully, rather appearing to be disturbed by their duty. Then let her be released again at someone's earnest request, and taken on one side, and let her again be persuaded; and in persuading her, let her be told that she can escape the death penalty [...] Others think that, after she has been consigned to prison in this way, the promise to spare her life should be kept for a time, but that after a certain period she should be burned.*

You might argue that this only proves how much we have progressed, and I agree (since we no longer burn witches), but this is not the point. My concern is how wrong beliefs lead to suffering and bloodshed.

Perhaps we are not any better now, just different. Let me quote for you a contemporary paragraph about torture: *We conclude that for an act to constitute torture as defined in Section 2340, it must inflict pain that is difficult to endure. Physical pain amounting to torture must be equivalent in intensity to the pain accompanying serious physical injuries, such as organ failure, impairment of bodily function, or even death. For purely mental pain or suffering to amount to torture under Section 2340, it must result in significant psychological harm of significant duration, e.g., lasting for months or even years. We conclude that the mental harm also must result from one of the predicate acts listed in the statute, namely: threats of imminent death; threats of infliction of the kind of pain that would amount to physical torture; infliction of such physical pain as a means of psychological torture; use of drugs or other procedures designed to deeply disrupt the senses, or fundamentally alter an individual's personality; or threatening to do any of these things to a third party. The legislative history simply reveals that Congress intended for the statute's definition to track the Convention's definition of torture and the reservations, understandings, and declarations that the United States submitted with its ratification. We conclude that the statute, taken as a whole, makes plain that it prohibits only extreme acts.*

This is how the infamous 50-page torture memo[21] begins, sent to Alberto R. Gonzales, Counsel to then-President Bush. We torture, all legally OK'd; who cares about ethics and human rights?

About witches (a satanic conspiracy), people believed things that are unbelievable to most of us. Accusations and trials highlight the following events that allegedly occurred in meetings with the devil at the Sabbath, at night lit by "a great bonfire, sinister and frightening."

- Night flight as a way of transport (on a stick, a chair, or animals).
- Ointments that allow or facilitate transport.
- The killing of children and cannibalism.
- The real presence of the devil in the form of an animal associated with his figure: The Great He-Goat.
- The devil's worship included the *osculum*, the infamous kiss (kiss on the ass — and we still use this expression).
- Blasphemy and sacrilege (imitation of the sacraments, insult of the consecrated host).
- The dance and the banquet.
- Indiscriminate intercourse of the attendants with each other and with the devil.
- The narration of the evils carried out since the last assembly.
- The delivery of powders or poisons that would allow the assistants to continue carrying out evil acts (curses).

Today, it is worthy of a cheap horror and sex movie, but many believed these stories of incestuous orgies, infanticide, and cannibalism. Even today, many believe in witches and witchcraft and make appointments with all sorts of paranormal operators using their cellphones. Weird.

Norman Cohn[22] tells us that this fantasy has ancient origins and that already in the first century, the "pagans" accused the Christians of these things. Minucius Felix (born, Africa? died c. 250, Rome), one of the earliest Christian Apologists to write in Latin, recounts these horrible secret cults: *just the secrecy of this perverse religion proves that all these things, or almost all, are true.* An excellent example of a fallacy used even today by believers of conspiracies. Cohn explains the phenomenon: *the urge to purify the world through the annihilation of some category of human beings imagined as agents of corruption and incarnations of evil.* Thus, began the fall of Europe into a millennium of darkness. A similar story was part of the myths used by European anti-Semites of the nineteenth century.

Cohn summarizes: *The great witch-hunt can, in fact, be taken as a supreme example of massive killing of innocent people by a bureaucracy acting in accordance with beliefs which, unknown or rejected in earlier centuries, had come to be taken for granted, as self-evident truths. It illustrates vividly both the power of the human imagination to build up a stereotype and its reluctance to question the validity of a stereotype once it is generally accepted. Until our own century. the operation and consequences of demonization have never been more horrifyingly displayed.* Unfortunately, it seems that the US has not been able to get rid of its inner demons: white supremacy and increasing xenophobia, recently aided and abetted by the pathetic figure of a second-rate reality show star, elected president.

The list of incredible (and false) things that people believe is long because there are infinite false things and very few true things. Some are so obviously false that it is difficult to understand how it is possible to believe them (say, the spontaneous combustion of humans or that the Shoah never happened). In contrast, others are false for less obvious reasons (such as flying saucers — which sometimes occur at a couple's kitchen). According to Michael Shermer,[23] people maintain these beliefs because they want to believe; they make them feel good and comfort them.

On the other hand, it is possible to have good reasons to believe a falsehood. Thus, the past belief that the motionless Earth was the center of the Universe was based on the thought that if it did move, you should feel it.

Today we understand that these objections were not valid. The observations of the aberration of starlight by James Bradley (1693–1762) in 1728 demonstrated the Earth's orbital motion. However, it was not until 1838 that Friedrich Bessel (1784–1846) published the results of the first measure of the parallax angle (0.3 seconds of arc) for the star 61 Cygni (caused by viewing a star from opposite places on Earth's orbit). The pendulum of Léon Foucault (1819–1868), (that today can be seen in many science museums), erected initially at the Paris Observatory in 1851, left little doubt about the reality of Earth's rotation.

Note something important: these facts (the movement of the Earth) were established as the result of several different and *independent* observations, which constitute a network of observations (or experiments) that together support the notion of the mobile Earth. Thus, our knowledge of the world is not based only on one crucial observation or experiment (although this might be initially so). In the same vein, global warming is not obtained from one thermometer stuck who knows where; it is based on millions of measurements of different variables by scientists in various countries.

It is wrong to maintain a belief, even if it is true, if it is not based on evidence, and you cannot later say: "ah, but I was right after all"; since you were not. After an event, someone might say, "I knew it was going to happen," and may very well have indicated it before the event as if it were a prophecy. But she did *not* know; there was no way to know or verify it before the event. It was an opinion, possibly supported by previous experiences. But many things that we think will happen do not happen, and we do not say: "I knew it was going to happen, but it did not happen." For example, we may think: "This marriage will not last." When the divorce happens, we will say: "I knew it would not last." But suppose that despite everything, they last "till death do us part." Then it makes no sense to say: "I knew it was not going to last, but I was wrong." The most we can say is: "I believed it wouldn't last, but I was wrong."

The maxims of Clifford and Russell are then not merely philosophical principles. The US (to give a recent but not unique example in history) has caused death, suffering, and destruction to other people based on a proposition without any foundation (that this attacked nation possessed weapons of mass destruction and harbored the terrorists of Al Qaeda). The belief in the superiority of one group over another is behind genocide that found its maximum expression in the tragedy initiated by the Nazis. The belief in a supreme being, whether earthly or celestial, has caused suffering to many, from the witches and heretics tortured, burned, or drowned without mercy under ecclesiastical supervision to those who could not escape from the earthly hell of the burning twin towers.

Article 10 of the Declaration des Droits de l'Homme et du Citoyen of 1789, one of the important documents of the Enlightenment states: *No one should fear for their opinions, including religious ones, as long as their manifestations do not alter the public order established by law.*

In the meantime, the invention of Dr. Joseph-Ignace Guillotin (1738–1814) went up and down without consideration, a mechanism of "justice" used in France until 1977. There remains a shuddering question: After the cut, did the victim have a few seconds of awareness about what had happened?

Today, religious beliefs of remarkable infantilism alter public order. Just look at what is going on in the Middle East, Africa, and increasingly in what we call the west. The intrusion of antiquated beliefs in government matters alters public order. Interference with the civil code alters public order. Opposing contraception, same-sex love, and abortion for religious reasons alters public order. Opposing vaccination alters public order. The list is long and varied.

It is totally unacceptable that pseudoscience, myths, superstition, and religion permeate professions and government and contaminate society. The potential damage is enormous since those unfounded beliefs lead to wrong action. It is alarming in these times in which we need to act with knowledge to solve the significant and challenging problems that we face. Opposition to measures (meager as they are) to mitigate climate change caused by our activities does not change its reality, a criminal position, since it will cause havoc and death. This is not a matter of (expletive deleted) opinion.

As formulated by Physicist Alan Sokal[24]: ... *Clear thinking, combined with a respect for evidence — especially inconvenient and unwanted evidence, evidence that challenges our preconceptions — are of the utmost importance to the survival of the human race in the twenty-first century.*

Not a few groups (from sects to dictatorships) do not allow their followers to communicate with those who do not belong to the group. That is the reason for the demand for intellectual freedom made by Sakharov.

The following was proposed by Hungarian mathematician George Polya (1887–1985)[25]: First, *we should be ready to revise our beliefs.* Second, *we should change a belief when there is a compelling reason to do so.* Third, *we should not change a belief capriciously.* These points sound trivial, yet one needs rather unusual qualities to live up to them. The first point needs "intellectual courage." You need the courage to revise your beliefs. Galileo, challenging the prejudice of his contemporaries and the authority of Aristotle, is a good example of intellectual courage. The second point needs "intellectual honesty." To stick to a conjecture that has been contradicted by experience just because it is my conjecture would be dishonest. The third point needs "wise restraint." To change a belief without serious examination, just for the sake of fashion, for example, would be foolish.

Believing is not the same as knowing. We believe many things that are not true, but we cannot know something that is not true. (You may *believe* it is raining when it is not raining, but you cannot *know* it is raining when it is not raining — that is the difference.) All we know is ultimately a belief expressed through a proposition that responds to a criterion — that of being true or false — so that given some proposition, the necessary question is: Is it true?

I believe that I see a tree, but I might be hallucinating, and in the same way, I claim to have seen a ghost. The important thing is the degree of independent confirmation of a belief — the evidence. We have all had the experience of believing something to discover later that it was false. Thinking something does not make it true. Believing is subjective, whereas knowing is objective. Toulmin[26] sums it up as follows: *We know something (in the full and strict sense of the term) if and only if we have a well justified belief; our belief is well justified if and only if we can produce good reasons in its support; and our reasons are really "good" (in the strictest philosophical norms) if and only if we can produce a conclusive or formally valid argument, which relates the belief to a starting point that is not questioned (and is preferably unquestionable).*

Before considering the justification of a belief, it must seem relevant and credible. It will be relevant if its truth value impacts what we think or how we act. It will be credible if it seems to us that we cannot rule it out from what we already know. The proposition: "There is life on another planet" is relevant: in this case because of its philosophical implication concerning who we are. But the proposition: "Aliens visit us every night" is not credible, and it can be discarded without further ado since all we have are some dubious anecdotes and blurry images, and good reasons to doubt this (as I will explain). The goal of the critical thinker is to acquire knowledge — *justified true belief* — increase the amount of knowledge, increasing the ratio (in the mathematical sense) of true beliefs to false beliefs. A good recipe to test a belief is to answer the following question: Is the foundation of the belief enough to justify it if we remove all possible reasons for wishing the belief to be true? (Not easy to do). We may have reasons to believe that something is true or false without knowing why it is true or false. I may have good reason to believe that Pythagoras's theorem is true (I learned it at school, my mother taught it to me), but I may not know the mathematical proof of the theorem, which is why it is true (for plane geometry).

Information and data are necessary to develop new ideas such as Darwin's, who spent a few years traveling the world on the HMS Beagle gathering it. Information is necessary but not sufficient, and that is the tricky bit, especially in this "information age," as I will discuss further. New ideas must arise from well-used reason since it is very common to reason badly, as I have discussed, and often act by instinct or emotion, like any other animal. Let us briefly look at some of the most prevailing beliefs.

8.1 Miracles

For miracles are so-called not because they are the work of God but because they happen seldom and for that reason create wonder. If they should happen constantly according to certain laws impressed upon the nature of things, they would no longer be wonders or miracles but might be considered in philosophy as a part of the phenomena of nature notwithstanding that the cause of their causes might be unknown to us.

— Isaac Newton[27]

When you keep putting questions to Nature and Nature keeps saying "no," it is not unreasonable to suppose that somewhere among the things you believe there is something that isn't true.

— Jerry Fodor[28]

Against stupidity, the very gods themselves contend in vain.

— Friedrich Schiller

The belief in miracles is pervasive. People talk about miracles, pray for a miracle, go to multitudinous events to request miraculous healing mediated by a pastor, and the Catholic Church requires it as proof of sanctity. The term "miracle" indicates a seemingly inexplicable event accepted as supernatural, generally understood as divine intervention. The Oxford English Dictionary defines it as: *An extraordinary and welcome event that is not explicable by natural or scientific laws and is therefore attributed to a divine agency* Merriam Webster defines it as: *an extraordinary event manifesting divine intervention in human affairs.* (Note the difference in "welcome.")

It is common before an unusual event to think: "It cannot be a coincidence," but it can be, and it probably is. Things of extremely low probability happen every day. It is quite improbable that a day will pass with nothing improbable happening somewhere. Percy Diaconis writes[29]: *With a large enough sample, any outrageous thing is likely to happen. The point is that truly rare events, say events that occur only once in a million [as the mathematician Littlewood (1953) required for an event to be surprising] are bound to be plentiful in a population of 250 million people.* If there is a sufficiently large number of opportunities, very low probability events will occur.[30]

A "Black Swan" defined by Nassim Taleb[31] describes low probability unanticipated events with vast consequences. If the consequences are good (say cancer remission), people call it a "miracle." If they are bad people, think of a conspiracy. Events that are perceived as significantly connected without an apparent cause leave us perplexed. Two or more coinciding events will attract our attention if we perceive the coincidence as improbable and significant, especially if it happens to us personally. We often look for something in common when we meet someone; We ask questions and are happy if there is a coincidence: "I also know Jack," or "I also am Libra," or whatever. We can all tell stories of coincidences that seem to be more than mere coincidences. Some events seem so unlikely that they make us think of mysterious connections, that there is something about the universe that we do not understand. (And by the way, there is, and that is what science is dedicated to.)

Here is a personal story. In June 1986, during a sabbatical in Germany, I had the opportunity to travel with a small group of colleagues to attend a congress in Armenia. I joined the group after a colleague decided not to travel (for fear of fallout from the Chernobyl accident on April 26 of that year). The flight first took us to Moscow, and we spent the night at the Soviet Union's Academy of Sciences hotel to continue the next day with another flight to Yerevan. I was waiting for one of the two elevators to go up to the hotel's fourth floor after receiving the key to my room, and when the first elevator arrived, I lined up, waiting for those inside to exit. Finally, one who came out exclaimed: "Altschuler!" and I stared at the face of an Indian-looking gentleman in a suit and tie, who was looking at me with a smile. Then, noticing my amazed look, he said: "I am Ram Subramanian, don't you remember?"

As my face showed that I did not remember, he continued: "I was a graduate student at Brandeis University when you started studying there 15 years ago, and once, I invited you to my apartment for dinner." "Oh yes!" I said, lying, and as his group left for their bus, he told me that he was passing through, that he lived in India, and had come to Moscow to participate in a congress on atomic energy. He gave me his card and left. I was stunned, and if I were someone else, I would think about cosmic vibrations, predestinations, and who knows what else. But consider:

millions of people meet every day, the vast majority being strangers who go unnoticed. But the huge number of opportunities means that a few of these meetings will be unusual, and one of them occurred to me. I tried but was later unable to find him.

Although some people (they are not few) say that they "do not believe in coincidences" for reasons that are not at all clear, the world is full of them, that is to say, two or more events that occur randomly, but give the impression of being connected. Winning the lottery is a coincidence: the number you bought at random (or even if you chose it according to certain numbers with a special meaning for you) coincides with another random number. The probability that a six-digit number printed on a paper in your pocket matches a number produced by a machine in the lottery draw is very low: one in a million. But if all the tickets were sold, then somebody must win.

When you look at the map on the wall at a train station, you have reached for the first time, and see a red dot that says: "You are here," you don't ask yourself: How did they know? It is not difficult to calculate that in a group of 23 people (people on a football match) there is a 50% chance that two of them will have their birthday on the same date. We get confused thinking about 23 people and 365 days of the year but think that there are many more possible pairings than people. Note that this is not the same as one of the players having the same birthday as the referee. In a meeting with 30 people, the probability increases to 70%, and you can place a bet with that probability of winning. The occurrence of a specific event or match can be very low (such as a particular person winning the lottery next Wednesday), but an unspecified unlikely event will likely happen (someone wins the lottery). That is why extraordinary coincidences are noticed after the fact, and the "seers" never say something very specific (and if they do so, they are almost always wrong). The mistake we often make when we are amazed by a coincidence that has happened to us is to ask: What is the probability of that happening to me? But it is the wrong question. The right question is: What is the probability of that happening to someone?

Coincidences might mean nothing, as well put by Austrian writer Marie von Ebner-Eschenbach (1830–1916): *A clock that is not working will show the correct time twice a day and after years can look back on a long series of successes.*

David Hume wrote about miracles[32]: *No testimony is sufficient to establish a miracle, unless the testimony be of such a kind, that its falsehood would be more miraculous than the fact which it endeavors to establish.*

Hume continued: *When anyone tells me, that he saw a dead man restored to life, I immediately consider with myself, whether it be more probable, that this person should either deceive or be deceived, or that the fact, which he relates, should really have happened. I weigh the one miracle against the other; and according to the superiority, which I discover, I pronounce my decision, and always reject the greater miracle. If the falsehood of his testimony would be more miraculous, than the event which he relates; then, and not till then, can he pretend to command my belief or opinion.*

In the face of testimony from one who claims to have seen a Guru levitating, one wonders what is most likely: That the Guru levitated, and a large part of Physics is wrong, or that it is a trick, not unlike those done by illusionists or a special effect.

Consider a renowned case: On May 13, 1917, and every month until October 13, Lucia dos Santos, a peasant girl, and her two cousins Francisco and Jacinto Marto, had a vision that presented herself as the Virgin of the Rosary. This happened at Fatima, a small village in Portugal. In her August appearance, the Virgin promised a miracle for October. On the 13th of that month, a crowd reported as 70,000 people gathered at Fatima, waiting for the miracle. The Virgin of the Rosary presented herself to the children and pointed to the Sun. Lucia did so, and everyone looked at the Sun. Those present were terrified to see that the Sun was falling on them. But before embracing them, the miracle ceased, and the Sun returned to its usual place in the sky.

Years later, on October 13, 1930, the bishop of the nearby city of Leiria, after years of research by the Church, accepted the children's vision as the apparition of the Virgin Mary and approved the cult of Our Lady of Fatima. The great basilica of Fatima was consecrated in 1953. On the 50th anniversary of the first apparition, on May 13, 1967, a crowd estimated at one million gathered at Fatima to listen to the mass from the mouth of Pope Paul VI. Lucia, a Carmelite nun, was at his side.

The secret message, presented by the Congregation for the Doctrine of the Faith[33] (established in 1542 as the *Supreme Sacred Congregation of the Roman and Universal Inquisition*), in a document signed by Joseph Cardinal Ratzinger (which among other inanities says: *Throughout history, there have been supernatural apparitions and signs which go to the heart of human events and which, to the surprise of believers and non-believers alike, play their part in the unfolding of history,* written by Sister Lucia only in August 1941 when the Germans were invading Russia (many years after the vision of 1917, year of the Bolshevik Revolution) recounts her vision of hell and then: *You have seen hell where the souls of poor sinners go. To save them, God wishes to establish in the world devotion to my Immaculate Heart. If what I say to you is done, many souls will be saved and there will be peace. The war is going to end:* (referring to the first World War) *but if people do not cease offending God, a worse one will break out during the Pontificate of Pius XI. When you see a night illumined by an unknown light, know that this is the great sign given you by God that he is about to punish the world for its crimes, by means of war, famine, and persecutions of the Church and of the Holy Father. To prevent this, I shall come to ask for the consecration of Russia to my Immaculate Heart and the Communion of reparation on the First Saturdays. If my requests are heeded, Russia will be converted, and there will be peace; if not, she will spread her errors throughout the world, causing wars and persecutions of the Church. The good will be martyred; the Holy Father will have much to suffer; various nations will be annihilated. In the end, my Immaculate Heart will triumph. The Holy Father will consecrate Russia to me, and she shall be converted, and a period of peace will be granted to the world.* The content is not surprising, considering what was happening at the time it was written and the deplorable attitude of the Vatican, which did nothing while the Nazis murdered millions. In this sordid episode, there is nothing divine and much that is very human. A prophecy it is not. Like the occasions in which something dramatic happens (earthquakes, murders, accidents), and then someone says: "I had a premonitory dream of that." Yes, so why did you not say it before rather than after?

Anyway, consider Hume. There are two options: A miracle happened, and the Sun really approached Earth, or possibly it was a case of mass hysteria. If the Sun moved, then an even greater miracle occurred since some three billion inhabitants of the planet did not notice the phenomenon, without considering the problem that such a movement of the Sun would mean the end of the world. Following Hume, we must choose the second alternative. It would remain as a research topic to understand how the phenomenon was generated in Fatima.

We can apply Hume's rule to other alleged phenomena. Thus, in the face of the testimony of one who has seen extraterrestrial spacecraft, following Ockham's principle, it is much more likely that they are mistaken perceptions than real aliens.

Miracles are the product of ignorance and statistics, and often ignorance of statistics. Those who prayed for a cure and did not heal for obvious reasons began begging for a priest and did not give testimony. The following questions can be asked: Why are miracles limited to childish events like the appearance during the night of colored tears on the face of a religious figure? Why, in a severe accident or natural disaster with hundreds of deaths, is one saved "by a miracle" and not all without a scratch (which would be a miracle)? Why does the healer only heal some chosen ones and not all the attendees of the event? Or better yet: Instead go to a nearby hospital and heal everyone?

Imagine that in these sad days, just when a group of jihadists are about to kill 100 Christian hostages, all the executioners suffer a fatal infarction. That would be miraculous, and that's why it does not happen. Or, more generally, it is necessary to ask: If God intervened so many times in human affairs during antiquity (according to the Bible), why did She not do so during the Shoah? Why do not all despots of this world get uncontrollable and constant diarrhea? If during a walk through an old town, a flowerpot falls from a balcony and bursts on the floor one meter away from a person we say, "He was miraculously saved," then why do we not say: "He was miraculously killed" if instead, the pot smashes his head?

8.2 Souls

Unless I am wrong it will first be necessary to understand the soul, and investigate the laws which he observes in these operations of the spirit, the powers and actions of the electrical spirit which pervades our bodies.

— Isaac Newton[34]

Again, again, I say confess we must,
That, when the body's wrappings are unwound,
And when the vital breath is forced without,
The soul, the senses of the mind dissolve,-
Since for the twain the cause and ground of life
Is in the fact of their conjoined estate.

— Lucretius[35]

When it comes to immortality of the soul, whatever that may be precisely, I can only say that it seems to me to be wholly incredible and preposterous. There is not only no plausible evidence for it: there's a huge mass of irrefutable evidence against it, and that evidence increases in weight and cogency every time a theologian opens his mouth.

— H.L. Menken[36]

Given the general uncritical acceptance of this idea, it is worth considering it in more detail. As I mentioned, in ancient times, it was believed that life resulted from a "vital force" (the Hindu prana, the Chinese ch'i). It was assumed that this vital force was the source of life and was associated with the soul or spirit, immaterial and separable from the material body. It, therefore, could survive death (*ontologically* different). The word "spirit" comes from the Latin "*spiritus*," which means breath. Thus, we understand what is meant when in Genesis 2.7 we read[37]: *And the Lord God formed man of the dust of the ground and breathed into his nostrils the breath of life, and man became a living soul.* Compare this with the Babylonian inscriptions of the *Enûma Elish* (from the 12th century BCE), one of the oldest stories we know, where we read of the god Marduk: *He made mankind ... creatures with the breath of life ... creator of all people,* an antecedent of the biblical story.[38] These ideas have survived to the present, even though there is no evidence to sustain them.

Science (particularly neuroscience) understands mental processes as neuronal interactions and, paraphrasing Laplace, "we do not need that hypothesis." The philosophers of the Renaissance were religious believers and understood the study of nature as the study of God's creation. Newton devoted a lot of time studying the Bible, and the quotation of the epigraph reminds us of his feeling. But science has no ulterior goal beyond understanding our world and ourselves in relation to it. Instead, science leads us where it may, often to discover the contradictions between our beliefs and the world as it is.

According to the catholic encyclopedia,[39] *The soul may be defined as the ultimate internal principle by which we think, feel, and will, and by which our bodies are animated.* It continues, demonstrating a high dose of ignorance: *The belief in an animating principle in some sense distinct from the body is an almost inevitable inference from the observed facts of life. Even uncivilized peoples arrive at the concept of the soul almost without reflection, certainly without any severe mental effort. The mysteries of birth and death, the lapse of conscious life during sleep and in swooning, even the commonest operations of imagination and memory, which abstract a man from his bodily presence even while awake-all such facts invincibly suggest the existence of something besides the visible organism, internal to it, but to a large extent independent of it, and leading a life of its own.*

It is strange to argue that since "uncivilized peoples" believe in a soul (*almost without reflection*), then it is true. The only thing that can be concluded about past people's beliefs (and some of the present) is that these people did not have the knowledge we now have. The mysteries of birth and death are superstitions, there is no mystery about these events, and the *almost inevitable inference from the observed facts of life* is different. What we know about the brain is not consistent with the idea of *an animating principle in some sense distinct from the body.* (the "in some sense" means I do not know what we are talking about).

In a classic book, New York University psychologist George B. Vetter writes[40]: *The prototype of all non-material forces or agencies conceived to exist is the subjective thought or volitional process that is the common experience of all mankind. It is this subjective process that provides the*

Daguerreotype Originally from the collection of Jack and Beverly Wilgus, and now in the Warren Anatomical Museum, Harvard Medical School.

Figure 8.2. Phineas P. Gage.

terms and symbols of a communication process that have been elaborated into a category of reality in metaphysics. What begins as an adjective for the covert, the symbolic, the private, the mental, becomes Mind, Spirit, or Soul by a process of reification, the conversion of adjectives into nouns of dubious validity.

If we were to accept the idea that the soul is the conductor of the body, then, as behavior and mental functions are altered by alcohol, we would have to conclude that somehow the intangible soul can be drunk as well. Nor is it consistent with the fact that brain damage can affect the functions of a soul that has nothing to do with the brain.

The classic case is Phineas P. Gage (1823–1861), a railway worker. In 1848 an accident while working with explosives caused a long-pointed metal rod about one inch thick and three feet in length to fly through his brain (Figure 8.2).

The rod penetrated behind his left eye and exited through the top front of his skull. Gage did not lose his life, nor fell unconscious after the horrible accident. But on recovering 2 months after the accident, although he did not present clear impediments, except for the blindness in his left eye, a drastic change in his personality was documented. His friends noticed that "Gage was no longer Gage." It represents the first case of personality alterations because of brain damage. The incident sowed the seed for the idea of the functional modularity of the brain. We must be careful here, this happened a long time ago, and many stories have distorted the facts.

Nevertheless, modern studies of brain function lead us to conclude that we are not an incarnate spirit but are our brain's activity (as remarked above by Francis Crick); unacceptable to many. Nevertheless, survival after death matters so much to us that despite the total lack of evidence and logic, we insist on the idea that "there must be something else." Well, no, there does not have to be anything, and everything we know indicates that this is so, however painful this may be.

It can be argued (and many do) that since there are many things we do not understand, it is not reasonable to discard something that many people believe to be true merely because of a logical or physical problem. But any hypothesis about the world can and should be tested with all available means if we want to get closer to the truth.

8.2.1 *A minor mathematical problem*

If, as they say, each of us is the recipient of a spirit that leaves us when we perish and then reincarnates into a newborn, then if we consider population growth that has multiplied the number of humans by a large factor, accounting gets complicated. Since it is necessary to supply this growing population with spirits, an increasing population of spirits is needed that must come from somewhere.

Another option is that with each new human being born, a spirit is born (*the spiritual soul is created immediately by God*, as indicated by the encyclical *Humani generis* of Pope Pius XII). Then there would now be about 100,000 million disembodied unemployed spirits since that is the estimated total number of humans who have ever lived.[41]

You can always invent a system that solves part of the problem; you can have this great reservoir of spirits that fight to reincarnate (or perhaps not to do so) in a "war of the spirits," but you can see that this quickly leads to the absurd. It is not really of great importance.

8.3 The Nature of the Supernatural

Christian thought that emerged in the early centuries often gave irrationality the status of universal "truth" to the exclusion of those truths to be found through reason. So, the uneducated was preferred to the educated and the miracle to the operation of the natural laws.

— Charles Freeman[42]

It is common to think about and invoke the supernatural, and indeed entire volumes are written to discuss such thing as the existence of *a superhuman supernatural intelligence who deliberately designed and created the universe and everything in it, including us* as defined by Richard Dawkins in his delightful and influential book[43] "The God Delusion" or as discussed by Daniel Dennett in his excellent "Breaking the Spell: Religion as a Natural Phenomenon"[44] where we read about *a social system whose participants avow a belief in a supernatural agent or agents whose approval is to be sought.*

According to Webster, the supernatural is

1: of or relating to an order of existence beyond the visible observable universe; especially: of or relating to God or a god, demigod, spirit, or devil.
2a: departing from what is usual or normal specially to appear to transcend the laws of nature.
2b: attributed to an invisible agent (as a ghost or spirit).

According to the Dictionary of the Spanish Royal Academy *sobrenatural* is something that exceeds the terms of Nature (*Que excede los términos de la naturaleza*).

But what does the supernatural mean? What is the meaning of exceeding the terms of nature, or have a cause outside of Nature? It is implicitly assumed that it is possible to be beyond or outside nature, but these are empty words, just magical thinking.

At any time in history, some phenomena appear to transcend the known laws of nature. Before Newton, it was not understood why the planets moved, and before quantum mechanics, it was not understood how an atom was held together. Were the motion of planets and the structure of atoms mediated by the supernatural? One of the currents "mysteries" in biology is the emergence of life. We simply do not know how it happened, and it is easy for some to invoke a "supernatural" cause, a "breath of life" which does not explain anything. When we discover it, it will be natural, and it may require a revision of our understanding of the way nature works, as has happened in the past.

Are ghosts (whatever they are) supernatural? How can anyone claim to see, feel, or detect them if they are not part of nature? If ghosts exist, they are natural and subject to all the rules of nature and evidence. We might need to revise our understanding in the light of what ghosts might tell us (in a metaphorical sense), but they will not be supernatural. If they are immaterial, as some say, then how do they interact with matter?

It makes as much sense to talk about something supernatural as to something outside the Universe. By using the word "supernatural," we unwittingly accept the possibility of such a thing. The objection of something outside of nature is like the one expressed by Princess Elizabeth of Bohemia to Descartes about how two ontologically excluding entities could interact. They cannot, as already mentioned.

8.4 The Hereafter

He wished and prayed that there would be no life after death. Then he realized that there was a contradiction and merely wished that there was no life after death.

— Douglas Adams

Anyone stupid enough to worry about how he will be remembered deserves to be forgotten.

— Gore Vidal

Do not let the following depress you and divert you from your chosen path. Do not get anxious, do not worry about it (it will be of no use) since it is not really something new or anything you do not already know. You may decide to change something here or there in your behavior, in your love life, but it is not my desire to torment you with this fact.

However, the fact is inescapable: you will die (as I already said), and it will happen very soon on a historical scale. That is the fundamental problem of life: *it is a terminal disease.* It may seem harsh, you may not like to think about it, but that's reality. In the end, worms and bacteria will consume you, and in a short time, you are not even a memory.

Melvin Konner writes about unhappy birthdays[45]: *Each birthday a little death' each rite de passage another passage toward it. There is a sense in which life is a continual condition of bereavement, during which we mourn the loss of ourselves. Think of the anger, the affectless acceptance, the denial, the shock, the colossal, ineffable sadness. Is it any wonder that we become more selfish with age, abandon certain youthful ideals, and come to believe that living well is the best revenge? Is it strange that the world's religions, great and small, with their venerable, contradictory narratives, their competing gods and spirits, and their occasional taste for holy war, have a grip on so many minds?*

Of course, this sad reality is not so if you believe an "afterlife" is waiting for us in the Hereafter, as was believed in ancient times, and many still do. The origin of religions is found in the grave. The graves of Neanderthals suggest that they already believed that when they buried a loved one, there was something that endured somewhere. Or, as is said, "they go to heaven."

And where is heaven? For many years, an increasing number of astronomers have studied all the space accessible to our instruments. All kinds of detectors located on Earth and in space have scrutinized the most remote places of the Universe and found many interesting things. With each passing decade, new technologies have improved our instruments. As a result, we have managed to see further to the threshold of our Universe, which occurred an unimaginable 14 billion years ago (0.000000005 of what you live).

But after many years of exploration, centuries of reasoning, and millennia of curiosity, no matter how much we see further, we have not found that place some call Heaven or the Hereafter or any evidence of its existence. Five hundred years ago, daring explorers sought to find the Garden of Eden, they never found it.

It is a good story, with Heaven and Hell, so that we behave and observe specific rules so that we do not think that we die and there is nothing else so that we do not lose hope so that everything does not seem so trivial. The thing is that many believe in a naive children's tale. Plato, the thinker, said it very well: *We can easily forgive a child who is afraid of the dark, the true tragedy of life occurs when men fear the light.* (Although there are doubts about this attribution.)

Brutal battles planned by adults for the youth of this world to kill or die, true butcheries of pain, atrocious episodes to get more quickly to where we all finally go. "Life is very short, and there's no time for fussing and fighting, my friend" (but we can't work it out) There is so much to do, so much to know, and at any moment, the lights turn off. Worse still, all those things of yours, the most precious memories, that old album of photographs, or the new one which resides on flash memory, prepared with so much love, this book written with care will end up sooner rather than later in the dump of oblivion. We indeed have children and friends who love us, appreciate us, and these things will last a while, but in no more than a 100 years, I assure you that there will be nothing left. And if we go on as we have, perhaps there will be really nothing left at all. Yet we insist.

Do you know your great-grandfather? "Know," I say, more than a faded photograph that can be hung on a wall in your home, "know" in the sense of knowing about him, his life, his feelings, what he was, what he did every day. I doubt it.

The belief that the afterlife is somewhere surrounds even the now-defunct Arecibo Observatory, where I worked for a while. (Now, on December 1, 2020, we mourn its loss, as a combination of wrong moves caused it to collapse and be destroyed.)

A few years ago, two very well-dressed gentlemen came to my office at the Arecibo Observatory to request my help to clarify a matter that concerned them. They had doubts about what one of the brothers of the Church was preaching. To their great relief, I assured them that it was not true that from Arecibo we received the sounds of last judgment's trumpets. Starting with the fact, I told them that although it is called a *radio* telescope, the instrument does not receive sound waves of any kind.

I further explained to them that if sounds were emitted from somewhere in Orion, as the brother maintained (although Orion is not even a place), such sounds would not be able to propagate between the stars since the interstellar medium is almost a vacuum. Sounds need a medium (like air) to propagate. They left reassured. I do not know what happened to the brother.

Some people claim they can communicate with the inhabitants of the Hereafter (at a very long-distance rate, of course). Sure, they have clients, who would not like to catch up with their dearly departed, tell them that they love and remember them and that we are well, that the grandchildren already have children. We could inquire if they are well, although it may be an inappropriate question, and finally say goodbye with an innocent "see you." It would be fabulous; it would be incredible, and it is.

I would ask the "medium" to communicate with Aristarchus of Samos, the Greek philosopher of the Pythagorean school who, almost 2,000 years before Copernicus, postulated that it was the Earth that orbited the Sun. We know of this indirectly since, unfortunately, his work has been lost. Could he dictate it to me? And then, could you contact Franz Schubert and ask him to complete his beautiful unfinished symphony number eight? It would be fabulous, truly amazing. Or perhaps dear grandfather can tell us where he hid his testament?

Not so long ago, it was believed that nature was governed by the will of divine beings (and many still do). This gave human beings a certain sense of security because the world was understandable in human terms. We were part of that eternal world, and in some religions, we were the reason the world was created. With the gods, there was a possibility of extending our ephemeral existence since there was the Hereafter.

There was reason for hope and to believe in a better life after this one, which was (and is) not very good for the vast majority. In recent centuries science has discovered that we do not need angels to move planets. Thus, the gods cut themselves off from the world, although many still believe in miracles.

The mistake is to think that by removing the angels to move the planets, science has also taken away spiritual values. This is not the case; in fact, the opposite, but only if we clearly understand what science says about us and our relationship with the world. Science fills us with feelings of beauty, wonder, and humility towards nature, feelings of high spiritual content. The problem is that the majority do not know what the new "Bible" of science says with its latest version of Genesis and our relationship with the Universe. (It also seems that the majority does not know that there are two contradictory stories of the creation in Genesis.)

There is a lot of mystery and fascination in this new vision that says that, although we are not special, in the sense of being chosen, we are the product of a wonderful process revealed by science. As scientific progress has allowed us to see further, to understand better, the more precarious the Hereafter has become. The most common conception of a god is a supernatural entity that created the Universe and can influence us and the world. In other words, it (she, he) can sometimes intervene in human affairs, and this is because it can do so or in response to our prayers.

Depending on the creed, this supreme being goes from something concrete with a human figure to something abstract and without material form. His role can be limited to nothing more than the creation of the Universe or meddling in even the most intimate affairs of people. But think: if a supernatural entity can influence the world, it must be considered a hypothesis about the world subject to scientific study. Therefore, it would cease to be "supernatural," as already discussed. It all boils down to the already mentioned impossibility of interaction between exclusive ontologies.

The saddest thing is that we forget about life here with so much concern for life in the Hereafter. With so much concern about moving to that mythical other world where everything is lovely (few consider the alternative that hell will be their destiny), we do not mind destroying our home, which is the only one we have. Some ask God to solve their problems instead of taking responsibility for them. Others, so convinced of something for which there is not the slightest evidence, can commit the most horrendous acts with the idea that in the Hereafter, their god will reward them, which does not speak very well of Her. God save us from those people!

Chapter 9

The Growing Ecological Footprint

Over billions of years, on a unique sphere, chance has painted a thin covering of life — complex, improbable, wonderful, and fragile, suddenly we humans (a recently arrived species no longer subject to the checks and balances inherent in nature) have grown in population, technology, and intelligence to a position of terrible power: we now wield the paintbrush.

— Paul B. Macready[1]

We . . . must avoid the impulse to live only for today, plundering for our own ease and convenience the precious resources of tomorrow. We cannot mortgage the material assets of our grandchildren without risking the loss also of their political and spiritual heritage. We want democracy to survive for all generations to come, not to become the insolvent phantom of tomorrow.

— Dwight D. Eisenhower[2]

The thought had somehow never ever even entered my head before, but, well, it goes exactly like this: we, on the Earth are constantly walking over a bubbling, crimson sea of fire, hidden there in the belly of the Earth. But we never think about it. But what if suddenly the fine crust of Earth under our feet became glass, and suddenly we could see... I became glass, I saw into myself, inside.

— Yevgeny Zamyatin[3]

It is true, as Zamyatin thought, that about 1,800 miles beneath our feet, you encounter the outer liquid iron-nickel core of the Earth, at a temperature of 4,200°C, increasing with depth to 7,700°C before reaching the hotter but solid inner central core. But even at a small depth of about 4 (1.86 miles), it is already about 100°C. So, we better stay on the surface.

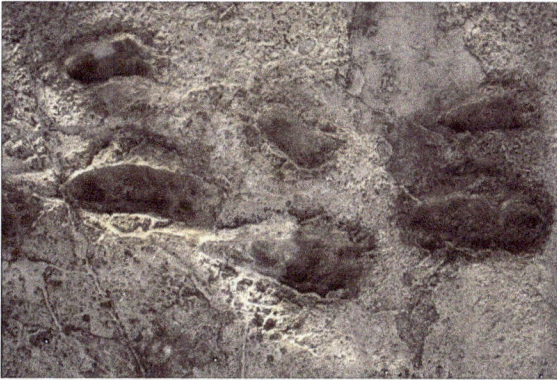

Figure 9.1.   The Laetoli footsteps.

About 3.6 million years ago, some hominids belonging to *Australopithecus afarensis*, the species from which *Homo sapiens* eventually evolved, went for a walk in Africa, in what is now northern Tanzania, at a place called Laetoli (Figure 9.1). They did this when a volcano, now called Sadiman, erupted and covered the ground with a thick layer of gray ash which preserved the footprints until they were discovered in 1976 by paleoanthropologist Mary Leakey (1913–1996) and her team. These footprints were not erased by sheer luck, as are those left on the beach, but buried under new soil were preserved for posterity. We do not know where our ancestors were headed, but it is not difficult for me to mentally connect their footsteps with the ones that we recently left on the gray surface of the Moon (Figure 9.2). We have traveled far, but perhaps in the wrong direction.

Our footprints are everywhere, on the most remote places of the planet, on the highest peaks of mountain ranges, the deepest ridges of oceans (not exactly footsteps there), and increasingly in areas populated by wild animals. As we intrude them, pathogens for which we are defenseless (and we see what that brings, as I write), can attack us.

Some footprints we associate with names such as Mattias Zürbriggen (the first to ascend Aconcagua in 1897) and Jorge Juan Link and his wife Adriana Bance (who lost their lives on Aconcagua in 1940). On the summit of the highest mountain in the world, in the Himalayas, Mount Everest

Source: NASA.

Figure 9.2. Footstep on the Moon.

perhaps carries the footprints left by Tenzing Norgay and Sir Edmund Hillary, who first managed to reach the summit at 8,848 m in 1953.

In 1930, on his 50th birthday, a German meteorologist, Alfred Wegener (1880–1930) and his guide Rasmus Villumsen (1907–1930), returning from a grueling trek to resupply their Greenland ice research station (Eismitte), succumbed to the inclement weather. Wegener formulated the idea of continental drift (now plate tectonics), Villumsen carefully buried Wegener and marked his grave with erect skis so that it could be found. Unfortunately, he disappeared without a trace, but as the ice keeps melting, perhaps we shall eventually find his frozen remains.

The stories of Robert Falcon Scott (1868–1912), explorer of Antarctica where he died, and of Norwegian Roald Amundsen (1872–1928), who beat him in the race to reach the south pole (on December 16, 1911) are fascinating, footprints left only to be erased by the next snowfall.

But the most profound trace is our *ecological footprint* which makes for not such a good story, an indicator of our impact on the planet, a concept developed in 1990 by Mathis Wackernagel and William Rees.[4] The ecological footprint measures the average amount of natural resources used by a population expressed as the area of productive land and water (using average values) needed to produce the resources consumed and to absorb the waste generated, including the area of forest required to absorb the CO_2 produced by energy use and the area of land used for housing and infrastructure to maintain a specific standard of living.

However, the ecological footprint does not include all possible categories; for example, it does not tell us anything about biodiversity, public health, or quality of life. Furthermore, the calculations do not include degenerative effects that diminish the global production capacity, such as loss of fertility and desertification of lands, deforestation, and loss of aquifers. The previous omissions make the calculated ecological footprint an approximation that underestimates the actual value. Nevertheless, it is a valuable quantity to analyze the state of health of the biosphere. It is agonizing.

It turns out that for a person in an undeveloped country, the ecological footprint is equivalent to about one hectare (10,000 m^2) and is ten times higher, about ten hectares, for those in developed countries (see Figure 9.3). The dubious distinction of being among the top ten corresponds to the US, with Singapore and the United Arab Emirates with about 8 hectares per inhabitant. The lowest correspond to Haiti, Afghanistan, Eritrea, and Yemen, which is not surprising.

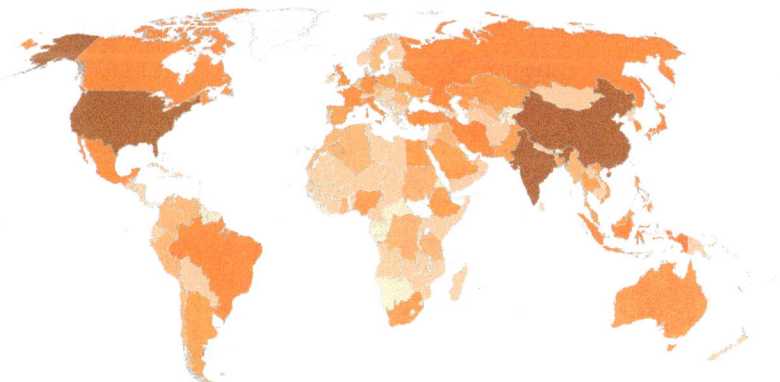

Source: https://data.footprintnetwork.org/?_ga=2.220618584.1554399153.1607641413-391021616.1607641413#/.

Figure 9.3.   World map with country's ecological footprint. It is measured by the number of global hectares for each country. Lighter shades denote countries with a lower ecological footprint and darker shaded for countries with a higher ecological footprint. Note that this is not a per capita representation, which would be different.

Fifty years ago, when the human population was "only" about 3 billion, Georg Borgstrom (1912–1990),[5] geographer, ecologist, and an authority on hunger, warned: *Words like limitless, inexhaustible, and boundless have figured prominently in the debate about the Earth and its resources. They have been persistently used despite the fact that they would not exist in a Dictionary of Nature. The propagandists utilize them with increasing frequency, as if anxious to prevent the human masses from realizing the truths about Man's own existence and the earth's limitations. But facts are stubborn things. Sooner or later, they break into the open and assert themselves, forcing man to make harsh accommodations to reality. The circus acts then get stripped of their luster. The bluff is called, and the swindle unveiled. Rather than a rich, well-fed and abundant world, excruciating poverty, widespread hunger, and debilitating diseases emerge.*

And here we are with almost 8 billion and counting. Climate change, to which we shall turn next, was not even an issue then.

The ultimate problem is overpopulation multiplied by increasing consumption, the unstoppable avalanche of mouths to feed (each year, we grow by about 70 million). And all are persuaded to use mobile phones, eat meat, and drive cars, among countless other things offered by the ever-growing hunger for material things promoted by greedy corporations.

Natural resources are voraciously consumed as if they were inexhaustible, corporations are on a rampage[6] competing for what is left,[7] and the waste is discarded into the air we breathe and the water we drink as if nature had an unlimited capacity to absorb it. Although the world seems gigantic, the atmosphere cannot withstand the injection of toxic gases. Neither do our lungs. It is estimated that millions die and more suffer from contaminated air and water, many in crowded mega-cities such as Mexico, Sao Paulo, Beijing, or Delhi.

The seas cannot cope with the effluents of our industries, fish suffer from unsustainable fishing or choke on plastic, and flora and fauna cannot endure our insatiable appetite. The level of deforestation is the highest in history, and drinking water is scarce in large parts of the planet and is further depleted by the giant bottled (mostly plastic) water industry. The consequence is a colossal wound afflicting the world, all forms of life, including us. We are painting ourselves into a corner.

Most of the world's arable land is already under cultivation. Its productivity is diminished by the effects of erosion, salinization, and soil degradation, each year causing the loss of several million hectares. As the climate changes, so will the availability and productivity of arable land.

The total area of biologically productive land is approximately 12,000 million hectares[a] (it is difficult to plow in water and irrigate deserts), so a simple division tells us that the present population of the planet (7,800 million) has on average about one hectare and a half per inhabitant. This limit, defined by the size of the earth, reveals that the idea of unlimited growth is a delusion and a cruel lie by those who propose impossible development models. In addition, there are hectares, and then there are hectares; all the lands are not equally productive.

The average ecological footprint has increased significantly despite certain gains in productivity which come at the expense of external resources (water, energy, and fertilizers). In 2017 we already needed about 1.7 earths, and unless we change drastically, we will need two planets by 2030 (Figure 9.4). Unfortunately, we only have one, so something must give.

If we do not *radically* change (and we will not do it despite all the unrest and protests), it will take several Earths for everyone to live at a consumption level of developed countries. The US is the first to block any country from seeking a different path, contradicting the established system. Cuba comes to mind.

In the long term (not very long), an unsustainable activity leads to ruin, just as if you only live off your savings: at some point, you run out of money. In the personal case, you could resort to a loan, but there is no such loan in the ecological case since the natural capital once consumed cannot be restored.

To understand the consequences of so-called development, a new accounting is necessary so that the true cost of all activities are considered, including the needed energy and natural resources consumed and the damages caused to the ecosystem and global society. For example, the

---

[a] One hectare is a square with sides of 100 meters (328 feet) in length and is about 2,47 acres.

How many Earths does it take to support humanity?

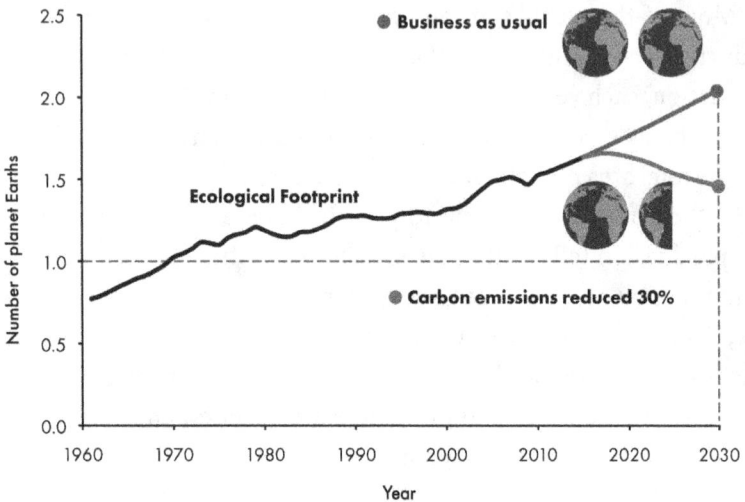

Figure 9.4. World Ecological Footprint. We surpassed one planet in the seventies.

cost of solar panels (I am the first to support their use) should include all activities associated with their production, such as the mining of scarce materials needed to produce them (which emit CO_2, etc.) and the ecological cost of producing a growing number of batteries and the environmental cost of disposal. It should also include the social cost of all those children who could not go to school because they are part of a cheap labor force in some countries.

Much has been written about this, excellent books indeed, and completely inconsequential, even if written by a Nobel Prize winner. The other book that many read does not serve for much either, except as an inopportune consolation to what seems inevitable.

It is practically impossible for the "undeveloped" world to develop following the model of developed countries. Although some do not understand it or perhaps do not want to understand, it is evident that given a finite number of resources, it is only possible to keep a finite number of activities for a limited time. This does not prevent large corporations from selling all kinds of illusions, as they say: "opening new markets," which only creates unattainable expectations and blocks any idea of alternative development.

The world's population is growing and will continue to do so for quite a while and is part of the catastrophic convergence. Before 1500, we were less than 400 million, and in no more than 20 years, we will be about 9 or 10 billion, and that difference is crucial. (As with all these studies, there are uncertainties in these projections which do not affect the discussion.)

With the scientific and technological advances of the last century, the global population went from 1.65 billion in 1900 to 6.1 billion in 2000, a growth by a factor of 3.7 in 100 years (Figure 9.5). This has gone hand in hand with our global consumption of energy. As it is for all things that have a measure far from our usual "middle world" values, 10 billion means as much as 5 billion because we have difficulties grasping very large (or small) quantities. Think that if you started now and counted 8 billion seconds, you would not be done until the year 2274.

Total population

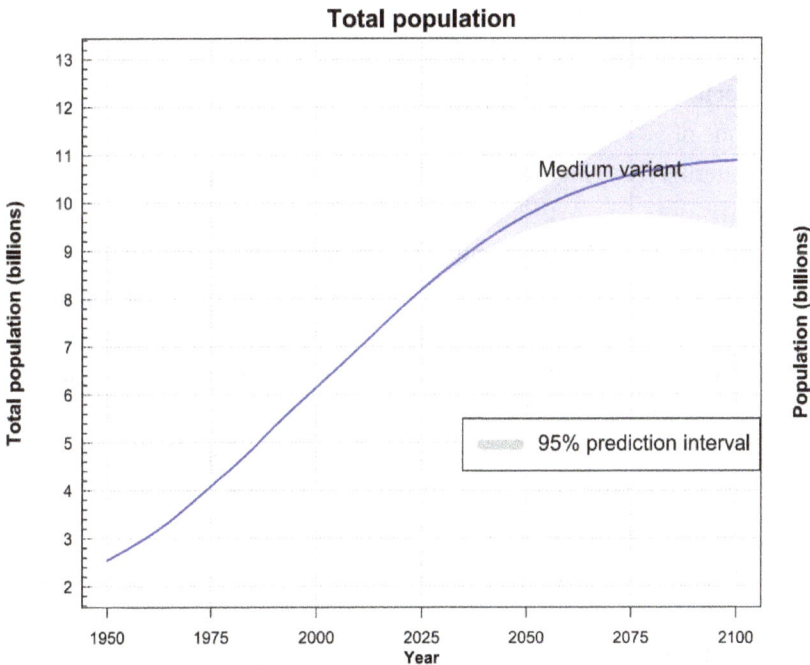

Source: United Nations.

Figure 9.5.   World total population.

Source: By Chensiyuan, CC BY-SA 4.0. https://commons.wikimedia.org/w/index.php?curid=10279326.

Figure 9.6.   Rocinha, a neighborhood in Rio de Janeiro, Brazil, is one of the largest and densest slums (favela) in Latin America.

Population growth, something we were warned about by Paul Ehrlich[8] and others many years ago, is the grey eminence of our troubles. Although several critics of Ehrlich have pointed out that many of his specific predictions of 50 years ago were inaccurate and perhaps overstated, in the end, it remains true (Figure 9.6). We would not be facing this crisis (or we would have more time to find workable solutions) if the world were inhabited by "only" 1 billion. Renowned author and columnist for The Nation, Katha Pollitt recently wrote[9]: *Does the world need more people? Not if you ask the glaciers, the rain forests, the air, or the more than 37,400 species on the verge of extinction thanks to the relentless expansion of human beings into every corner and cranny of our overheated planet. There are now 7.9 billion of us and growing — 50 years ago there were fewer than half as many. I'd say we've more than fulfilled the biblical injunction to be fruitful and multiply.*

Most future population growth will occur in poorer, underdeveloped counties; the population of sub-Saharan Africa, for example, is projected to double by 2050. Just imagine. It's a matter of time, and the present refugee "crisis" will look like a kindergarten outing. Even then, most will just have to suffer the worsening conditions in their own countries, surviving

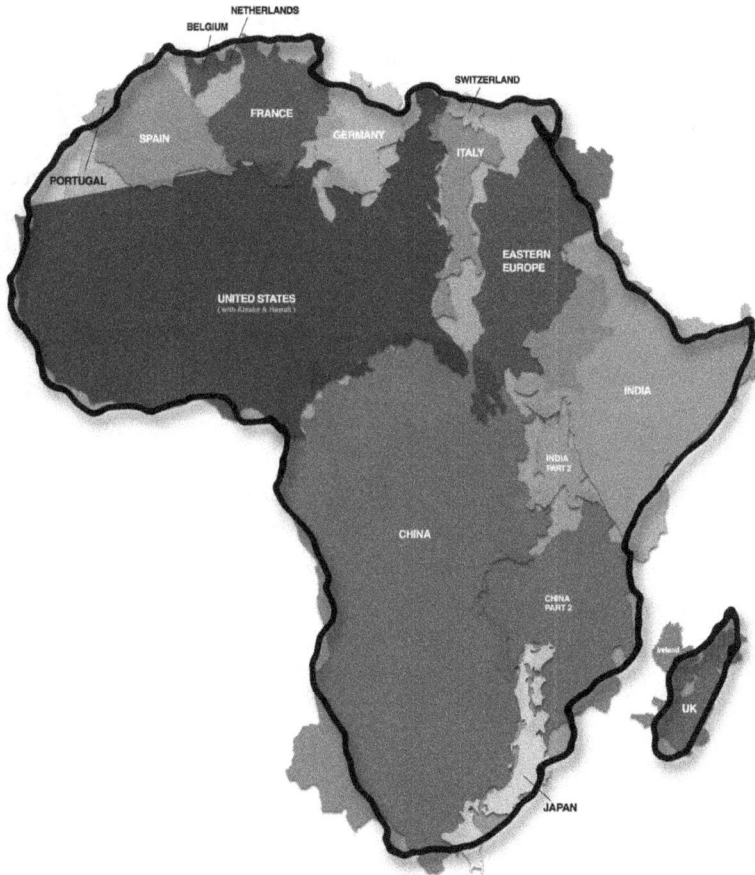

Figure 9.7. Africa is much larger than what it seems on usual maps (Kai Krause).

in filthy, unhealthy, and violent slums governed by ever more corrupt individuals. As a byproduct (collateral damage), we populate unsafe places, encroach on wildlife (catching viruses), and get to see more dramatic scenes on the news as fire, wind, and water decimate people living at the edge of safety. Violence will be inevitable, and some countries might resort to measures that will remind us of Hiler's Germany.

Being used to a normal world map, you get the impression that Africa is relatively small, but this is an illusion produced by representing the spherical earth on a flat surface (Mercator projection) which will distort areas, making them seem larger as you move away from the equator. Greenland (which is still primarily white) looks larger than the US on usual maps but, it is five times smaller. Africa, as you can see from Figure 9.7, is huge, and you can fit China, India, the US, Europe, and then some into it. It is not surprising that the US has expanded its military presence in Africa and engaged in secret operations.[10]

Chapter 10

Ozone and CO$_2$

The climate is a common good, belonging to all and meant for all. At the global level, it is a complex system linked to many of the essential conditions for human life. A very solid scientific consensus indicates that we are presently witnessing a disturbing warming of the climatic system. In recent decades this warming has been accompanied by a constant rise in the sea level and, it would appear, by an increase of extreme weather events, even if a scientifically determinable cause cannot be assigned to each particular phenomenon. Humanity is called to recognize the need for changes of lifestyle, production, and consumption, in order to combat this warming or at least the human causes which produce or aggravate it. It is true that there are other factors (such as volcanic activity, variations in the earth's orbit and axis, the solar cycle), yet a number of scientific studies indicate that most global warming in recent decades is due to the great concentration of greenhouse gases (carbon dioxide, methane, nitrogen oxides and others) released mainly as a result of human activity.

— Pope Francis[a]

Climate change is a negotiation between human beings and physics and physics doesn't compromise. Past a certain point there's no more room for maneuver.

— Bill McKibben[1]

The main reason to be careful when you walk up a flight of stairs is not that you might slip and have to retrace one step, but rather that the first slip might cause a second slip, and so on until you fall dozens of steps and break your neck.

— Robin Hanson[2]

---

[a] Encyclical letter *Laudato Si'* of the Holy Father Francis on Care for our Common Home.

In what follows, I will not get into the details. That would lead me to an entire book on the issue (and there are plenty). So, all I wish to do is tell you the essentials, and perhaps you already know, but this might help you explain it to your skeptical neighbor. And although it is common to express that "the devil is in the details," in this case, disregarding the details and considering the essentials might let us understand how not to follow the devil.

If our planet did not have an atmosphere, things would be very different. Not only would we have an obvious problem with breathing and air transport, among other things, but in geophysical terms, the surface of the planet would be much colder, so cold that the oceans would freeze, and our planet would be a ball of ice. This is because life on the planet (any planet) is intimately linked to the climate and its surface temperature. Too hot or cold, and there will be no life.

Carbon dioxide, which constitutes a tiny part of the atmosphere (presently 0.042%), methane, and water vapor — greenhouse gases — are transparent to sunlight but not to infrared or ultraviolet radiation. As a result, sunlight can reach the surface of the Earth, where the soil absorbs its energy and heats up. In *equilibrium*, the Earth emits the energy received from the Sun back into space, since otherwise, it would continue to heat up. As already presented, the Sun emits energy mostly as light and ultraviolet radiation (at a temperature of about 5,500°C) the Earth (at a temperature of about 15°C) emits in infrared (See figure 6.7).

But the atmosphere is *not* transparent to infrared radiation; the greenhouse gases absorb much of this radiation. They then cool by emitting this radiation in all directions, part of which then returns to the Earth's surface. In other words, the surface of the Earth is heated by direct solar radiation and by radiation emitted by atmospheric greenhouse gases. This causes the average temperature of the Earth's surface to be 15°C (59°F), while without the atmosphere, it would be –18°C (0°F), and there would be no life on the frozen planet (although this latter number is an approximation depending on how much solar radiation the surface would reflect (the "albedo")).

Figure 10.1 describes the various kinds and amounts of energy that enter and leave the Earth's system. It includes both, radiation components (light and heat) and other components like conduction convection and

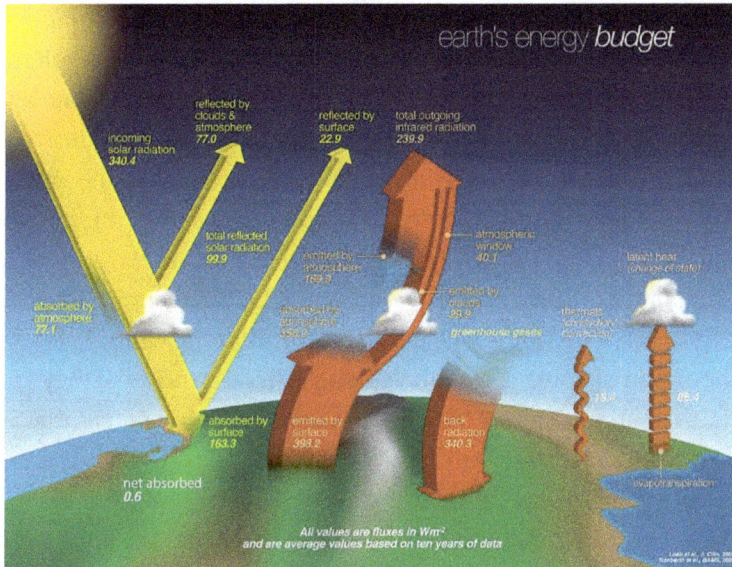

Figure 10.1. The Earth energy budget (NASA).

evaporation which also transport heat from Earth's surface. On average, and over the long term, there is a balance at the top of the atmosphere. The amount of energy coming in from the sun is the same as the amount going out from reflection of sunlight and from the emission of infrared radiation.

To reach equilibrium under this increased received radiation from the greenhouse gases, the planet must emit more energy into space, and this it does by increasing its average temperature. So, increasing greenhouse gases means a catch-up game where the planet keeps getting warmer. Figure 10.1 illustrates the various energy flows starting with what arrives from the Sun (a nuclear fusion plant), about 1,360 W/m^2, of cross-section outside the atmosphere (this number is called "solar constant" although it varies slightly). Since only the illuminated part of the spherical earth receives solar energy, the averaged value received is 340 W/m^2, (the factor of 4 is the difference between the area of a sphere and that of a circle of equal radius), of which approximately one third is reflected.

The atmosphere quickly becomes thinner with height. About 90% of the air is in the lowest thirty kilometers, and half of it is concentrated in the first five kilometers, below the top of the highest mountains. Air contains

78% molecular nitrogen (two nitrogen atoms) and 21% molecular oxygen (two oxygen atoms) with carbon dioxide, methane, water, and argon traces.

The cultural, political, economic, and technological development of *Homo sapiens* went hand in hand with the terrestrial climate, determined by natural processes related to the orbital configurations of the Earth and complex processes on its surface such as volcanism and the circulation of huge marine currents. The latter transport large amounts of water that affect the temperature of some regions, as does the Gulf Stream in the present, affecting the climate (for good) in the North Atlantic region.

There is an ongoing discussion among expert scientists (if they are not experts, their arguments seldom matter) about technical details that apply to the results of measurements and climate models one way or another. But let us not lose sight of the forest for the trees; all agree that *we* are causing unprecedented changes in Earth's climate, and the uncertainty is only about when (and not if) hell will arrive.

Let me elaborate a bit on this crucial point. Arguments between experts point out disagreements on the results of a study (empirical or theoretical), which usually are of the form $X \pm x$, where x denotes a measure of the uncertainty in the value of big X. What we want is a small x, but for complex systems, this is often difficult. But even if x is larger than we wish, it is better than nothing and a good reason to continue research. It is, however, not a reason to reject X outright. For example, we cannot (yet) predict with a week's anticipation where a hurricane will make landfall (X) and can only tell $X \pm x$, x getting smaller as the thing approaches. But if you live within x from X, it would be foolish not to prepare.

Our consumption of energy is estimated globally to be about 150 PWh per year and growing constantly. (The prefix Giga is 10^9, Tera is 10^{12}, and Peta is 10^{15}), (I clarify that power (Watts) is the rate at which energy is used, that is, energy per unit of time). This energy is extracted mainly from fossil fuels (coal, gas, and oil), producing waste carbon dioxide (CO_2) that we discard into the atmosphere (Figure 10.2).

Physicist Varum Sivaram, in a book[3] that looks carefully at the technologies and policies needed to tackle the climate crisis writes: *Our star produces 2 billion times as much energy as a sliver of light that eventually*

Global primary energy consumption by source

Primary energy is calculated based on the 'substitution method' which takes account of the inefficiencies in fossil fuel production by converting non-fossil energy into the energy inputs required if they had the same conversion losses as fossil fuels.

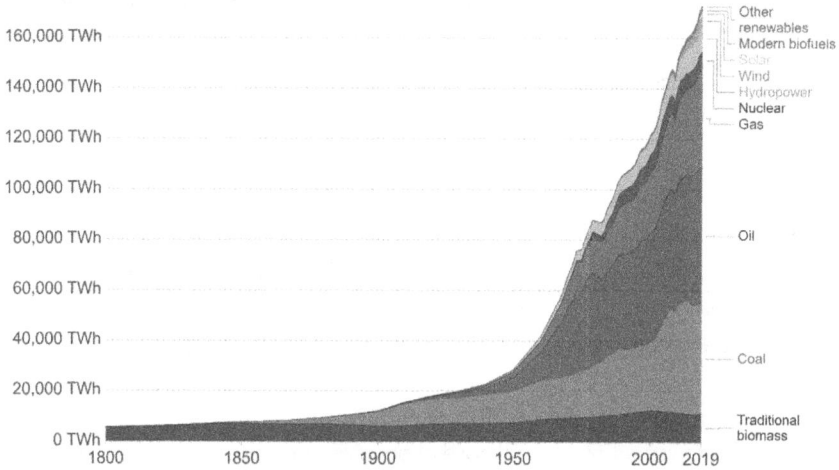

Source: Vaclav Smil (2017) & BP Statistical Review of World Energy

OurWorldInData.org/energy • CC BY

Source: https://ourworldindata.org/grapher/global-primary-energy?country=~OWID_WRL.

Figure 10.2.   Global energy consumption.

reaches the Earth, yet that sliver delivers over 10,000 times the power that the world needs.

The big problem resides and how to harvest this energy (say with photovoltaics (PV) — solar panels) and to deal with the fact that PV only work when there is sunlight and that 70% of the planet is covered by oceans. Converting water to hydrogen (with solar energy) might be a solution. Part of this energy is the source of the photosynthetic processes that feed us. The Sun, and windmills seem the most viable sources for the future, but in terms of the generation of electricity, which runs through the veins of technological civilization, they produce a small percentage of the total, although it is growing (According to International Energy Association over the next years about 70,000 solar panels *per hour* will be added: most in China.)

We need an intense global project akin to the Manhattan project, or Apollo, to capture solar energy. Instead, during the Trump administration, funding for alternative energy research was cut drastically, ending the

ARPA-E program, which gave $300 million a year in grants for research in technologies aimed at reducing fossil-fuel consumption and improving energy efficiency. Expressing that *the private sector is better positioned to finance disruptive energy R&D and to commercialize innovative technologie,* we lost four precious years.

10.1 A Terrifying Story

> We find ourselves, one way or another, in the midst of a large-scale experiment to change the chemical construction of the stratosphere, even though we have no clear idea of what the biological or meteorological consequences may be.
>
> — Frank Sherwood Rowland

> If we don't take action, the collapse of our civilizations and the extinction of much of the natural world is on the horizon.
>
> — David Attenborough[b]

Ozone is a triatomic form of oxygen (O_3) discovered by the German Christian Schönbein in 1840 (normal oxygen we breathe has two atoms — O_2). The ozone layer resides in the stratosphere at a height between 15 and 30 km. This thin layer forms a shield that protects us from the dangerous ultraviolet radiation emitted by the Sun. O_3 is formed when O_2 molecules absorb solar ultraviolet radiation, dissociating them into two oxygen atoms that can then be combined with another oxygen molecule to form O_3. In the absence of this shield, ultraviolet radiation is lethal to all forms of life, causing genetic mutations, eye injuries, skin cancer, and damage to plants. It is no coincidence that life on Earth began underwater where UV radiation does not reach far and only ventured to the surface about a billion years ago after the ozone layer formed.

At the beginning of the seventies decade, there were intense debates about the possible effect on ozone caused by oxides of nitrogen produced by the planned fleet of supersonic aircraft that would fly in the stratosphere. These planes would allow an executive to attend a meeting in London

---

[b] Speaking at a U.N. climate summit in December 2018.

or Paris and return to his office in New York on the same day (I do not see what was so great about that, but it does not matter). This problem stimulated the interest to investigate the chemistry of the atmosphere, the effect of several compounds generated in increasing quantity by us, and the behavior of chlorofluorocarbons (CFCs).

CFCs began to be used approximately 80 years ago, and by the end of the 1970s, the industry used about 1 million tons per year. CFCs and other chemical products containing chlorine and bromine were used because they were non-toxic, non-flammable, and very stable, not reacting with other compounds. These were ideal properties for refrigerants, electronics, and aerosol propellants (spray cans). For example, it would not be helpful to spray your hair with a product propelled by another that burned it.

But in 1974, two critical studies were published. Richard Stolarski (born 1943) and Ralph Cicerone (1943–2016)[4] explained how chlorine atoms could destroy ozone. And Mario Molina (born 1943) and F. Sherwood Rowland (1927–2012) demonstrated that CFCs *could* reach the stratosphere[5] where solar ultraviolet radiation decomposed them releasing chlorine atoms. In a catalytic reaction, a chlorine atom can destroy a 100,000 ozone molecules.

In 1984 a British research group led by geophysicist Joseph Farman[6] (1930–2013), collecting atmospheric data since 1957, detected a drastic drop in the concentration of ozone over Antarctica that occurred each October (Figure 10.3). This caused quite a stir and a controversy with NASA scientists whose satellite measurements did not show such a depletion (because of faulty software, it turned out). This time coincides with the beginning of austral summer when the first rays of sunlight illuminate the days after 4 months of darkness, the so-called "ozone hole." But most alarmingly, measurements made in successive years revealed that the minimum amount of ozone in the hole decreased each year. Although the effect was more dramatic in the South Pole, around 10% ozone depletion was found at all latitudes.

It is estimated that a decrease of only 1% in ozone levels causes a 2–3% increase in the incidence of skin cancer.

These investigations caused great consternation in an increasing number of scientists and culminated after many debates in several

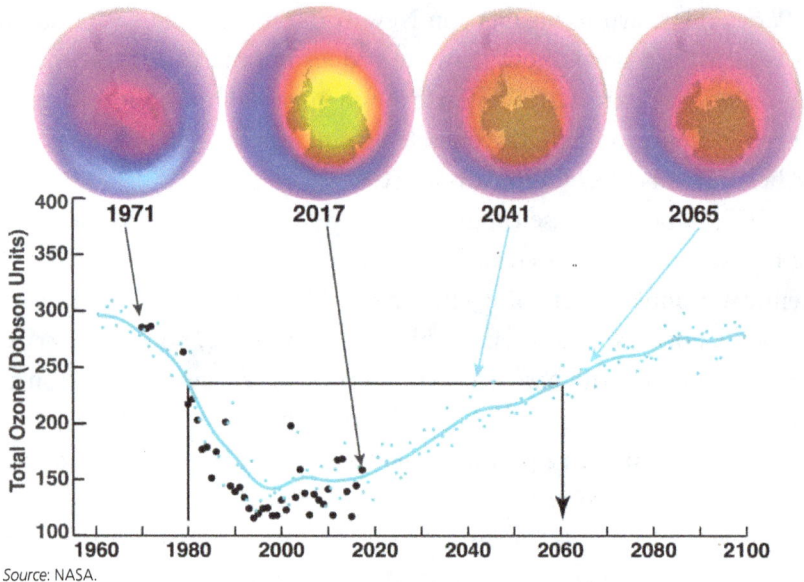

Figure 10.3. The Antarctic ozone level over the years.

international laws and treaties (Montreal Protocol of 1987) that prohibit the industrial production of CFCs. However, over time the ozone layer will slowly recover (as shown in Figure 10.3).

Rowland, Molina, and Paul Crutzen (1933–2021) received the 1995 Nobel Prize for chemistry (Crutzen for his work related to nitrogen oxides). The citation states in part[7]: *By explaining the chemical mechanisms that affect the thickness of the ozone layer, the three researchers have contributed to our salvation from a global environmental problem that could have catastrophic consequences.* (Do not ask why Farman and his team didn't, but they are mentioned in the award ceremony speech.)

This left all those who thought that there was no need to worry in a terrible light. In the seventies, there was a real "aerosols war"[8] between those who were concerned about the effect on ozone of these products and those who defended the industry, investing large sums of money in advertising campaigns claiming that scientists did not know what they were talking about, that the data were erroneous or had significant uncertainties and that "more studies were needed" before prohibiting or modifying a helpful technology.

Paul Crutzen[9] said that we were much closer to a global disaster of enormous proportions than we think. Suppose Farman and his group of investigators had not persevered in making ozone measurements in the harsh Antarctic environment for all those years, beginning in 1958. In that case, the discovery of the problem could have taken longer, and the damage would have been much greater. More alarming is the thought that compounds based on bromine instead of chlorine could have been developed. In this case, given that bromine is a hundred times more destructive of ozone than chlorine, the effect would have been more severe, generating an ozone hole everywhere and all the time before anyone knew what was happening.

This episode, in which humanity was on the verge of disaster without anyone knowing, demonstrates the need to be alert. Independent scientists supported by supranational funds should be allowed and encouraged to investigate and publish the results of everything that worries them, without intrusion by anyone. This time we squeaked by, thanks to science.

Everything sounds very similar to the current discussions on carbon dioxide emissions and climate change, another new war (the "climate war") between industrial interests and the public, with governments standing in the middle without knowing what to do, or rather, without wanting to do what needs to be done. Again, we hear that in the face of uncertainties, "more studies are needed." It seems that history is quickly forgotten, and we repeat the mistakes of the past.

Given uncertainty (inevitable in these cases), we must act with caution. In the case of environmental problems, this principle was enunciated as part of the United Nations Framework Convention on Climate Change, the result of the United Nations meeting in Rio de Janeiro in 1992, which in its third article says[10]: *The Parties should take precautionary measures to anticipate, prevent or minimize the causes of climate change and mitigate its adverse effects. Where there are threats of serious or irreversible damage, lack of full scientific certainty should not be used as a reason for postponing such measures, ...* This was 40 years ago and is the reason why I have previously referred to a "desperate lack of progress."

At that meeting, a speech was given saying in part[11]: *An important biological species — humankind — is at risk of disappearing due to the rapid and progressive elimination of its natural habitat. We are becoming aware of this problem when it is almost too late to prevent it. It must be said that consumer societies are chiefly responsible for this appalling environmental destruction. They were spawned by the former colonial metropolis. They are the offspring of imperial policies which, in turn, brought forth the backwardness and poverty that have become the scourge for the great majority of humankind [...]. Unequal trade, protectionism and foreign debt assault the ecological balance and promote the destruction of the environment. If we want to save humanity from this self-destruction, wealth and available technologies must be distributed better throughout the planet. Less luxury and less waste in a few countries would mean less poverty and hunger in much of the world.*

In 1994, the General Assembly of the United Nations agreed that September 16 would be the "World Ozone Day" to commemorate the signing of the Montreal Protocol on that day of 1987.

But we are not out of it yet. Many old refrigerators contain CFCs, and if improperly recycled, they release this chemical into the atmosphere. A rise in CFC-11 has been recently observed,[12] and the source is probably the careless recycling of refrigerators in China.

10.2 Planetary Emergency

What is the use of having developed a science well enough to make predictions if, in the end, all we're willing to do is stand around and wait for them to come true.?

— Frank Sherwood Rowland[c]

A simple calculation shows that the temperature in the arctic regions would rise about 8° to 9°C, if the carbonic acid increased to 2.5 or 3 times its present value.

— Svante Arrhenius[13]

The choice we now face in responding to the climate crisis is between a just transition and just a transition.

— David Judt[14]

---

[c] Reportedly said while accepting his Nobel Prize.

The greenhouse effect caused by trace gases in our atmosphere is not a myth nor a scam, as some declare. The influential US Senator James Inhofe, authored (perhaps ghostwritten?) a book: "The Greatest Hoax," and influencer Rush Limbaugh (1951–2021) poisoned the well, calling climate-change science *the biggest scam in the history of the world*. Scott Pruit, controversial director of the Environmental Protection Agency (2017–2018), had this to say: *I think that measuring with precision human activity on the climate is something very challenging to do, and there's tremendous disagreement about the degree of impact.* Director of the EPA?). There are many other deniers; sowing doubt about scientific results is standard procedure.

However, life on Earth depends on the good services of these gases. Without CO$_2$ no plants would grow, and as already mentioned, Earth would freeze. However, good things in excess can be fatal. For example, an increase in the concentration of CO$_2$ and other greenhouse gases in the atmosphere will increase the average temperature of the surface. Any astrophysicist studying planetary atmospheres knows this.

It is a well-understood mechanism, and you can go to Venus if you have any doubts. Due to its atmosphere being composed of 96.5% CO$_2$, the surface temperature is about 900°F (450°C). Without CO$_2$, it would be in the range of 100–150°F depending on other variables, such as how much solar energy the surface would reflect (the "albedo").

It is nothing new, but for a long time, few had worried about it. Forgotten scientist Eunice Newton Foote (1819–1888) experimented in 1856 with gases exposed to sunlight, concluding that[15]: *An atmosphere of that [CO$_2$] gas would give to our earth a high temperature.* Swedish scientist Svante Arrhenius knew this in 1895,[16] and as early as 1912, in *Popular Mechanics*, the problem was discussed as shown in Figure 10.4.[17]

Later, in 1956, a *New York Times Science Review* article by Waldemar Kaempffert warned of possible consequences[18]: *Warmer Climate on the Earth May Be Due to More Carbon Dioxide in the Air* was the title. Ten years later, in a 1966 article in the industry publication "Mining Congress Journal" by James Garvey, then President of *Bituminous Coal Research Inc.*, we can read[19]: *Among the gaseous materials discharged from the stack is carbon dioxide. This is not generally considered to be a pollutant*

POPULAR MECHANICS   341

The furnaces of the world are now burning about 2,000,000,000 tons of coal a year. When this is burned, uniting with oxygen, it adds about 7,000,000,000 tons of carbon dioxide to the atmosphere yearly. This tends to make the air a more effective blanket for the earth and to raise its temperature. The effect may be considerable in a few centuries.

Figure 10.4.   A page from popular mechanics.

inasmuch as it has never been demonstrated to have any adverse effects on plants or animals. However, to illustrate the far-reaching aspects of the air pollution problem, it should be noted that serious studies are underway to determine whether more restrictions should be placed on the emission of carbon dioxide to the atmosphere. There is evidence that the amount of carbon dioxide in the earth's atmosphere is increasing rapidly as a result of the combustion of fossil fuels. If the future rate of increase continues as it is at the present, it has been predicted that, because the CO_2 envelope reduces radiation, the temperature of the earth's atmosphere will increase and that vast changes in the climates of the earth will result. Such changes in temperature will cause the melting of the polar icecaps, which, in turn, would result in the inundation of many coastal cities, including New York and London.

That was 1956!

An internal Exxon technical review, CO_2 *"Greenhouse" effect*, prepared on April 1, 1982 (they did fool us), *restricted to Exxon personnel and not distributed externally*, included the Figure 10.5.[20] The review also said: *In addition to the effects of climate on global agriculture, there are potentially catastrophic events that must be considered. For example, if the Antarctic ice sheet which is anchored on land should melt, then this could cause a rise in*

GROWTH OF ATMOSPHERIC CO2 AND AVERAGE GLOBAL
TEMPERATURE INCREASE AS A FUNCTION OF TIME

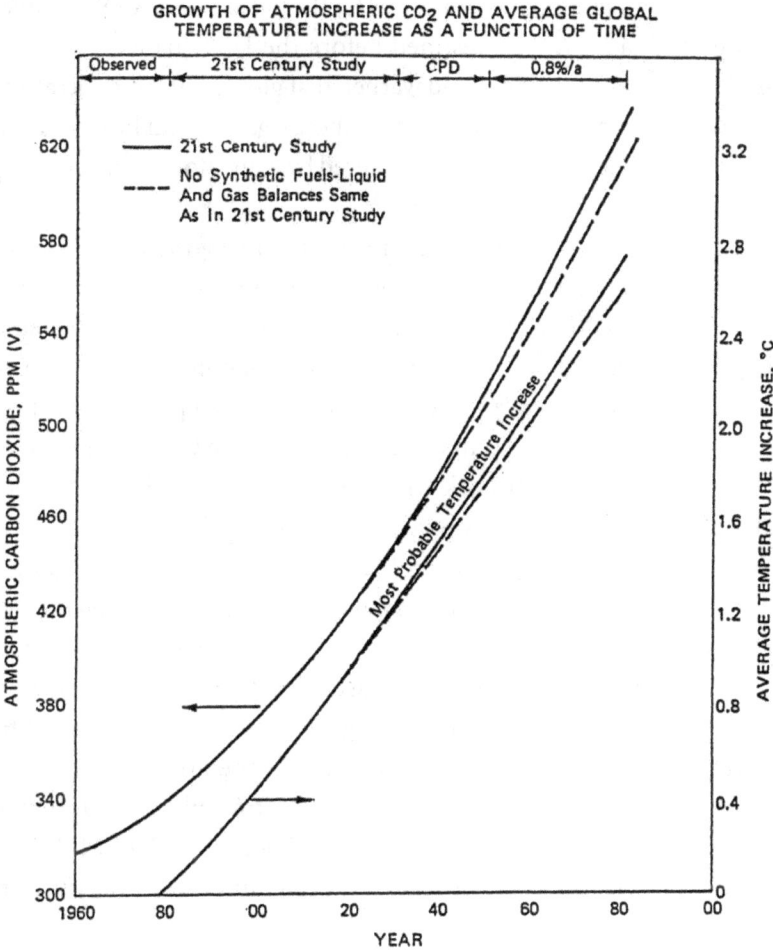

Source: Exxon via Inside Climatenews.

Figure 10.5.   Exxon 1982 prediction from a report.

sea level on the order of 5 meters. It is uncanny: Despite what must have been relatively primitive computations compared to modern climate models, the predictions of the effects of CO_2 on temperature are right on the mark. So yes, energy corporations knew, Exxon knew, and the US government knew very well in the 1980s what was coming and tried to hide and even deny it. You could label it as a crime against humanity.

In 1988, James Hansen (born 1941), then head of NASA's Goddard Institute for Space Studies, testified before the US Senate's Energy and Natural Resources Committee and warned that global warming was already happening due to human-produced greenhouse gases.[21] But little was done, and we lost 30 crucial years, a loss furthered by a disinformation campaign organized by the petroleum industry.[22]

In his 2009 book Hansen again warns: *During the past few years, however, it has become clear that 387 ppm (CO_2) is already in the dangerous range. It's crucial that we immediately recognize the need to reduce atmospheric carbon dioxide to at most 350 ppm in order to avoid disasters for coming generations.* We have now surpassed 420 ppm, and it keeps increasing. (There was a dip in emissions due to the Covid19 pandemic, but we are on track to get back where we were.[23]) Bill McKibben (born 1960), leader of the climate campaign group 350.org, writes[24]:

Other greenhouse gases, such as methane (CH_4), produced by bacterial fermentation and by animal digestion (farting cows which will increase as we consume more meat), or nitrous oxide (NO_2), a by-product of the use of fertilizers and the combustion of gasoline, also contribute to global warming. However, their long-term effects are less important than carbon dioxide because they remain in the atmosphere for less time (of order 10 years). On the other hand, atmospheric CO_2 remains in the atmosphere for a long time (estimated between 50 and 100 years) and accumulates. Thus, the problem will remain even if we could stop emissions right now.

Methane is much more potent than CO_2. Although burning natural gas (mostly methane) as a fuel produces about half as much carbon as burning coal, the problem arises with a small fraction that is not burned and escapes from drilling and pipelines into the atmosphere.

Many, like the already mentioned Inhofe, confuse meteorological weather with climate. Inhofe went as far as bringing a snowball he picked up to a hearing in Congress to argue that DC was cold, so no global warming. Do me a favor! Meteorological weather, which the news informs us every day is highly variable with the seasons and local

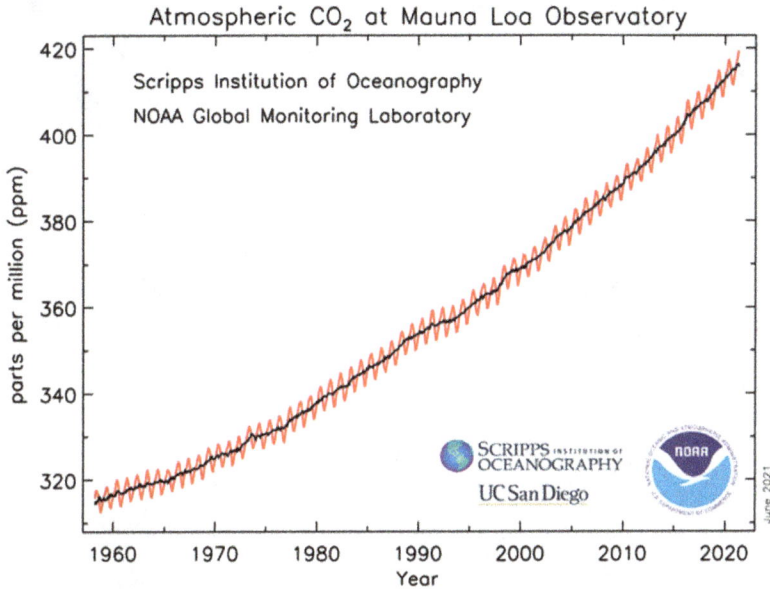

Figure 10.6.   Atmospheric CO$_2$ at Mauna Loa.

natural effects. The temperature changes from day to night, cloudy days and sunny days.

On the other hand, *climate* refers to changes in average long-term weather (over more than 30 years). We can predict that temperatures will rise in the future, and summers will be hotter on a global average, not necessarily in a particular place. We know the cause and the consequences (for example, water expands when warming, so sea levels must rise, independent of how much gets added through land ice cover melting, which is accelerating).

I consider the graph (Figure 10.6) to be one of the most dramatic in existence. It is the "Keeling curve" (Charles D. Keeling 1928–2005), showing the measurements of atmospheric CO$_2$ concentration over time, obtained from Mauna Loa, at an elevation of 3.4 km. They are

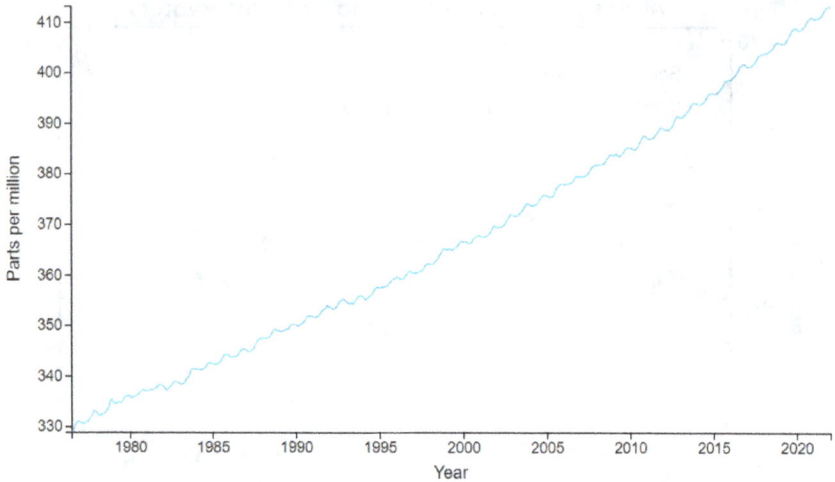

Figure 10.7.   Carbon dioxide measured at Cape Grim. (CSRIO)

very precise and even display the annual oscillation due to the growth of plants in the northern hemisphere, which absorb a fraction of the gas. The data (these are the *facts*, no alternative) show a progressive increase since 1960 that has recently surpassed 418 parts per million (ppm) and keeps increasing.

Some have objected that the Keeling graph shows a local measurement on a site near a volcano (that emits CO_2) and is therefore not representative of the global atmosphere. However, the atmosphere mixes on timescales of weeks to months. So, here is an independent measurement from a wholly different and far away site: Cape Grim, in Tasmania, showing the same result (Figure 10.7). Compared to the concentration of 200 years ago, carbon dioxide has increased by over 40%, methane has doubled, and nitrous oxide has grown by 15%. Although carbon dioxide constitutes only a tiny fraction of the atmospheric content, the total amount of *carbon* in the atmosphere is enormous: 875 gigatons (one gigaton (Gt) equals one billion tons), corresponding to 3,210 Gt of CO_2. (Note two different measures that appear in the literature: 1Gt of carbon (C) equals 3.67Gt of CO_2). It is estimated that human activities currently pump 10 Gt per year (36 gigatons of CO_2) into the atmosphere (including the effects of deforestation), of which about half remains in it.

Recently the Intergovernmental Panel on Climate Change (IPCC) of the United Nations) reported[25] that allowing the global temperature to increase above 1.5°C from preindustrial levels would be catastrophic. This will happen if we pump another 420 Gt of CO$_2$ into the atmosphere. Do the numbers and conclude that this will occur in less than 10 years at current levels. A recent IPCC[26] report concludes that to not surpass this limit, we need to curtail CO$_2$ emission by 45% by 2030. *It ain't gonna happen.* Being asked about this, Trump responded that *I want to look at who drew it — you know, which group drew it.* He had no clue and cost us years of inaction; he was, to quote the Beatles, *a real nowhere man, sitting in his nowhere land, making all his nowhere plans for nobody.*

Depending on the future use of fossil fuels, atmospheric carbon dioxide concentration could double in less than 100 years. It is estimated that this will cause an increase in temperature of at least 3°C. It does not seem like a big deal, and maybe you think it would not bother you if Montevideo or Chicago were a bit warmer. Still, we are talking about the global temperature, equivalent to a 20% increase in the planet's average temperature, that will have dreadful consequences. However, we still do not know them in detail. The last ice age resulted from a decrease in the average temperature of only about 5°C, so these seemingly small changes can have far-reaching consequences.

We have all seen the news of increasing deadly heatwaves and forest fires, and as temperature increases, we will see more of them, and they will last longer. However, there is a hard limit: humans cannot survive heat above 95°F and high humidity for long. This will affect all living things in wide swaths of the planet.[27]

It is possible to measure the concentration of greenhouse gases in the atmosphere of the past by analyzing small air bubbles trapped in polar ice by the accumulation of snow and retrieved by drilling. The deeper the ice, the older it is. The graph (Figure 10.8) shows the variation of CO$_2$ over the past 800,000 years. (The quasi-periodic changes mark epochs of glaciation interrupted by warm interglacial periods.) What is remarkable is that the concentration of CO$_2$ never exceeded 300 ppm. There is a correlation between the climate of the past and certain parameters associated with *slow* variations in Earth's orbit (Milankovich cycles — named for the Serbian

Atmospheric CO2 concentration

Global average long-term atmospheric concentration of carbon dioxide (CO₂), measured in parts per million (ppm).
Long-term trends in CO₂ concentrations can be measured at high-resolution using preserved air samples from ice
cores.

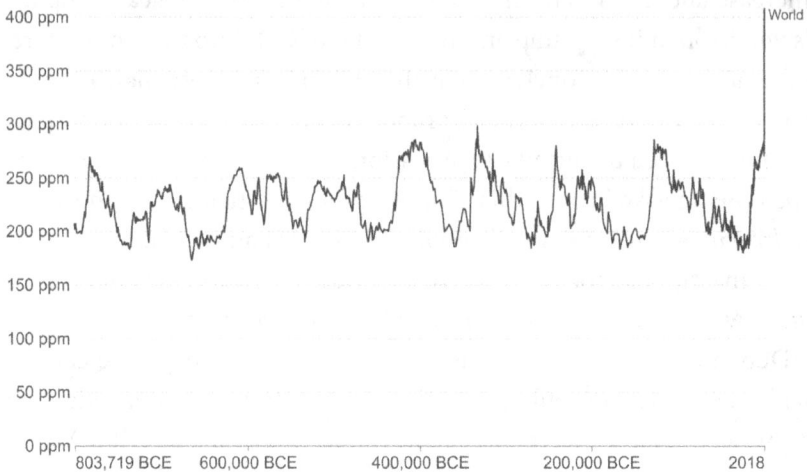

Source: EPICA Dome C CO₂ record (2015) & NOAA (2018)   OurWorldInData.org/co2-and-other-greenhouse-gas-emissions • CC BY

Figure 10.8.   About 800,000 years of past CO_2 concentration.

scientist Milutin Milankovich (1879–1958)), which have contributed to past dramatic climate changes, as shown in Figure 10.8. These changes happened on a timescale of thousands of years, ample time for life forms to adjust, migrate, or die.

Over the years, millions of temperature measurements have been accumulated, made by meteorological stations on land and ships on the seas.

Different research groups coincide. By averaging these data considering their geographical distribution and changes during the year, an increase of approximately 1.3°C (34.34°F) has been measured for the last 100 years. This growth is interrupted by natural highs and lows, with a more pronounced increase beginning at the end of 1970.

Although most discussions of climate change center on temperatures and their steady increase because it is easy to understand and felt by all, the evidence supporting climate change is much broader than just temperature measurements. They include studies in other *independent* fields outside climate science and atmospheric models (glaciology, geology, ecology, ocean

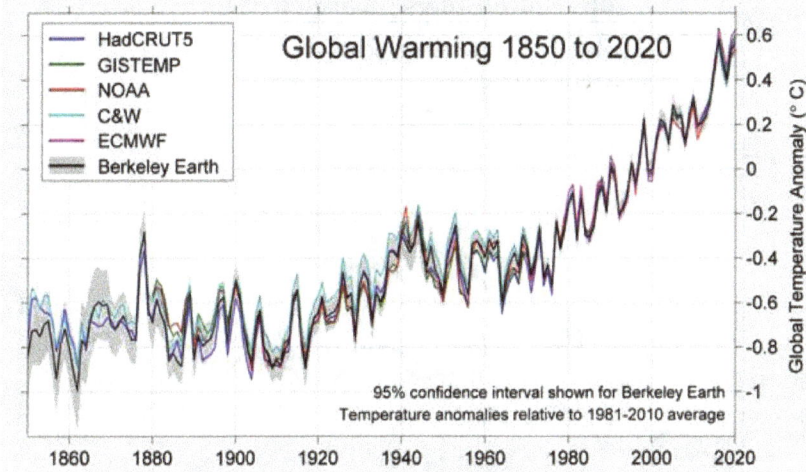

Figure 10.9.   Comparison of global warming trend from six different datasets. Credits: Berkeley Earth.

science), pointing in the same direction. So it is not, as infamous Sarah Palin once said, *doomsday scare tactics pushed by an environmental priesthood.*[28]

We observe a systematic shrinkage of the planet's glaciers, an accelerating decrease in polar ice mass, and an increase in sea level (which is still small). The rise in sea level is due to two causes: the rise in water due to the melting of continental ice and the increase in volume with increasing temperature. Note that the level of the sea does not increase due to the melting of *floating* ice. But if the entire Greenland ice sheet melted (3 km thick), sea levels would increase by about seven meters (goodbye Florida). And it *is* melting currently at a rate of about a billion tons of ice per day![29] If the ice of the Antarctic melts (South Pole), which contains 90% of all land ice, the sea level would increase by seventy meters (goodbye, most of us)! There is no doubt that the water will come, as indicated by an excellent book that explains the details.[30] Since a large proportion of our population and cities are located near coastlines, the damage will be monstrous. Meltwater from glaciers is a freshwater source for consumption and agriculture in many regions worldwide. As the glaciers go, so does this vital source of water, creating a collateral crisis.

Some, for whom it is difficult or inconvenient to accept that *we* are the problem, look for other possible reasons. One that was frequently

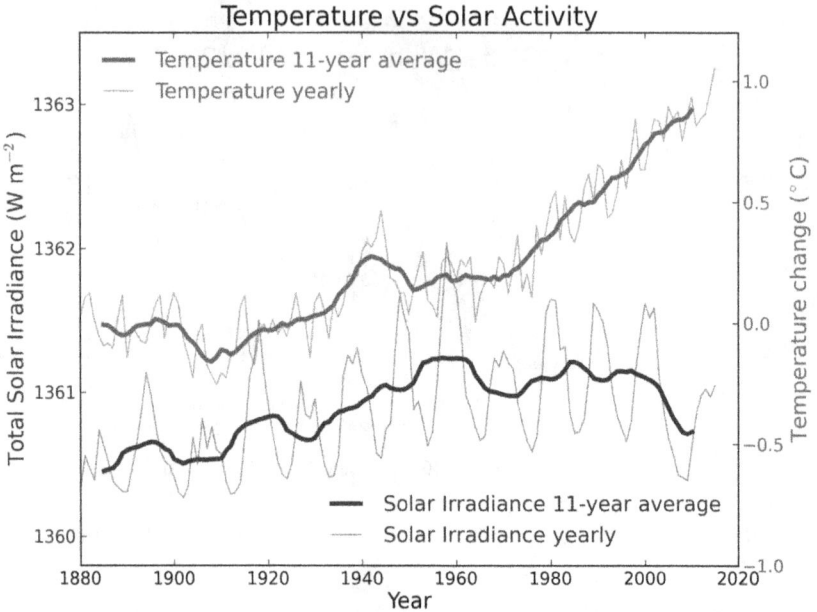

Source: https://skepticalscience.com/solar-activity-sunspots-global-warming.htm.

Figure 10.10. It is not the Sun.

cited in the past (hoping it were true) was that the increase in temperature was due to an increase in the luminosity of the Sun (solar irradiation). If the Sun increased its emitted energy, it would cause global warming, and as an alternative hypothesis, it was plausible. Over millions of years, the sun's luminosity has slowly increased and affected Earth's past climate. It is also true that solar radiation is not constant and varies in a cycle of 11 years, but not by much. But all hypotheses must be tested to consider whether the empirical data support it or not — that is how science works. As an example, look at the graph that shows the periodic changes in solar radiation (Figure 10.10) (Total solar irradiance is a measure of the total energy received from the sun at the top of the atmosphere being on average 1,361 W/m^2) and terrestrial temperature change on the right axis. Notably, the data, after 1960, show that the hypothesis of a solar cause, for which a correlation between both quantities would be expected, is falsified, which is why I wrote "was" above. So, sorry, it is not the Sun.

Life on Earth has adapted to changes in atmospheric composition and surface temperatures and has sometimes survived tremendous environmental upheavals. However, the fossil record bears the remains of myriads of species that did not survive these trials caused by natural events and shows that most species that ever lived are extinct, knowledge which 200 years ago caused a great commotion for those who believed that God had created everything, and nothing could change, let alone disappear. We know of five episodes of mass extinctions where over 75% of species vanished in a short time interval, never to return. Recent studies[31] indicate that the well-documented mass extinction, the "Great Dying" (at the Permian-Triassic transition) 252 million years ago, happened during a period of a few 1,000 years and was due to a huge increase in atmospheric CO$_2$ produced by an extended episode of Siberian volcanism. It warmed the planet (reducing the dissolved ocean oxygen) and wiped out almost all life on earth (like the trilobites).

A *sixth* mass extinction is underway. The difference is that *we* are the agents of destruction. We, who claim high moral standards and the highest status in the living world, *have become Death, the destroyer of worlds.*

It could, of course, be that atmospheric CO$_2$ concentration has increased and caused climate change, but that it is not our fault, nor a consequence of burning (although burning produces CO$_2$). Well, it happens that we have very good scientific evidence that it *is* us. It is a detail related to the fact that carbon atoms come in various forms called *isotopes*, differing in the number of neutrons in the atomic nucleus. The most common and stable ones in the atmosphere are Carbon-12 (^{12}C, 99%) and Carbon-13 (^{13}C 1%). But the precise ratio ^{13}C/^{12}C (δ^{13}C) is different in fossil fuels because, ultimately, these are derived from ancient plants. Plants prefer the lighter isotope (^{12}C), and so as we burn fossil fuels, that difference should show up with time in the isotope ratio of the atmosphere; that is, the ratio ^{13}C/^{12}C should decrease as ^{12}C increases. Several studies have shown that starting at about 1850 the ^{13}C/^{12}C ratios began to decline just as the CO$_2$ started to increase, as expected and shown in the Figure 10.11.[32]

The problem of climate change is no longer scientific[33] and has become political, but nature cares little about politics. In the US and other

Mauna Loa Observatory, Hawaii and South Pole, Antarctica Monthly Average δ¹³C Trends

Data from Scripps CO₂ Program   Last updated April 2020

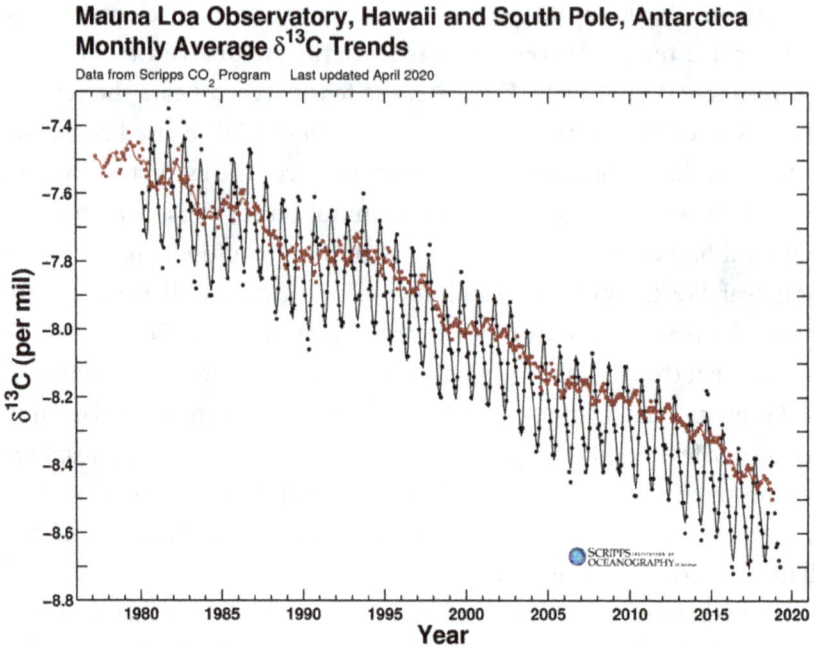

Source: Scripps CO₂ program.

Figure 10.11.   It *is* us.

industrialized nations, it is particularly political because the changes necessary to alleviate the situation will not come from industry but from radical and sweeping government regulations, which is anathema to corporate interests and the political right (which is wrong). Dale Jamieson writes in a must-read book[34]: *The climate change that is underway is remaking the world in such a way that familiar comforts, places, and ways of life will disappear on a timescale of years or decades. Over the next few centuries, climate change risks putting an end to a great deal that we value, including much of humanity and its Creations. Climate Change is not an isolated phenomenon but is occurring in concert with other rapid environmental, technological, and social changes.* (I do not think we need to worry about the next few centuries.)

The atmosphere is not only a vital resource, but it is also unique. Contrary to other resources or national assets such as minerals, oil, forests, etc. (belonging to nations which might then be subjected to the invasion by

Table 10.1. CO$_2$ production of selected countries.

Country	% of total 36 Gt CO$_2$ 9,8 Gt C	Tons per Capita CO$_2$
China	30	7,7
India	7	2
USA	16	16,1
Spain	0,7	5,7
Argentina	0,5	4.4

another nation to get them), the atmosphere and the oceans are a common good that belongs to all forms of life on the planet and all forms of life need it. A molecule of carbon dioxide does not require a passport or a visa to go where it goes, no matter where it was born, and a toxic substance poured into the ocean will eventually reach all shores.

In particular, the US emits approximately ten times more carbon dioxide per capita than the less developed countries. So, what right has a nation to unilaterally damage this resource with serious repercussions for all, and what responsibility has that nation has to repair the damage?

The global average CO$_2$ emission is approximately 5 tons per inhabitant (since we are about 7.7 giga-inhabitants to give a total of 36 Gt). Table 10.1 provides a sample. Of course, the numbers change every year, but they provide an idea of the situation. Some nations, such as China (1.5 billion, 19% of the world population), and India (1.4 billion, 18% of the world), contribute significantly to atmospheric CO$_2$ because they have huge populations, although their per capita production is relatively small (but increasing).

The chart shows the trends (Figure 10.12). Although we can "blame" the industrial revolution for starting it all, note that more than half of the CO$_2$ emitted into the atmosphere was in the last three decades.

The environmental deterioration is a product of the growing industrialization and the associated production of sometimes invisible toxic wastes. The beginning of the increase in the measured levels of atmospheric CO$_2$ coincides with the beginning of the Industrial Revolution. The idea of unlimited industrial growth is fundamentally incompatible

Annual CO2 emissions from fossil fuels, by world region

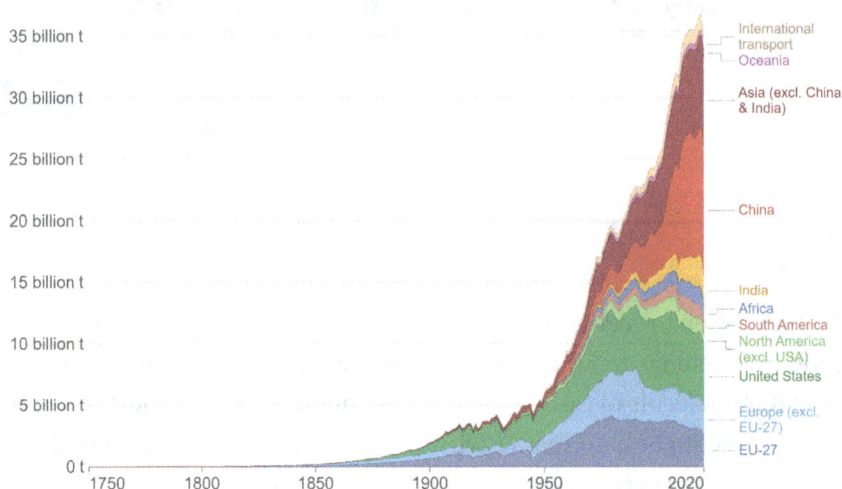

Figure 10.12.   Global CO_2 emissions by region.

with a viable future. The way to avoid what will undoubtedly be a global catastrophe would be driving social and economic changes that few of those who have the power to do so are willing to consider. The situation is not encouraging. Just look at the socio-economic development of China and India, nations that together contain almost half of the planet's population, to realize that we are utterly on the wrong track with little time and the will to change.

Approximately one-third of global energy is extracted from oil, a high fraction of this for transportation. Naturally, the easiest and best quality is extracted first. Over time, the extraction becomes more complex and less efficient, and lower quality products are extracted, which are more difficult to refine. Another third is coal, followed closely by natural gas. (Alternative energies increase rapidly but still do not reach 10% of the total and do not significantly reduce emissions).[35]

Approximately one thousand billion barrels (one barrel = 159 liters) of oil have been extracted in the last 100 years. Every day, 95 million barrels are used — 35 billion a year! (Apart from the other primary fuels, coal and natural gas). Estimates of the amount of oil still available are uncertain (also for gas and coal) and range from 1,000 to 2,000 billion barrels, which is approximately the same amount we have already used.

Here comes the arithmetic: If we divide the estimate of available oil (a number with large uncertainty, but that is the best we can do) — say about 1,500 billion barrels — by the present annual consumption of 35 billion gives us 1,500/35 = 43 years. That easy! *In about 40 years, there will be no more oil* (Maybe new reservoirs will be found, but what is important here is that we should not exploit them). You might object that this is an oversimplification (and it is), but as I mentioned at the beginning, it helps us focus on the essentials, and it tells us that the more we wait, the more it will hurt. Similar calculations indicate that natural gas will be finished in 70 years and coal in 200 years. But consumption will increase, mainly due to demographic increase and the industrialization and development of the "developing" countries, so these resources will run out sooner. Recently (May 2019), the US Department of Energy (DOE), announcing an increase in US export of natural gas, referred to it as "freedom gas." It takes nerve.

But that is not what matters most; the problem is that *if we use them, we will have increased the concentration of atmospheric CO$_2$ to such a point of causing a global catastrophe.* The inertia of the system, and that of human societies, already brings us closer to the inevitability of an increase over 2°C above pre-industrial levels, a limit agreed by consensus at the UN meeting in Copenhagen in 2009, which anyway, is not mandatory, and just happens to be a nice round number. It gets worse.[36]

As the world warms and polar ice melts (in mid-2020, artic temperatures 80°F were recorded and Siberia reached 86), artic resources hitherto unreachable become reachable. Northern countries see an opportunity to claim these resources for themselves as if it were not understood that getting at any gas and oil discovered as the ice recedes will do more harm in the long term than any short-term gains they might bring. It is deranged.

Consider the following example of positive feedback: The temperature rise affects vast regions of northern Canada, Alaska, and Siberia. Their frozen soils (permafrost — no longer "perma"nent) contain organic matter,

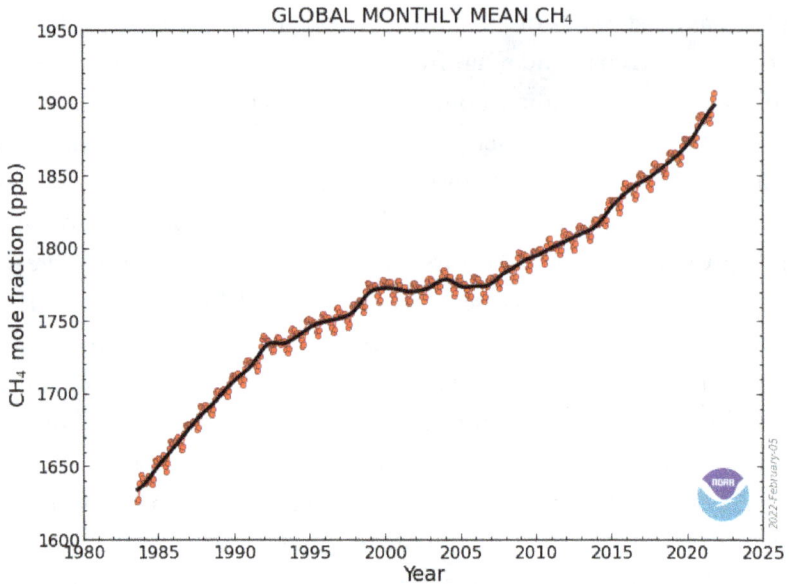

Figure 10.13.   Growing methane concentration.

which can be metabolized by methane-producing bacteria after defrosting. The increase in temperatures will increase the emission of methane and CO_2 by these soils, reinforcing the increase in temperatures, setting up a hellish feedback loop.[37] Permafrost can also contain long gone viruses and bacteria, which will wake up as the ice melts and threaten public health, and for some of them, we may lack immunity, as is the case for SARS-Cov-19.[38] Recent studies indicate that permafrost is thawing sooner than predicted, and this is alarming and is acting as the proverbial canary in a coal mine. Indeed, methane emission has increased, as is shown in Figure 10.13. It is no small matter; permafrost covers about one-fifth of the Northern Hemisphere. People are abandoning the towns they live in when the ground turns into a swamp, arable land diminishes, houses sink, and forests "get drunk."[39,40]

To make things worse, the decrease in polar ice that we observe means that less sunlight is reflected. Therefore, more is absorbed, causing a potential accelerated effect in the polar thaw. Deforestation, as mentioned above (to make room for increasing need of space to live, methane farting cows, and harvest) also increases CO_2 concentration since forests absorb CO_2 as trees grow.

The possibility of abrupt climate changes due to a sudden loss of equilibrium in the complex system that is our home, known as "tipping points," cannot be ruled out and represents a critical threat over and above the progressive increase in temperatures. If you stand on a canoe, you can lean to one side and return to the vertical, but if you go beyond a specific limit, you lose equilibrium and to the water you go. The climate system is much more complex than a canoe, and a tipping point could appear before we understand it. This is indicated by a US National Academy of Sciences report that says[41]: *Abrupt climate changes were especially common when the climate system was being forced to change most rapidly. Thus, greenhouse warming, and other human alterations of the earth system may increase the possibility of large, abrupt, and unwelcome regional or global climatic events. The abrupt changes of the past are not fully explained yet, and climate models typically underestimate the size, speed, and extent of those changes. Hence, future abrupt changes cannot be predicted with confidence, and climate surprises are to be expected.*

A study[42] published in the Proceedings *of the National Academy of Sciences* alerts about the possibility that feedback loops will bring the earth to a threshold that would take the planet into a new state: *Hothouse Earth.* It would be impossible to recover, causing a severe upheaval to ecosystems, societies, and economies. That is what awaits after the Anthropocene Transition: Hell.

In a publication in the journal Nature,[43] the authors note: *In our view, the evidence from tipping points alone suggests that we are in a state of planetary emergency: both the risk and urgency of the situation are acute. The historical record of the climate shows such abrupt changes.*

Science alerts, politicians pushed by the FIRM complex do not dare to do what needs to be done, and the people suffer the consequences. It is not that they do not know. A 2014 pentagon report (so the M of the FIRM knows) says[44]: *Among the future trends that will impact our national security is climate change. Rising global temperatures, changing precipitation patterns, climbing sea levels, and more extreme weather events will intensify the challenges of global instability, hunger, poverty, and conflict. They will likely lead to food and water shortages, pandemic disease, disputes over refugees and resources, and destruction by natural disasters in regions across the globe.* We already see it, especially in the equatorial zone that Christian Parenti has baptized as the Tropic of Chaos.[45]

In the fifth Climate Science Special Report (of 2017) of the US federal government,[46] we can read: *This report concludes that it is extremely likely that human influence has been the dominant cause of the observed warming since the mid-20th century. For the warming over the last century, there is no convincing alternative explanation supported by the extent of the observational evidence.* To say it mildly, it is regrettable that the 45th POTUS (I write in 2019) reduced the funds for these investigations and limited research on climate change and other environmental issues, wishing to kill the messenger since he and his supporters did not like the message. It is a characteristic of fascist politics to attack academia, science, and experts, mocking and devaluing them.[47]

The current Pandemic has given us a chance to *see* what we have done to the planet. One month of shutdowns and for the first time in a long-time, people can see the Himalayas in India, clear water in Venice, and the LA skyline with a blue sky. But, of course, governments want to go back to normal, "open as soon as possible," so enjoy the view while it lasts. But remember that "normal" is what caused the US to be unable to control the virus and avoid hundreds of thousands of deaths. On the other hand, aside from all the individual suffering that COVID-19 is causing, and even if it should kill millions, it is less than what we have done to each other, and perhaps it is also sending us a message we should heed. But the FIRM will not hear. It is sad and ironic: the country with by far the mightiest military on earth could not produce enough cotton swabs to keep its population safe.

Of course, there is talk about technological fixes for the climate crisis: carbon taxes, sequestration, new ways to remove carbon from the atmosphere, and even pilot projects showing what could be developed. But one of the problems with all these ideas is that to implement them, a huge enterprise, one needs time and money, we do not have time, and money we squander on military gadgets. These ideas are seductive, accustomed to fixing most problems with better technology and fanciful geoengineering ideas. (And no, Congressman Gohmert, we cannot change Earth's orbit[48]). Geoengineering treats our planet as if it were a test tube without knowing the result, a pretty dangerous endeavor because if the test tube breaks, we have no other. In the words of Charles Mann[49]: *In effect, the human race has entered into a great wager. We are, so to speak, betting the planet.* It gets worse since a geoengineering project without understanding the outcome will not be a global endeavor; it will be unilateral but affect everyone, which is unacceptable.

Raymond Pierrehumbert,[50] an atmospheric physicist at Oxford University, puts it this way in a recent article: *There is simply no good fix if we fail to stop pumping carbon into the atmosphere. We are already suffering some of the harms due to human-caused climate disruption. The only question is how much we will ratchet up the toll of human suffering, and the destruction of the ecosystems with which we share the Earth before we finally achieve net zero carbon emissions.*

So, why do so many people not believe it or deny it, and politicians oppose it? (There is no need to explain the reason for the opposition by particular industry sectors, since a restriction or prohibition of the use of fossil fuels means their ruin, including gas stations (about 125,000 in the US) and all businesses which cater or depend on that industry). Here are some reasons:

- Psychological — we do not perceive slow changes, often difficult to grasp until it is too late. Think that in 1960 we were 3 billion. One hundred years later, we will be over 9 billion, a huge change at the root of our problems, but we barely noticed. Between day and night, or winter and summer, the temperature changes are larger, and we have not yet realized that we are paying for the colossal accident of the Industrial Revolution. At least the ozone hole was tangible, and a solution to the problem was viable. Trump, for example, said in a Washington Post interview on November 27, 2018: *As to whether or not it's man-made and whether or not the effects that you're talking about are there, I don't see it.* Now you see?

- Self-transcendence — related to our ability to really care about the distant future, in which we will not exist as individuals. The selfishness of capitalistic thought does not help. Others are not worried, thinking that the world is supposed to come to an end, to be replaced by a perfect one ruled by the redeemer (Millennialism).

- Pragmatism — the necessary changes to completely decarbonize the world's economy are urgent and huge, even sacrificial, and it is not even understood if they are viable. It is simply impossible, for example, to build a nuclear plant per day for the next 40 years.[51] Decarbonizing the global economy without an alternative implies shutting down 80% of energy production. Furthermore, it is a multifactorial problem, and we cannot solve it by focusing on only one of the factors, as we usually do with other problems. Some argue that solving the problem is just too costly (Senator Bernie Sanders proposes a $16 trillion over 10 years plan) without thinking that not solving it will cost much more. Furthermore, it calls for collective global action, something challenging to achieve even for minor issues.

- Predatory Delay — a term coined by Alex Steffen[52] referring to the *blocking or slowing of needed change, to make money off unsustainable, unjust systems in the meantime.* The problem is the inevitable passage of time. Some of the problems we face can be resolved given enough time, but other issues discussed in this book must be solved *now*, and although we know it, we cannot do it. The longer we keep thousands of nuclear weapons, the sooner one will explode in our face.

- Political — necessary changes which imply a severe upheaval in production and consumption, a radical reorganization of our societies, and a change in living conditions require strong regulatory intervention by governments at a local and global level. This is anathema for many (Republicans in the US and the political right in general), but also goes against the gigantic profits generated by the energy industry. There is no global body that has the needed implementation power. Although we elect governments to protect us from harm, and they are the only institutions with the power to do so, they have failed due to the control by the FIRM.

- An overarching point, which applies more generally to the public understanding of science, is that science has become quite difficult and complex. Whereas in the past, anyone could understand intuitively how a mechanical clock worked, this is not the case for modern technology, like cell phones or genetic engineering, which are like magic. Otto[53] writes: *Science and technology become more a matter of belief than know-how, making people more vulnerable to disinformation.* A few greedy scientists well paid by industrial allies to lie to the public (through many front institutes and foundations with nice-sounding names. The *Competitive Entrepreneur Institute*, the *Heartland Institute*, or the *American Legislative Exchange Institute* (all funded with millions from Koch enterprises) are enough to sow doubt among the public who then elect the wrong person or the wrong treatment.[54]

- Depravity — we will not control the climate; it is too late, and the cost is too high. Instead, the world's oligarchs will foster the building of walls, militarize the police, and close ports to save themselves from an "invasion" of climate refugees who do not come from Norway. Therefore, let us enjoy the ride and place ourselves in God's hands, and move to higher ground and air-conditioned mansions. To Hell with humanity, free markets *über alles*. And that is exactly where we are headed.

The inhabitants of the countries that emit the least, the poorest, will be the ones who will be unable to defend themselves from the consequences caused by this pollution. It is estimated that globally millions *die each year* from pollution produced by coal-burning, and the more general contamination of our living space. There is no reason, in principle, to accept this unequal use of the atmosphere, although there are historical reasons that explain this inequality.

As is pointed out by Amitav Ghosh,[55] one cannot discount the looming problem posed by the fact that a large part of the world population lives in Asia — about 60%, mostly in India and China– and in time will become the largest emitters of CO$_2$ and its greatest victims. Tens of millions will have to migrate from flooded coastal areas, cities along the coast will be wiped off the map. In addition, melting glaciers and changing weather patterns will adversely affect this large population's crop yield and food supply, leading to great upheavals in what Ghosh calls the Great Derangement.

He writes: *Inasmuch as the fruits of the carbon economy constitute wealth, and inasmuch as the poor of the global south have historically been deprived of this wealth, it is certainly true, by every available canon of distributive justice, that they are entitled to a greater share of the rewards of that economy. But even to enter into that argument is to recognize how deeply we are mired in the Great Derangement: our lives and our choices are enframed in a pattern of history that seems to leave us nowhere to turn but toward our self-annihilation.*

The much-admired Mohandas Gandhi (1869–1948) was aware of this a long time ago, in 1928 when he said: *God forbid that India should ever take to industrialism after the manner of the West. The economic imperialism of a single tiny island kingdom is today keeping the world in chains. If an entire nation of 300 million took to similar economic exploitation, it would strip the world bare.* By the way, an author will choose the quotations that serve him well (we cherry-pick) and just to be honest, I also quote Gandhi saying in 1940: *I do not want to see the allies defeated. But I do not consider Hitler to be as bad as he is depicted. He is showing an ability that is amazing and seems to be gaining his victories without much bloodshed. Englishmen are showing the strength that Empire builders must-have. I expect them to rise much higher than they seem to be doing.*[56] I guess we all say stupid things, occasionally (and that is OK as long as we don't persevere).

Although in the past there have been conflicts between scientific discoveries and other interests (religious and industrial), think about Galileo, Darwin, or the disputes about tobacco, DDT, or Ozone, in the long run, science prevailed. But this time the outlook is gloomy because the dimension of the conflict is enormous, and we are driving things to the limit. In the words of Shawn Otto[57]: *The industrial clash with environmental science came into adulthood over the issue of human caused climate disruption. It is an issue that has profound existential stakes for the world's most powerful industry (energy), as well as for the environment, environmental science, the world economy, and democratic governments. The showdown is propelling an industrial war on science the likes of which the world has never seen.*

Those who have the power and resources to protect the future do not have the ethical vision or courage to act. Some years ago (in 2000),

President Bush expressed what sums up the US ethical posture[58]: *We will not do anything that harms our economy. Because first things first are the people who live in America* (without doubt he referred only to the US when he referred to "America", but then he referred to "Africa" as a country). And the people whose economy he defended are the top 1%. Indeed, "America first" has been a much-used rallying slogan with a dark history, as pointed out by Sarah Churchwell.[59]

Peter Singer,[60] professor of Bioethics at Princeton University, points out that the difference between the tragic events of September 11, 2001, and the deaths caused by air pollution, particularly carbon dioxide produced by the industrialized countries, is only one of perception. The first event was shocking and had great mediatic coverage. The second is barely noticeable, measured by instruments that slowly collect data. But billions will suffer the consequences. Globalized ethics entails that it is not possible to give different values to human life depending on whether the unfortunate ones are ours or not. (This was the thought when the first bomb was dropped on Hiroshima.) This is far from what is observed, where, for example, the US soldiers killed in the invasion of Iraq were carefully counted (several thousand) but the Iraqi dead were not; that is, they were not counted because they did not count. Several estimates give figures above one hundred thousand, with a large proportion of civilians.[61] The number of wounded or displaced, many of whom will have to live the rest of their lives with the consequences, is much higher. Like all wars, it is a senseless carnage in which the innocents suffer since those who make the decisions and send others to kill are safe.

The sleep of reason creates monsters, said Goya as shown in the frontispiece, and we are feeding one with our fumes as if it were a holocaust as if it were a god. We do not see it coming because it advances at the speed of a huge lazy bear, and when it arrives, it will be too late. I return to what Günter Anders said: *either we change, or nature will take care of the changes — without us.*

I end this section with the epilogue of a book written by one of the nation's leading climate scientists[62]: *We look back now with revulsion at the corporate CEOs, representatives, lobbyists, and scientists-for-hire who knowingly ensured the suffering and mortality of millions by hiding their knowledge of tobacco smoking's ill effects for the sake of short-term corporate profits. Will we hold those who have funded or otherwise participated in the fraudulent denial of climate change similarly accountable — those individuals and groups who both made and took corporate payoffs for knowingly lying about the threat climate change posed to humanity, those who willfully have led the public and policymakers astray, and those politicians and media figures who have sought to intimidate climate scientists using McCarthyite tactics? Though the impact is more subtle and difficult to measure, their recalcitrance may end up costing more lives than cigarette smoking ever has. And, unlike the case of tobacco and cigarette smoking, many of those who will suffer the worst impacts of climate change will have played little or no role in creating the problem.*

Attending to the problems posed by the catastrophic convergence have been delayed by the pandemic just at this critical time. When we get back to them, precious time will have been lost, getting closer to an uninhabitable Earth.[63] Our house is falling apart. It is time to panic. At the 2021 Davos meeting 18-year-old Greta said: *We understand that the world is very complex, and that change doesn't happen overnight. But you've now had more than three decades of bla bla bla. How many more do you need? Because when it comes to facing the climate and ecological emergency, the world is still in a state of complete denial. The justice for the most affected people in the most affected areas is being systematically denied.*

Chapter 11

Escape

Especially since this world of ours was made
By natural process, as the atoms came
Together, haphazardly, quite by chance
Quite casually and quite intention less
Knocking against each other, massed, or spaced
So as to colander others through, and cause
Such combinations and conglomerates
As form the origin of mighty things,
Earth, sea and sky, and animals and men.
Face up to this, acknowledge it, I tell you
Over and over — out beyond our world
There are, elsewhere, other assemblages
Of matter, making other worlds. Oh. ours
Is not the only one in air's embrace.

— Lucretius[1]

We are running out of space, and the only place we can go to are other worlds. It is time to explore other solar systems Spreading out may be the only thing that saves us from ourselves. I am convinced that humans need to leave Earth.

— Stephen Hawking[2]

On Thursday morning in Campo di Fiore, the dastardly Dominican friar from Nola was burned alive, of which one writes in the past: obstinate heretic and having of his caprice formed different dogmas against our faith, and in particular against the Holy Virgin and the Saints, he obstinately wanted to die in that way the scoundrel; and said he was martyred willingly, and that his soul would rise to paradise with that smoke. But now he will know whether he was telling the truth.

— AVVISO DI ROMA IL 19 FEBBRAIO 1600
(public edict to the people of Rome)

Since a world without us is likely, let us consider other worlds, so our disappearance might be less dramatic. Some imagine we might find one to emigrate to (if there are no walls) when ours becomes hell with high water. This story is also a good example of how science proceeds and advances.

Unidentified flying objects — UFOs — and their alleged extraterrestrial origin is a common belief, the theme of dozens of films, hundreds of books, and numerous programs in the media and websites, and currently reignited by the Pentagon who knows why. Nevertheless, there is absolutely no evidence that this is happening despite the many instances of alleged visits by people who have witnessed something or pilots looking at a blurry splotch on a screen. The idea that extraterrestrial beings exist is not new, and already in antiquity, it was thought that the sky was the residence of a great variety of mythological beings. However, these ideas were suppressed during the European Middle Ages. They disagreed with prevailing religious dogma, which posited a universe with our unique Earth at its center, created for us by one God.

Four hundred years ago, on February 17, 1600, to be precise, a 52-year-old man was stripped with his "tongue imprisoned because of his wicked words" (a nail was driven through his tongue), tied to a stake, and burned alive in a place in Rome called *Campo Dei Fiori* (a market square). Pope Clement VIII had ordered this (who was not clement) for being a heretic "impenitent, obstinate and stubborn." This man had written[3]: *Thus there is not merely one world, one earth, one sun, but as many worlds as we see bright lights around us, which are neither more nor less in one heaven, one space, one containing sphere than is this our world in one containing universe, one space or one heaven.* The man was Giordano Bruno (1548–1600).

This undoubtedly contained a high dose of heresies and speculation that stars were just distant suns, not at all evident. (Indeed, I know of people today who do not know that the Sun is a star, and why should they, given the evident differences.) Moreover, his belief in the plurality of worlds was not consistent with the conception of the universe and the place occupied by humans, promulgated by the ecclesiastical authorities. Consequently, they burned him on a sad day in God's name. To the judges who sentenced him, he said: *Perhaps you pronounce this sentence against me with greater fear than I receive it.*

One wonders why so many people are willing to believe that *they* visit us frequently. With a little knowledge and critical thinking, it is easy to realize that a visit is problematic. The basic fact here is that nothing can exceed the speed of light (300,000 kilometers per second). As I have mentioned, it is not a technological limit but a limit imposed by nature, discovered by physics, and established in thousands of experiments. A visit is problematic and even less credible because these visitors are usually portrayed as humanoids (an accidental product of evolution on Earth).

Even a conversation is difficult since the star closest to the Sun (Proxima Centauri) is about four light-years away. So, 8 years will elapse between posing a question and receiving an answer. And this is only for the nearest star since most are thousands of light-years away. So, think that between our message: "Here Earth calling, how are you?", and the possible answer: "Here planet X, we did not copy well, repeat the question," tens or hundreds of years would elapse, and if it happened, nobody on Earth would remember what the question was if there is anyone left. Not to mention the problem of translation.[4]

Some say they communicate by "telepathy," but this does not prevent the limitations mentioned above. Then there is the issue of distinguishing telepathic messages from hallucinations (and no, there is no evidence for telepathy). Those who believe interpret the denials of scientists and government officials as confirmation of their suspicions (in a typical conspiracy theory reaction — see below).

There are also those, including renowned physicist Stephen Hawking (1942–2018), who think that other habitable planets will be the salvation for humanity when we have finished with this one. Therefore, let us briefly consider the most pertinent issues, which also threatens our conception of ourselves and the notion of our place in the universe.

11.1 Ecospheres

When people thought the earth was flat, they were wrong. When people thought the earth was spherical, they were wrong. But if you think that thinking the earth is spherical is just as wrong as thinking the earth is flat, then your view is wronger than both of them put together.

— Isaac Asimov[5]

It is out there in the starry sky that whatever is more than animal within us must find its solace and its hope. And so, in hope and solitude my story ends.

— H.G. Wells[6]

When we think of life on another planet, we refer to a process with the general attributes of life on Earth but not its unique features (accidental). Although in the public's imagination, nourished by films, fiction novels, and websites, an alien is portrayed as a humanoid or a reptilian, you do not have to look far to understand that it will not be like that. It is enough to observe the astonishing diversity of life forms on Earth, despite their common origin, to be able to conclude that any extraterrestrial, with an evolutionary history different from ours, will resemble us as much as we do a sofa.

It is a good first reason to discard all the stories that those who claim to have seen them want us to believe. Our appearance (that we present a bilateral symmetry with two eyes and not three and four legs and not seven, for example) is a product of fortuitous factors combined with some that have to do with physical laws. So, naturally, the shapes must adapt to the local environment (most marine creatures are shaped to move efficiently in water). In addition, organisms need to function effectively in the local gravitational field, adapt their eyes to the atmospheric transparency and character of the star's light, and their shape and weight to the air's density (if they are going to fly). Therefore, the whole alien visitor's thing, including Roswell, is silly, to say it kindly.

Living beings' most common chemical elements (about 95% of the mass) are carbon, hydrogen, oxygen, and nitrogen (CHON). There are other essential minority elements without which life (as we know it) would not work. For example, phosphorus is part of the fundamental nucleotides in the information macromolecules of every organism (DNA). Sulfur is a component of two of the twenty amino acids (cysteine and methionine) present in all proteins. Calcium is an ingredient of skeletal structures and bones, and potassium is necessary for nerve conduction, and so on for a few others.

The fact that we are built mainly with CHON could be considered a particular property of life on Earth, and some think that instead of carbon, you could find silicon-based life (perhaps artificial robotic life). But there are good reasons to believe that these elements would also form life in other places starting with the fact that they are by far the most abundant in the universe, along with helium that does not react with anything since it is a "noble" gas. Furthermore, these elements form covalent chemical bonds (bonds that share electrons) between them that are stable and varied, allowing for a wide range of compounds with various important structures for any biology.

However, the most common elements on Earth are oxygen, silicon, aluminum, and iron, three of which are not abundant in living beings, which suggests that life, regardless of what the most prevalent local elements are, will be made with CHON. Note that one of the most important compounds for life consists of two of the four main ones: water — H_2O.

An ecosphere or habitable zone is the space around a star where planets and moons can reach stable climatic conditions long enough to develop carbon-based life. It must maintain an appropriate temperature for liquid water and a suitable environment for the necessary chemical reactions. Life, as we know it, developed in water (without ozone, the land was dangerous), and we do not know about life as we do not know it.

Several factors determine these conditions. First, there is the nature of the star, the amount of energy it emits (its luminosity) since this energy determines the temperature on the surface of some planet in orbit. This will also depend on the distance from the planet to its star since, at greater distances, less energy is received. For water-based biology, the temperature should be between approximately 0°C and 100°C to have liquid water under an earthlike atmosphere since all the vital processes occur in an aqueous solution (although some bacterial extremophiles exist outside these limits). The greater the mass of the star, the more energy it emits, but for less time. On the other hand, the composition of the planet's atmosphere (if it has one) determines, through the greenhouse effect, the temperature on its surface (as we have seen).

Whether or not a planet has an atmosphere depends on its mass because, if it is low, its gravitation will not prevent it from escaping, as was the case for Mars (atmospheric pressure 1/100 of the terrestrial one) and the Moon. On the other hand, if it is too hot, an atmosphere will escape if it ever forms.

The Sun's ecosphere is relatively narrow, place the Earth at the orbit of Venus and it will become hell, move it to the orbit of Mars, and hell will freeze over.

The ecosphere is as stable as its star. But, sooner or later, the star dies, and the ecosphere collapses — for the Sun, there are a few billion years to go, the actual end of the world, nothing to lose sleep about.

The fact of seeing many stars at night (a tiny fraction of the total) is the reason many imagine that around some of them, on some planet, there must be life, like Giordano Bruno, and Lucretius already surmised. Therefore, it is easy to *imagine* that aliens visit us. But the distances that mediate between stars are huge, making it very unlikely that they would visit us. (I guess there are better places to visit if they could travel that far.)

The beautiful image of the great Andromeda galaxy presents a huge star system (a trillion stars) similar to but larger than the Milky Way (Figure 11.1). It is about *two and a half million light-years* away, and its diameter is about 220,000 light-years. All the points of light that you see in the image are stars that belong to the Milky Way and are in the foreground of Andromeda's image. We appreciate here its disc shape, the increasing luminosity towards the central nucleus (more stars, perhaps a black hole), the clouds of interstellar gas, two small satellite galaxies (one at the bottom left and another as a luminous globule above the nucleus).

The Milky Way, with a diameter of about 170,000 light-years, harbors about *two hundred billion* stars, more than the number of grains of sand contained in a beach several miles long. Other galaxies are millions and billions of light-years from us. What can I say? The universe is a place of inconceivable dimensions, impossible to imagine, and guarantees our isolation. And our planet is not ours and cares little about us, in the words of Jacques Monod[7]: *Man must at least wake out of his millenary dream and discover the total solitude, his fundamental isolation. He must realize that, like a gypsy, he lives on the boundary of an alien world; a world that is deaf to his music, and indifferent to his hopes as it is to his sufferings and crimes.*

The Sun is in what could be called the suburbs of the Milky Way, about 30,000 light-years from its turbulent nucleus. The Sun orbits the nucleus due to the gravitational force of the entire Galaxy at the incredible speed of some 900,000 km/h (yes, we do not feel anything, but at that speed, it

Figure 11.1.   The beautiful Andromeda galaxy (Hubble Space Telescope).

moves). So, it takes some 250,000 years to complete an orbit. This could be called a "galactic year." The Earth is 16,000 years old in these units, and our species arose just a year or so ago. We are newcomers.

Although all stars are essentially the same thing (huge spheres of mostly hydrogen and some helium held together by their own gravity), they differ in mass. This determines their central temperature, luminosity, and longevity, as shown in Table 11.1, which summarizes the different properties of the stars of what is called "the main sequence," which are the great majority, about 90%. (Other stars are white dwarfs or red giants, not relevant to the present discussion.)

There are more massive and luminous stars than the Sun. Those of types O, B, and A are of little interest since they represent only 1% of all

Table 11.1. Main sequence stars (typical properties).

Spectral type	O	B	A	F	G	K	M
	Produce supernovae		A	F	Sun	Red Dwarfs	
Temperature (Degrees Kelvin)	40,000	20,000	8,500	6,500	5,700	4,600	3,200
Radius (Sun=1)	10	5	1,7	1,3	1	0,8	0,3
Mass (Sun=1)	50	10	2	1,5	1	0,7	0,2
Luminosity (Sun=1)	100,000	1,000	20	4	1	0,2	0,01
Life (million years)	10	100	1,000	3,000	10,000	50,000	200,000
Abundance (%)	0,00001	0,1	0,7	2	3,5	8	80

and have short lifespans. But they are of great importance because they are the ones that explode as supernovae and produce and disseminate the heavier chemical elements of the periodic table. For example, a star with ten solar masses lasts "only" 100 million years and is a thousand times more luminous than the Sun. It is thought that this is not enough for a slow evolutionary process (although we cannot be sure that what happened on Earth is a good example). Also, they emit large quantities of ultraviolet radiation, not good for life as we know it. Other stars are red dwarfs with smaller masses than the Sun, are much less luminous and colder (that is why they are red), and more abundant (80% of the total). A star with 70% of the solar mass lasts fifty times the life of the Sun, but its luminosity is only 20% of that of the Sun. Thus, stars like the Sun represent only 5% of the total, meaning that the Sun is not an average star as commonly believed.

Besides hydrogen, stars have some helium and produce heavier elements, such as oxygen, carbon, calcium, manganese, and iron, which are a byproduct of the nuclear fusion reactions in their high-temperature centers. Without these heavy elements, there would be no water, there would be no fundamental structures for life (carbon-based molecules) there would be no planets like Earth.

In our galaxy (without thinking of the billions of other galaxies), if we just consider the 5% of stars that are like the Sun, and even if we eliminate half of them for being part of multiple star systems (which leads to possible instabilities for a planetary system), and then arbitrarily say that only *one in a thousand* of the rest has a planet located in the ecosphere, we still obtain of the order of a million possibilities.

The question is whether they are a million gigantic bacterial cultures, brown spheres coated with cockroaches, or green ones covered with mold or something else — something intelligent. Hopefully, we are not the only conscious beings in this vast universe. As some say: "they are out there." If they are, we will be one of many, and our disappearance will be less dramatic, and it would be a kind of solace if at least we knew we were not alone.

Let me suggest why I think most are lonely planets. We know that life on Earth developed relatively early and produced millions of species (most now extinct) over time. We also know that only once and very late in the game, one species developed a large and complex brain to be conscious of all these facts, namely *Homo sapiens*. So, although we only have one sample (and it is hazardous to generalize), it suggests that life might be easy but intelligence difficult, maybe a fluke, an aberration that could explain a lot.

Furthermore, in hindsight, the industrial revolution was a colossal accident, with its unintended consequence of ending our existence or throwing us back to the Middle Ages as we change Earth's climate. We can expect this to happen elsewhere, where sentient and intelligent beings (in whatever form) develop. For purely logical reasons: they will be macroscopic beings, denizens of the middle world, composed of atoms and molecules just as we are. You need a lot of neurons (or equivalents) before you have enough processing power to start to understand and manipulate the environment. And when this happens, they will act to the best of their abilities, not waiting to evolve superintelligence that could foresee (as our computers try to do) much more than they. And their middle world will be ruled by the same universal natural laws that we have discovered here on Earth. That also means that they will discover fire and fuels before other means of energy generation (you need quantum physics to build nuclear reactors, solar panels, computers, and cellphones). And so, they too will fall into the trap of converting heat to work and start an industrial revolution. As a result, they will be just as unable as we were to foresee what would come 200 years down the road.

Of course, this idea of a universal tragedy is my speculation, but it seems reasonable from all we know (I would not propose it otherwise). So, life may flourish in this huge universe, billions of times on billions of planets. Still, as soon as some being discovers a way to generate energy by combustion, it leads in due course to the end of that lineage (greenhouse gases operate in any atmosphere). Thus, life in the universe is more like slow-motion fireworks than a cozy warm hearth. Well, perhaps this is too harsh. Perhaps there are planets where fossil fuels never form, or they somehow skipped this tragic step and started using their star's energy, and their story will not be as sad as ours. Let us hope so.

Finally, if by chance we find that we are not alone, we will disturb some theological doctrines, as was already observed by Thomas Payne in 1794[8]: *From whence, then, could arise the solitary and strange conceit that the Almighty, who had millions of worlds equally dependent on his protection, should quit the care of all the rest, and come to die in our world, because, they say, one man and one woman had eaten an apple? And, on the other hand, are we to suppose that every world in the boundless creation had an Eve, an apple, a serpent, and a redeemer? In this case, the person who is irreverently called the Son of God, and sometimes God himself, would have nothing else to do than to travel from world to world, in an endless succession of deaths, with scarcely a momentary interval of life.*

The first step is to find planets — exoplanets.

11.2 Exoplanets

> Not that there are any nights on Ursa Minor Beta. It is a West zone planet which by an inexplicable and somewhat suspicious freak of topography consists almost entirely of subtropical coastline. By an equally suspicious freak of temporal relastatics, it is nearly always Saturday afternoon just before the beach bars close.
>
> — Douglas Adams[9]

Michael Mayor and Didier Queloz, two Swiss astronomers, reported in 1995[10] that observing from the *Observatoire de Haute Provence* in the French Alps, they had managed to determine that the star 51 Pegasus, a star like the Sun at about 42 light-years away, had a planet. The measurements indicated that its mass was like that of Jupiter (which is 318 times the mass of the Earth) and followed an orbit very close to the star (eight times closer to it than Mercury is to the Sun). The planet takes a fantastic 4 days to complete an orbit — its year. They were awarded the 2019 Nobel Prize in Physics for this discovery.

Since then, the list of stars with planets around them has increased, and we have discovered thousands of strange worlds of great variety waiting for better telescopes that will allow us to know their properties in detail.

All the discovered planets are too faint and distant to observe directly with the telescopes available against the bright background of the light of their stars. In the range of visible light, the brightness of a star is a billion times greater than that of a planet which reflects a tiny fraction of it. Thus, capturing and analyzing a distant planet's light is like detecting the light of a candle in front of a powerful beacon a 1,000 km away. We expect that the recently deployed Webb Space Telescope, observing in the infrared, will discover smaller planets, and directly analyze their light.

Most planets discovered to date are of high mass, greater than the mass of Jupiter. This does not mean that extrasolar planets like the Earth are scarce, but it results from a selection effect: they are just harder to find.

The planets cause a slight oscillation of the star they orbit due to their mutual gravitational interaction. This manifests itself as a tiny cyclic change in the star's speed relative to us, which can be measured as a few meters per second, and causes, a tiny change in the frequency at which certain absorption lines are observed in the spectrum of light from the star. The more massive and closer to its star, the greater the effect, which is why most of the planets found are much more massive than Earth and tend to be close to their stars. For our solar system, for example, the Sun is mainly affected by Jupiter and Saturn. Also, planets further away from their star will take many years to complete an orbit so that the star's oscillation will only be detected after a long time, something that limits research.

Another method of discovering these distant planets is detecting small eclipses caused if the planet places itself between the star and us. NASA's Kepler mission, a space telescope launched in 2009, sought these eclipses by continuously observing some 145,000 stars. It has confirmed over 2,000 planets, a small fraction of which are comparable in size to the

Earth. Another satellite[11] (TESS — Transiting Exoplanet Survey Satellite launched on April 18, 2018) finds hundreds of them. Giordano Bruno has been redeemed.

So, should we start packing? The distances that separate us from a potential cosmic neighbor are enormous, as we have seen, and a potentially habitable planet will be tens if not thousands of light-years from us. Yet, despite this, some cling to the absurd idea that the aliens visit us after crossing these distances. They adhere to the notion that we are not alone, I imagine because they know the consequences of extreme isolation that leads to madness. And the truth is that if someone on a distant planet could observe ours, it would conclude that we were all condemned to solitary on this planet and have lost our minds.

Even so, we search to see if we are not so alone and imagine that we will emigrate in some sort of huge cosmic ark populated by those who can afford a seat (leaving behind "telephone sanitizers"). We will abandon our "shithole" to establish a new civilization, to colonize a pristine planet that awaits with open arms, with their version of the statue of liberty.

But that planet, even if it were to have water and an adequate atmosphere and temperature, would not be ideal for visitors. If it does not already have a biology, it would not be a place to visit (potatoes might not grow), and if it does have biology, it will most likely not be compatible with ours, which implies sure death to newcomers. Should biology be advanced and technological, the death of the visitors will be even more certain, or maybe they will put them in a zoo until they die of hunger since they cannot feed on a biology different from ours. We can eat and digest cows and bananas due to our evolutionary kinship, which is not the case on another planet. Our future is either here or nowhere. Others, more pragmatic, are buying real estate in New Zealand.

Michael Mann ends his excellent book with: *We will not, we cannot, wreck this planet. There is no Planet B. Earth is a rarity of literally cosmic proportions. It is an overflowing treasure chest of life-forms of unimaginable variety and beauty. It is perfectly fitted to us as humans because we evolved to fit it. It would amount to the gravest criminal act of irresponsibility in human history were we to throw it into fatal imbalance because of a wanton addiction to carbon.*

11.3 The Anthropic Fallacy

When I consider the short duration of my life, swallowed up in the eternity before and after, the small space which I fill, or even can see, engulfed in the infinite immensity of spaces whereof I know nothing, and which know nothing of me, I am terrified, and wonder that I am here rather than there, for there is no reason why here rather than there, or now rather than then. Who has set me here? By whose order and design have this place and time been destined for me?

— Blaise Pascal[12]

Most of humanity seeks to escape the terrifying reality of human history, to make some sense of events, to hope for something better (an afterlife? a redemptive life? remembrance?) than what we have. And most of all, a majority of humans refuse to accept that the universe, the world, god(s), are utterly indifferent to our plight.

— Teofilo Ruiz[13]

The anthropic fallacy refers to a question about the teleology (the purpose) of the universe. It has been argued[14] that if a set of physical parameters had different values, the universe would be quite different and could not be the seat of intelligent life, perhaps not of any life. (The density of the universe, the universe's age, and fundamental physical constants such as the magnitude of the nuclear force that binds atomic nuclei, the value of the Planck constant that determines the chemical bonds, and others.)

Accordingly, it seems as if we live in a fragile and special world; a universe *finely tuned* for our existence, which some take to the extreme of postulating a designer, since, according to them, it is highly unlikely that all these parameters have the values they have by coincidence.

It is argued that someone must have adjusted these values (The mouth of whoever asks where that someone came from will be washed with bitter soap.) There is no reason why these parameters have the values they have (that we know of). They could have been others, and therefore the reason why they have them is our existence. So it is argued, which is putting the cart before the horse.

A couple of examples: (a) If the Earth's orbit were not what it is (its distance from the Sun and its almost circular shape), then our planet would probably not have a biology. Change the composition of the Earth's atmosphere (as we are doing) or change the distance from Earth to the Sun (in either direction) by a small percentage, and we would not be here to tell the story; (b) If the nuclear force which keeps the protons and neutrons of the atomic nuclei in place despite the electrical repulsion between the protons were greater by 2%, then the di-proton (a nucleus with two protons, an isotope of Helium) would be stable. Stars would not only explode if they formed, but stable stars composed of hydrogen would not have formed since all available hydrogen would have become helium at the beginning of the universe. Again, we would not be here to tell the story.

Much has been written about this, but in the end, it is a circular argument saying that if the universe were not appropriate for life, we would not be here to think about it. Furthermore, it is in the wrong direction. The universe is not finely tuned for us; we are finely tuned to the universe, or at least to a tiny part of it: Earth. Nothing can discard the idea of multiple universes (although it is not strictly a scientific statement) with different cosmological parameters and different physical constants (an eternal Multiverse) in which, for the vast majority, something we would call life could not develop. We are in a universe (or an environment of the universe), where life is possible because we could not be anywhere else.

To be astonished and seek an explanation of why the universe is so good for life is not very different from being astonished and search for an explanation of why the earth's atmosphere is transparent to sunlight precisely in the band of light visible to our eyes (it is the other way around, well explained by natural selection). Or to be amazed and seek an explanation of why someone won the lottery, or more prosaically, why the mold fits so well to the cake when it was baked.

Paul Braterman observes,[15] that if we were to accept the argument that the complexity of the universe and life can only arise from an intelligent designer, then we fall into an infinite regression: A complex system arises from intelligence that must be even more complex, which in turn had to arise from something more complex still ... Moreover, we know of many complex systems that occur without the need for the intervention of intelligence. Think of a drop of water, which, when frozen, becomes a complex crystal or a neuron that generates consciousness when connected with billions of others.

Cause and effect are confused. And if the environment changes and the life form in question cannot adapt in time, it will not survive.

In the end, all that remains, despite the many attempts proposing a teleological universe designed by someone — a disguised religion — is: The universe is as it is because it is what it is. The purpose of your life is what you make of it.

Chapter 12

Blowback and Nuclear Weapons

The unleashed power of the atom bomb has changed everything except our modes of thinking, and thus we drift towards unparalleled catastrophes.

— Albert Einstein[1]

A society that has lost its life values will tend to make a religion of death and build up a cult around its worship — a religion not less grateful because it satisfies the mounting number of paranoiacs and sadists such a disrupted society necessarily produces.

— Lewis Mumford[2]

The first point which comes to my mind is the possibility that it is the destiny of historical man to be annihilated not by a cosmic event but by the tensions in his own being and in his own history.

— Paul Tillich[3]

Let's get back to Earth. In this chapter, I will walk you through another reason we are in a lot of trouble, and I begin by reminding you of the last part of Frost's poem cited above: *I shall be telling this with a sigh. Somewhere ages and ages hence: Two roads diverged in a wood, and I — I took the one less traveled by, and that has made all the difference.* We took one path, that of war rather than peace, and it has made all the difference. It is disconcerting that this issue no longer seems to be newsworthy.

Furthermore, the following subjects are less amenable to scientific scrutiny because data is not always known or are uncertain. (Either because they are kept secret or because they are not known.) And also due to the varied premises that go into the argumentation, often about what is right or wrong. Still, logic leads the way, and certain things are not a matter of opinion. As already discussed, the fallacy of hasty generalization must be kept in mind. If it is true that we wish to live in peace, then we must fight two powerful enemies. One the slowly approaching monster which I already discussed. The other is *us*, to which I turn.

Sixty years ago, in September 1961, at age 89, philosopher Bertrand Russell was jailed for 7 days for a "breach of peace" in an anti-nuclear demonstration. Authorities offered to exempt him from jail if he pledged himself to "good behavior," to which Russell replied: "No, I won't," and here we are with thousands of nuclear weapons, many much more powerful than those dropped over Hiroshima and Nagasaki.

With nuclear weapons, we have acquired the power to obliterate ourselves. Thus, we have become the mythical God, much more real than the fantastic biblical tale in the Book of Revelation, with its catastrophic end of times. It is distressing that the distinction between the biblical story and the real immolation fades away for many. Some even think that the nuclear immolation caused by our sublime stupidity is really God's desire. Ignorance, fanaticism, superstition, and ambition will lead us to Armageddon.

No God will forgive us for this possible mega-terrorist act. A delirious being who could not get rid of his anachronistic monsters. Who knows if another planet with a different life form lives in peace in some other corner of our galaxy? They will likely not find out about our tragedy, but were they to do so, they surely would put us into the hall of universal infamy, and the ruler with a cat will not shed a tear. The tragedy of intelligence is that it enables stupidity because, without intelligence, the concept of stupidity has no meaning. German philosopher Günther Anders[4] referred to our apocalyptic blindness when he questioned how we were not afraid. It amazes me how we live in denial.

As it increasingly affects the dominant countries, the catastrophic convergence will foster the temptation to resort to these weapons to resolve what some may consider a threat to "national security." However, the spark could also be an attack by a disaffected group of the periphery to the metropolis, with a nuclear bomb obtained by contraband or homemade.

Total nuclear disarmament is an inescapable imperative. If not, "Blowback"[5] will be the last of the blows. "Blowback" is a term used to refer to the retaliation resulting from a military operation that only leaves destruction and anger in its wake, without the affected population understanding the reason. The attacks on the twin towers of New York and the Pentagon on September 11, 2001, were an example of Blowback. This was well understood by former Secretary of State Hillary Clinton. In remarks before a subcommittee of the House Appropriations Committee on April 23, 2009, she said[6]: *We also have a history of, kind of moving in and out of Pakistan. Let's remember here, the people we are fighting today, we funded twenty years ago. And we did it because we were locked in this struggle with the Soviet Union [...] but let's be careful what we sow because what we will harvest.*

Before that tragedy, Chalmers Johnson wrote[7]: *Right now terrorism by definition strikes at the innocent in order to draw attention to the sins of the invulnerable. The innocent of the 21st century are going to harvest unexpected blowback disasters from the imperialist escapades of recent decades. Although most Americans may be largely ignorant of what was, and still is, being done in their names all are likely to pay a steep price — individually and collectively — for their nation's continued effort to dominate the global scene.* In the second edition (2004) Johnson relates that: *the book was largely ignored in the US [...] and the house organ of the Council on Foreign Relations, Foreign Affairs, wrote that 'Blowback reads like a comic book'.* That is how particular views once entrenched are difficult to change; it took the tragedy for the comic book to become not funny at all. Keep on bullying others, and at some point, you will get blowback.

Some European countries are paying a deferred blowback. Their past overseas colonies were hindered in their development, subject to exploitation, violence, and genocide.[8] That is behind all those who today cross the Mediterranean trying to escape African violence and hunger, some being washed up on sandy shores, as do dead fish. European racism is still alive, perhaps not as shameless as that of the US, but enough to develop vigorous protest against immigrants, not bothered nor understanding that it is in great measure a consequence of their shady colonial history.

Most Europeans and Britons thought (and many still think) that non-whites were inferior and treated the people in their African and Asian colonies accordingly. King Leopold II[9] (1835–1909) of Belgium is just one example, being behind killing an estimated ten million people in the Congo. Few know this terrible history of white brutality against black people, even those who see his statue in Brussels, proudly mounted on a horse. It should be dynamited, and if not, and perhaps even better, surrounded by other bronze figures of Africans without hands (as such was the punishment for not delivering enough rubber to the masters). As I write, there is a lot of talk about statues, and some would be better placed in a museum with appropriate explanatory plaques.

The US has learned little from this blowback, where black lives do not seem to matter even today. Its military might and increased national security apparatus are not perfect and will not solve the problems of an increasingly unstable world. Rejection and outright hate arise from its exploits and acts of war in foreign lands. The best way to avert this self-inflicted wound is to change the idea of domination, the idea of US exceptionalism and its superiority complex, by a more humble and friendly stance of collaboration, spending less on endless wars destroying distant countries and more on constructing peace. The international view of "us" and "them," including the internal racist idea of "us" and "them," is utterly destructive for the world and the nation.

But this has been said before with little effect. Something is very wrong in the US collective mind. Ruben Andersson writes[10]: *As new technologies are supposed to bridge the geographical divides, as global risks expand, and as the climate is heating up, peoples and governments need to be more connected, not less. Yet, instead of deepening cooperation among the world's rich and poor, we are being torn apart. We are seeing the emergence of a global geography of fear: a parsing up of the world map in which the dirty work in distant crisis zones is left to middlemen and advanced technology while borders are reinforced and contact points severed.*

According to the Bulletin of the Atomic Scientists[11]: *as of mid-2017, there are nearly 15,000 nuclear weapons in the world, located at some 107 sites in 14 countries. Roughly, 9,400 of these weapons are in military arsenals; the remaining weapons are retired and awaiting dismantlement. Nearly 4,000 are operationally available, and some 1,800 are on high alert and ready for use on short notice* (14 countries because some US bombs are in European countries) (Table 12.1).

Table 12.1. Countries with nuclear weapons (2021).[a]

Country	Number (total inventory)	Year of the first test
Five countries of the NPT		
United States	5,600	1945 ("Trinity")
Russia (ex USSR)	6,257	1949 ("RDS-1")
United Kingdom	225	1952 ("Hurricane")
France	290	1960 ("Gerboise Bleue")
China	350	1964 ("596")
Other countries with nuclear weapons		
India	160	1974 ("Smiling Buddha")
Pakistan	165	1998 ("Chagai-I")
North Korea	~45	2006
Non-declared nuclear weapons		
Israel	90	never

[a] Numbers from FAS. https://fas.org/issues/nuclear-weapons/status-world-nuclear-forces/.

There is no justification for maintaining these weapons unless some deranged minds think they might be helpful in the future. (Which would be one way to quickly enter the Anthropocene transition.) The threat of 15,000 nuclear weapons or 1,500 nuclear weapons is almost the same, although larger numbers also mean a higher probability of something going wrong. But, their use is not acceptable under *any* circumstances.

If we do not eliminate them, we risk that someone without reason finds a reason to use them. If not, then at some point, the world will end in fire and then ice. We tend to confuse something improbable with something impossible, and that is where the danger lies. But it is illogical and hypocritical to pursue a policy that demands that other nations "denuclearize" while keeping thousands of bombs, with the implicit definition of who is "good" and who is "bad." At the end of World War II, Robert Jackson, the American chief prosecutor at the Nuremberg Tribunal, expressed: *If certain acts of violation of treaties are crimes, they are crimes whether the United States does them or whether Germany does them. And we are not prepared to lay down the rule of criminal conduct against others which we would not be willing to have invoked against us.*

Instead, the US is building new nuclear missiles, spending billions (and the aerospace industry is happy) to deploy a system by 2030. *$100 billion to replace machines that would, if ever used, kill civilians on a mass scale and possibly end human civilization is just another forgotten subscription on auto-renew* as written in the Bulletin of the Atomic Scientists.[12]

On August 9, 1945 (the day Nagasaki was bombed), President Harry Truman informed citizens by radio from the White House about the Hiroshima bombing. He said[13]: *We must constitute ourselves trustees of this new force — to prevent its misuse and to turn it into the channels of service to mankind. It is an awful responsibility that has come to us. We thank God that it has come to us, instead of to our enemies; and we pray that He may guide us to use it in His ways and for His purposes..*

Use it in His ways and for His purposes? Really?

Although Truman later became ambivalent, he always defended his decision to drop the bomb and authorized the development of the much more powerful hydrogen bomb against the views of some of the leading atomic scientists, including Oppenheimer.[14] Increasingly often, politics trumps science (no pun intended). In fairness, he said in his 1952 farewell broadcast to the nation: *Starting an atomic war is totally unthinkable for rational men.* (Unfortunately, there are irrational "men".)

Nuclear weapons are very different from conventional weapons; they represent a discontinuous change. The most noticeable immediate effects of a nuclear detonation, which cause the most deaths in the first few minutes, are the emission of thermal radiation (heat) and the production of a shock wave with hurricane-force winds that wipe out everything in their path. Not very different from the effects of a conventional explosive, but much more powerful. It is estimated that within a mile of the hypocenter of Hiroshima (the point on the ground directly below the aerial explosion), the heat pulse had a temperature of a few thousand degrees, like placing the Sun right there. It flattened the city.

Nuclear explosions also generate large amounts of radioactive isotopes, high-energy gamma rays, and neutrons, producing more radioactive nuclei. The radioactive products (some with long half-lives — the time it takes for half of the material to decay) such as Strontium-90 (half-life 30 years).

They spread to the four winds, where they can remain for months causing biological damage. This is known as fallout.

The ingestion of contaminated food continues to attack the body's cells from within, causing genetic damage and a high incidence of cancer. This is why the Partial Test Ban Treaty (PTBT) was signed in 1963 (it allows underground tests).

I had the opportunity some years ago to visit Hiroshima with my wife and visit a home where those kids that survived (the *hibakusha*) are taken care of. Although I had nothing to do with it, to see those old sad faces brought tears to my eyes, looking at me with polite smiles, many were sitting in wheelchairs. So I guess I cried for humanity.

In a nuclear conflagration, with hundreds of detonations, after the initial chaos, the long-term effects are those that could ultimately lead to the biosphere's demise. Several studies indicate that combustion products from bombed cities (dust and smoke) will accumulate in the stratosphere, staying there for months or years, depending on the number of nuclear weapons used, blocking sunlight, and destroying ozone.[15]

This causes a rapid and persistent decrease in the planet's average temperature in what is known as "nuclear winter,"[16] more acute than the one that characterized the last glaciation. As a result of the decrease in temperatures, the global hydrological cycle would be affected, drastically reducing rainfall, which, together with the loss of sunlight, would cause the collapse of vegetation and the death of a large part of the biosphere with devastating consequences. However, these computer models have not yet been verified experimentally.[17]

It is hard to trust our leaders, who are minions of the FIRM complex whose power determines the political leadership. They have shown that they are not up to the task of solving once and for all our fundamental problems and divert our attention with lesser but more emotionally charged issues (abortion, terrorists, immigrants, etc.).

A nuclear weapon is a testimony of the power of the mind, of the surprisingly physical reality of some mathematical formulae written in a scientist's notebook. In the words of one of the participants in the first test, physicist Emilio Segré: *it was one of the greatest physics experiments of all time*. But it was also a testimony of our idiocy since we were not intelligent enough not to build it or continue after Japan surrendered. Some argue that it was out of fear that the Germans would make one, but that fear was not substantiated since they were not even close, in part because the imbecile sent his best scientists (including Einstein) into exile.

So why does any country keep hundreds or thousands of nuclear weapons, spending many resources to maintain them and risking an accident when they should not be used? What would happen if the US unilaterally and for everyone's safety eliminated all but 100 (to give a number) nuclear warheads? *Nothing.* It might inspire others to follow suit to finally get rid of them all. It is called leading by example. Call me a dreamer, but think about it. Instead, the US is embarking on a costly program to modernize its nuclear capabilities: *Meanwhile, the exploding price tag of the National Nuclear Security Administration's long-term plan to sustain and modernize the nuclear warheads and production facilities — now an exorbitant $505 billion — flies under the radar.*[18]

Lloyd Dumas[19] writes: *Because all systems that human beings design, build, and operate are flawed and subject to error, accidents are not bizarre aberrations but rather a normal part of system life. We call them accidents because they are not intentional, we do not want them to happen, and we do not know how, where, or when they will happen. But they are normal because, despite our best efforts to prevent them, they happen anyway.*

Note that nuclear weapons' danger is no longer a cause for protest; we have accepted it as normal. But if it is proposed to generate energy with nuclear plants, some will scream bloody murder, preferring to continue feeding the invisible monster burning coal and oil. Few seem to be bothered by a nearby military base storing nuclear weapons. That is how we are. Daniel Ellsberg writes[20]: *The hidden reality I aim to expose is that for over fifty years, all-out thermonuclear war — an irreversible, unprecedented, and almost unimaginable calamity for civilization and most life on earth — has been, like the disasters of Chernobyl, Katrina, the Gulf oil spill, Fukushima Daiichi, and before these, World War I, a catastrophe waiting to happen on a scale infinitely greater than any of these. And that is still true today.*

Suvrat Raju writes[21]: *In the nuclear era, the survival of humanity is closely tied to the abolition of war; this much has long been clear. But lasting peace is possible only in a just international order — where aggression by powerful countries isn't tolerated, international relations are guided by equality instead of exceptionalism, and science is guided by social rather than military objectives. On the 70th anniversary of the bombings of Hiroshima and Nagasaki, it is time for the world to acknowledge and act on these lessons.*

We urgently must disarm; we urgently must "change our mode of thinking" and get rid of a self-made menace to humanity. We must tell our leaders that their decisions are entirely wrongheaded and answer any attempt to silence us with "No, We Won't."

12.1 Pax Americana

There was a time when the American empire recognized limitations or at least the desirability or behaving as though it had its limitations. That was largely because they were afraid of somebody else — the Soviet Union. In the absence of this kind of fear, enlightened self-interest and education have to take over.

— Eric Hobsbawm[22]

Two centuries ago, a former European colony decided to catch up with Europe. It succeeded so well that the United States of America became a monster, in which the taints, the sickness, and the inhumanity of Europe have grown to appalling dimensions.

— Frantz Fanon[23]

The civilized have created the wretched, quite coldly and deliberately, and do not intend to change the status quo; are responsible for their slaughter and enslavement; rain down bombs on defenseless children whenever and wherever they decide that their "vital interests" are menaced, and think nothing of torturing a man to death: these people are not to be taken seriously when they speak of the "sanctity" of human life, or the "conscience" of the civilized world.

— James Baldwin[24]

"Manifest destiny," the idea that the US was destined by God to expand its territory across the entire North American continent, was first expressed in 1845. Native Americans paid the price. As written by Sirvent and Haiphong[25]: *The colonial genocide of indigenous people was the first root planted in the formation of America. That root remains a fundamental anchor of American society and one that permeates American culture.* It was the American quest for "lebensraum."

Belgians cannot and should not forget what happened in Africa during king Leopold's reign, and Germans cannot and should not forget what happened during Hitler's regime. Many other nations need to reckon with their past and their rationalizations explaining away bloodshed and genocides. Likewise, the US cannot and should not ignore what was done during the founding of the Republic to Native and African Americans. Only this *mea-culpa*, accepting the not so glorious past and making the necessary restitution to at least come to terms with this different narrative (as Germany has), will allow for peace. Unfortunately, the US has yet to act, starting with an admission of its bloody hands (past and present) that to this day have yet to come clean and, with a commitment to world peace, stopping all its foreign military activities.

Today, the spirit of the US's "original sin" lives on in the minds of the oligarchs controlling de FIRM complex who propose that the US is exceptional and seeks to make the world safer and better, lead and fight for liberty (especially that of the markets) and justice. And the rest of the world (and a good chunk of US citizens) pays the price. Have you ever asked safer from what and for whom?

These obsolete notions of domination instead of collaboration in the heads of fossils with a cold war mentality do not help. Bacevich writes in his latest book[26]: *Sadly, however, even today, the failed national security paradigm remains deeply entrenched in Washington. Its persistence testifies to the influence of the military-industrial complex, the lethargy of an officer corps that clings to demonstrably flawed conceptions of warfare, and the policing of mainstream discourse to marginalize critical voices. Enabling each of these is a pronounced apathy of the American people who, apart from ritualistic gestures intended to "support the troops," have become largely indifferent to the role this country plays in global affairs.*

Do you know what the US has done to other people all over the world because of this? Probably not; it is not part of the official narrative. So let me tell you a little.

I propose that one of the greatest threats to the peace that most of us yearn for is the US, without forgetting that we also have Putin's Russia,

Xi Jinping China, and some others, and without belittling the hyped and mediatic, but genuine although limited terrorist danger.

On first reading this, you might think I just crossed a red line, gone too far. How can I write this when the US did sacrifice the lives of its military personnel fighting against the fascists of WWII? But bear with me. Indeed, the US lost about half a million persons in WWII (and the Soviet Union, in defeating the Nazi's lost many more). But that was 80 years ago. Nonetheless, today we have idiots flying the Nazi flag. You can get further details about the alternative stories, which shine a very different light on US history, reading the revealing books by Sirvent and Haiphong,[27] Steven Bronner,[28] or Andrew Bacevich.[29] Try also that by Fareed Zakaria.[30] An important one was written in 1973 by Noam Chomsky and Edward. S. Herman (and initially suppressed), with a new expanded edition published in 1979.[31] Historian Gabriel Kolko (1932–2014) commented: *A brilliant, shattering, and convincing account of United States-backed suppression of political and human rights in the third world, it relentlessly dissects the official view of Establishment scholars in their journals. The best and brightest pundits of the status quo emerge from this book, thoroughly denuded of their credibility. By virtue of the importance of the subject and the excellence of the book, it is obligatory reading for any American seeking to comprehend the role of the United States in the world since 1946.*

And If you are honestly interested in getting a story that is closer to the truth, then try US Army major Daniel Sjursen's *A True History of the United States*.[a]

If I insist on reading these and other different narratives about the US and its relationship with the rest of the world, I do so because unless the premises and beliefs about the US are changed, we shall never find peace. It is not something new, but it has been systematically suppressed from the US public. Dirty laundry is not something you show.

The US aided and abetted and often acted directly as a terrorist organization, torturing, bombing, and subverting governments all over the globe, causing millions of innocent people to suffer the consequences. Often this is done by hiring private armies of mercenaries who become "freedom fighters." You might not like this harsh judgment, you might dismiss it as fake, but it is a fact. Countless documents attest to this.[32] The

---

[a] Daniel A Sjursen (2021). *A True History of the United States*, Lebanon, NH: Steerforth Press.

detail of this shady business is well laid out in Steve Coll's book: *Ghost Wars: The Secret History of the CIA, Afghanistan* (much in the news these days) *and Bin Laden, from the Soviet Invasion to September 10, 2001*. A reviewer in Amazon wrote: *Precise scholarship gracefully written and immensely valuable in the era of our historically deaf electorate who act as if the Afghanistan War came out of nowhere. Colls tracks the attempts of the CIA, Saudi Arabia, and the Pakistan Intelligence Services as they lose control of the Islamist extremism they invented while pushing the Russians out of Afghanistan. The Taliban and Osama Bin Laden are their legacies.*

And then some ask: "why do they hate us?"

In a speech to Congress just 9 days after 9/11, the question was answered by then-President Bush, who said: *They hate our freedoms, our freedom of religion, our freedom of speech or freedom to vote and assemble and disagree with each other.* Well, that was hubris. They hated US aggression. How would you feel if a drone fired a missile that killed and injured those at your wedding party? How would you feel if bombs destroyed your home? How would you feel if the US interfered in your elections or "changed your regime"? It has done so often and worse, but if we discover that others are doing it to us, we scream to high heaven. No one hates the "American way of life," and increasingly, many feel sorry for it. I quote from a recent Irish Times article[33]: *the United States has stirred a very wide range of feelings in the rest of the world: love and hatred, fear and hope, envy and contempt, awe and anger. But there is one emotion that has never been directed towards the U.S. until now: pity.*

Lest you think that all the CIA does is awful, there are, of course, actions that, especially during the cold war, were considered necessary.[34] Covert actions, lying, and spying are the CIA's bread and butter (and all other state's "intelligence" services). Espionage is as old as history. The early intelligence with satellites also led to a good outcome; a trove of early images now declassified led to environmentally relevant material that contributed to significant research in this area.[35] But taking pictures from satellites is quite different from killing and torturing people.

The US declares itself unilaterally the chosen nation to determine the world's destiny with "Pax Americana" interventions around the globe.

Although there was a bit of fresh air and different rhetoric with President Barack Obama, especially in contrast to President Bush's previous

8 years, the fundamental positions have not changed. The "national security state" continued, the drones flew, the tortured screamed, and the US society went down a slippery slope, at the bottom of which we saw an attempt to violently overthrow the government. The rhetoric hit rock bottom, and it made a difference as we witnessed rabble-rousing speech coming from the highest levels of government which energized and encouraged the lunatic fringe to act, *stand back and stand by*. It reminded me too much of the Germany of the late thirties.

The US is a nation dominated by a class that suffers from a superiority complex (white supremacy) and a warrior mentality that proposes that everything is resolved with military actions. It is also evident that there is a clear dichotomy in terms of national and international politics. US citizens vote according to their perception of what is good for them (as happens everywhere) and are mostly ignorant or do not care about what is done in foreign countries, and most could not place Afghanistan on a map. Of course, the rhetoric will improve with President Joe Biden (how could it not?) but remember it is rhetoric. We need revolutionary action, rebels who, in the words of civil rights leader and congressman John Lewis (1940–2020), can make "good trouble." But I fear that we shall get more of the same, especially on the international scene.

I have no doubt that, in the future, with the growing scarcity of resources important to maintain the industrial machinery, the US armed forces will be used to secure them, the real reason for keeping its military might (and Africa is fertile ground). Of course, it will be argued that if we don't do it, our competitors will, which may be true. This is the reason to change the rules of the global game.

Or do you really think that if the military were reduced to, say, one half, which would still make it by far the most powerful one and leave a lot of resources for what the US really needs, then it would be invaded by some enemy? Dreaded Cuba, perhaps? Or do you really believe we drop bombs and declare embargoes on other countries for the sake of human rights? (We would have to bomb half of the world). Once you spend over 800 billion a year on the military, things become difficult to change because a lot of money is to be made in this business, and millions live from it. The military is very much involved in our daily lives, more than what you imagine.[36]

We can place the beginning of this US vision of the world, an extension of manifest destiny, with the overthrow of Queen Lili'uokalani of Hawaii in 1893 by US sugar growers supported by the marines, culminating in the annexation of the territory in 1898. (It became a state in 1959.) Cuba, Puerto Rico, the Philippines, Nicaragua, Honduras, Iran, Guatemala, Indochina, Uruguay, Chile, Grenada, Panama, Afghanistan, Iraq, and many others followed.[37] According to William Blum,[38] since World War II, the US has intervened militarily in other states over 70 times, bringing suffering, torture, and death to citizens of foreign countries. You paid for it.

As mentioned, we can see antecedents of this predatory stance in the glorified (by the victors) "conquest of the west," a genocide perpetrated by white immigrants that snatched from the true Americans (which were *not* white), lives and lands, relegating them to inhospitable and miserable "reservations."

Spanish and Portuguese conquerors did the same in Central and South America (I mean the continent). But those were different times, times of colonization and slave trade belonging to the dark history of European actions, where white supremacy was born. Still, it seems that many cannot discard that inheritance, but we must.

The proclamation of General Nelson Miles during the US invasion of Puerto Rico in 1898 shows a rhetorical pattern that has been copied, with changes appropriate to the circumstances since then: *We have not come to make war upon the people of a country that for centuries has been oppressed, but, on the contrary, to bring you protection, not only to yourselves but to your property, to promote your prosperity, and to bestow upon you the immunities and blessings of the liberal institutions of our government.*

If you are aware of the present conditions in Puerto Rico, you might want to laugh (or cry). Not very different from the expressions of President Bush in 2003, at the start of operation Iraqi freedom: *We come to Iraq with respect for its citizens, for their great civilization, and for the religious faiths they practice. We have no ambition in Iraq, except to remove a threat and restore control of that country to its own people.*[b] (And what about Abu Ghraib)? The following caveat must be inserted here. When there is talk about a US invasion somewhere, the invaders are usually poor, brainwashed soldiers who are ordered to fight, but the real invaders are those who made the decision, who later eulogize the brave dead soldiers. Many are shocked, but few are awed. As mordantly told by Bacevich[39]: *As it turned out, however, shock and awe was to war what Donald Trump's promotion of hydroxychloroquine as a cure for coronavirus was to the pandemic of 2020: a con job the wreaked havoc on the lives of untold innocents.*

William Blum[40] calls them an Imperial Mafia and writes: *But these men are perhaps not so much immoral as they are amoral. It's not that they take pleasure in causing so much death and suffering. It is that they just don't care ... the same that could be said about a sociopath. As long as the death and suffering advance the agenda of the empire, as long as the right people and the right corporations gain wealth and power and privilege and prestige, as long as the death and suffering aren't happening to them or people close to them ... then they just don't care about it happening to other people, including the American soldiers whom they throw into wars and who come home — the ones who make it back alive — with Agent Orange or Gulf War Syndrome eating away at their bodies. American leaders would not be in the positions they hold if they were bothered by such things.*

Back in 1980, then-President Carter proclaimed that Middle Eastern oil was a vital interest for the US and would be defended by any means, including military force. *Let our position be absolutely clear: An attempt by any outside force to gain control of the Persian Gulf region will be regarded as an assault on the vital interests of the United States of America, and such an assault will be repelled by any means necessary, including military force.*[41] So, the US was not an "outside force?" I would think it would be much easier and less harmful just to buy what you need from their rightful owners.

Eric Hobsbawm[42] comments: *Few things are more dangerous than empires pursuing their own interest in the belief that they are doing humanity a favor.* And although I am referring to the US today, it applies well to

---

[b] President Bush Addresses the Nation. March 23, 2003. http://www.whitehouse.gov/news/releases/2003/03/20030319-17.html.

all past empires. Also, as is well said by Stiglitz[43]: *But for any country to exercise that (global) leadership, it has to be seen as not just serving its own interests, but as having a vision that sees the benefits of cooperation, without the use or threat of force.* Then it might become a city upon a hill. But, unfortunately, those in government, red or blue, have done all the contrary for many years. The question now is if the US can change course radically, a daunting challenge to undo the idea that the US is God's gift to the world. It must stop the entrenched view that Historian and retired Colonel Andrew Bacevich[44] calls the sacred trinity: *maintain a global military presence, global power projection, and a policy of global interventionism.*

There will not be much to be happy about unless Biden and Harris work hard to undo the misguided Pax Americana. Yes, he can try and perhaps succeed in "bringing us together," but the basic problem of US imperialism and the damage to the world will continue. I will eat my words; in fact, I will cook and eat this book if he would bring **all** the military and associates' homes and reduce nuclear weapons to a minimum. And, of course, stop meddling in other nation's affairs.

In the official 2002 document that presented the US security strategy and exposes the hubris of those who wrote it; you can read[c]: *The U.S. national security strategy will be based on a distinctly American internationalism that reflects the union of our values and our national interests. The aim of the strategy is to help make the world not just safer but better. Our goals on the path to progress are clear: political and economic freedom, peaceful relations with other states and respect for human dignity.* One problem with this statement is that national interests have mostly eclipsed values. The other problem is its hypocrisy. And though written 20 years ago, the spirit remains.

Though the Obama administration eliminated this shameful document as part of a supposed new vision (but the killing continued), let us not forget that the authors of this document are still around, resisting change. They did return to power for 4 years which seemed like an eternity, and are

---

[c] The National Security Strategy of the United States of America (September 2002). The White House, Washington, D.C. https://history.defense.gov/Portals/70/Documents/nss/nss2002.pdf?ver=2014-06-25-121337-027.

now back on the sidelines, waiting to return. But the damage they wrought will take a long time to repair. As noted by Max Blumenthal[45]: *Under the watch of Secretary of State Mike Pompeo, an evangelical zealot and anti-Iran hard-liner, neoconservative retreads were filling up mid-level posts. In just over a year, the Trump administration was beginning to resemble the second coming of George W. Bush's first term.* Nor let it be overlooked that no official has been prosecuted for these past abuses of human dignity.

And to respect human dignity, to respect the fifth article of the declaration of human rights, suspects were kept in the most wretched conditions to reappear, or not, and sent to be tortured to "black sites." (In what was called "extraordinary rendition.") Some complained bitterly when other countries did the same. More hypocrisy. So much for the hallowed and hollowed "rule of law."

As reported by William Blum in a chapter titled, *Torture — as American as apple pie*[46]: *The precise pain, in the precise place, in the precise amount, for the desired effect.* This was explained by a US instructor in the art of torture, one of many. The words are those of Dan Mitrione, who had served as Richmond's (Indiana) chief of police, and later became an adviser to the Uruguayan police and military. In this little-known event, he was killed by the Tupamaro movement in September of 1970, after being kidnapped.

New York Times reporter A J. Langguth[47] documented his story and the disturbing history of US complicity in destroying democracy and human rights in Latin America. The remains of this man who had killed several under torture were received in the US as a hero. Then-white House press secretary Ron Ziegler (1939–2003) (Nixon administration) solemnly stated that *Mr. Mitrione's devoted service to the cause of peaceful progress in an orderly world will remain as an example for free men everywhere.*

There is a sequel to this sordid period of Latin American right-wing dictatorships. José Alberto "Pepe" Mujica Cordano (born 1935), a Tupamaro leader imprisoned in squalid conditions for 13 long years during the military dictatorship in the 1970s and 1980s, became the 40th president of Uruguay from 2010 to 2015. He has been much admired by people worldwide and is considered one of the few presidents in the world who remained humble and uncorrupted. Quite a contrast with Nixon or with the 45th US president.

Further down, the document states: *To forestall and prevent such hostile acts by our adversaries, the United States will, if necessary, act preemptively*, which just means that to prevent attack we will attack (something which at the Nuremberg Trials was cataloged as "supreme crime"). Strangely, leaders of a country with by far the greatest military power in the world think another country will attack (the 9/11 attacks were not organized by a government). Absurd logic, but valuable for the FIRM. The US explicitly proclaims that they rule the world, and those who do not like the *Pax Americana* will suffer the consequences. As President Bush's letter of introduction stated: *The United States welcomes our responsibility to lead in this great mission*. I wonder what the "great mission" is and who gave them this responsibility.

John Bolton, the national security advisor to the 45th POTUS, wrote an article in the Wall Street Journal[48]: *The Legal Case for Striking North Korea First*, which reaffirms that right to a preemptive strike and shows a toxic way of thinking: *if it is legal it is OK*. Now we are back to a more "traditional" presidency, but will things change significantly? Will notorious Guantanamo be closed as Obama promised? It is twisted logic; North Korea, like it or not, could then use the same logic, scared as they are about a US intervention, and strike first. Sorry, Mr. Bolton, you are wrong, and aside from legal considerations, there are also ethical ones (if you know what I mean). And by the way, the people of North Korea are human beings. The idea that just because something is legal, it is also ethical pervades much thought in the US, leading to the absurdity of having other countries torture on behalf of the US because it is illegal in the US.

In Latin America, the Monroe Doctrine (established by President James Monroe in 1823), and initially not much more than words, turned into a guiding principle for US intervention in the affairs of nations, and with the excuse of confronting left-wing governments, resulting in the death, and suffering of uncountable people. Most US citizens did not care or know or believed the official narrative, although much was kept secret Outside the US, the "Cold War" was not so cold. Andrew Bacevich writes[49]:

As it unfolded across several decades, it produced ruinous consequences. It fostered folly and waste on a colossal scale, notably in an arms race of staggering magnitude. It bred hatred, hysteria, and intolerance, creating conditions rife with opportunities for demagogues. It warped political priorities, subordinating the well-being of people to the imperatives or state security. Whether directly or indirectly, it provided a pretext for murder and mayhem, even if the victims tended not to be citizens of the United States or the USSR.

Nothing to be proud of. We all recall the dreadful events of September 11 (meaning the terrorist attack in New York and Washington DC in 2001). I clarify this because few in the US know of another sad September 11 (in 1973), when the Chilean military bombed the presidential palace as part of a military coup with US complicity. (It was the time of Richard Nixon and Henry Kissinger.) As a result, the democratically elected socialist president Salvador Allende was overthrown and committed suicide at the palace.

You can read about this sad story in the fully documented book by Peter Kornbluh,[50] which despite the heavily redacted then-secret memos, shows enough to understand the shameful involvement of the US government. The Chilean dictatorship of General Augusto Pinochet (1915–2006), ruled by terror, torture, political persecution, and assassination. Orlando Letelier (1932–1976), a Chilean diplomat, ambassador to the US under Allende, released from Chilean prison camps and torture in 1974, went into exile to Washington DC. He was assassinated there with coworker Ronni Moffitt (1951–1976) by a car bomb. Two years earlier in Buenos Aires, Argentina, General Carlos Prats González (1915–1974), Commander in Chief of the Chilean Army in exile, was similarly assassinated with his wife, Sofía Ester Cuthbert Chiarleoni. Michael Vernon Townley (born in Iowa in 1942), working for the Chilean secret police (DINA), currently living under terms of the US federal witness protection program, was the organizer of both assassinations. Townley served 62 months in a US prison for the murders.

During the horrendous Pinochet dictatorship (ended in 1990), between 10,000 and 30,000 were killed (or disappeared) and more tortured or forced into exile. By the numbers and in proportion to its population of about ten million, it was far deadlier than the second 9/11 tragedy. Then-Secretary of State Collin Powell (1937–2021), responding to a question in February of 2003, stated[51]: *With respect to your earlier comment about Chile in the 1970s and what happened with Mr. Allende, it is not a part of American history that we're proud of.* (quite an understatement).

Similar reprehensible things happened with military dictatorships in other Latin American countries, with US complicity.[52] To remove any doubt, here is a quote from an internal CIA memo[53]: *On September 16, 1970, [CIA] Director Richard Helms informed a group of senior agency officers that on September 15, President Nixon had decided that an Allende regime was not acceptable to the United States. The president asked the Agency to prevent Allende from coming to power or to unseat him and authorized up to $ 10 million for this purpose…. A special task force was established to carry out his mandate, and preliminary plans were discussed with Dr. Kissinger on 18 September 1970.* At a high-level meeting related to the affairs in Chile, Kissinger had the nerve to say[54]: *I don't see why we need to stand by and watch a country go communist due to the irresponsibility of its own people.* Kissinger was awarded the Nobel peace prize in 1973, the year of the other 9/11, leading The Guardian to recall musician and mathematician Tom Lehrer's comment,[55] *Political satire became obsolete when Henry Kissinger was awarded the Nobel peace prize.* Argentinian miliary threw thousands of drugged "terrorists" into the Rio de la Plata and the Atlantic Ocean from aircrafts (vuelos de la muerte). I am not aware of any US protests about this and many other evil acts.

Those who condemn us to war, declaring that it is necessary to ward off some evil (real or invented) that it is waged to bring "freedom," are those who are part of the FIRM complex. They think that they have little to lose (even if wrong) since they send the sons and daughters of the poor to kill or die, not understanding why *endless columns of uniformed boys, white, black, brown, yellow, marching obediently toward the common grave,* in the words of Aldous Huxley.[56]

As said in Alfred Lord Tennyson's (1809–1892) tragic poem "The Charge of the Light Brigade"[d]: *Theirs not to make reply, theirs not to reason why, theirs but to do and die.*

All prepared by a previous disinformation campaign for the public to applaud the noble campaign and accept the body bags as they return home. As was well said by Orwell[57]: *All the war propaganda, all the screaming and lies, and hatred come invariably from people who are not fighting.*

Naturally, the atrocities committed by the glorious forces of freedom are not published, they are not investigated unless there is no choice, and then a minor scapegoat is processed and given a laughable sentence, even if it is not so funny.[58]

It gets worse with a lucrative growing market (worth hundreds of billions) for private military companies (PMC) that recruit mercenaries from all over the world, get juicy contracts from the Pentagon, respond to nobody, and get away with anything.

This dangerous privatization of the US military is convenient for the US government to wash its hands and lie about the number of "boots on the ground."[59] US soldiers' war crimes and torture seem not to bother military and government officials (what bothers them is if someone blows the whistle), sending the wrong message to the troops that anything goes, not very different from the attitude of US militarized police. Just look at a hair-raising video that shows a US Apache helicopter aircrew shooting down Iraqi civilians in 2007, including two Reuters journalists, released by WikiLeaks, a video provided to them by Chelsea Manning, a former United States Army soldier. (Look at the video and see for yourself.) Manning spent 7 years in prison, harassed and mistreated, after being convicted by court-martial to a 35-year sentence. It was commuted by President Obama. (Is it a crime to expose government crimes?). On the video of the "engagement," you can see as civilians are mowed down, and you can hear: "nice, good shoot" and another voice: "thank you" as if it were a Pickleball match. It is sickening, makes your stomach turn. It strikes me as insensitive that an attack helicopter used to kill innocent people be called "Apache." Others are called Blackhawk and Comanche.

---

[d] You can hear the poem read by Tennison himself in 1890, recorded on an Edison wax cylinder here: https://www.youtube.com/watch?v=YMl5c2gDoO8.

Noam Chomsky believes he sees the emergence of fascism in the US.[60] I have already used the term several times but not really defined it, mainly because we all think we know what fascism is until we are asked to say it.

Rob Riemen[61] tells us: *Fascism is the political cultivation of our worst irrational sentiments: resentment hatred, xenophobia, lust for power, and fear.* According to Jason Stanley,[62] the politics of fascism includes many distinct strategies: *the mythic past, propaganda, anti-intellectualism, unreality, hierarchy, victimhood, law and order, sexual anxiety, appeals to the heartland, and a dismantling of public welfare and unity.* Does this remind you of anyone or any place you know?

Mussolini famously said that if you consolidate power by stealthily plucking the chicken one feather at a time, people do not notice. Clinical Psychologist Elizabeth Mika tells us[63]: *Tyrannies are three-legged beasts. They encroach upon our world in a steady creep more often than overcome it in a violent takeover, which may be one reason they are not always easy to spot before it's too late to do much about them. Their necessary components, those three wobbly legs are: the tyrant, his supporters (the people), and the society at large that provides a ripe ground for the collusion between them. Political scientists call it "the toxic triangle".*

The President (from 1953 to 1961) of the United States (and five-star general) Dwight D. Eisenhower, a witness of the suffering and carnage of war, said in his Farewell Address to the Nation on January 17, 1961: *A vital element in keeping the peace is our military establishment. Our arms must be mighty, ready for instant action, so that no potential aggressor may be tempted to risk his own destruction...*

This conjunction of an immense military establishment and a large arms industry is new in the American experience. The total influence — economic, political, even spiritual — is felt in every city, every statehouse, every office of the federal government. We recognize the imperative need for this development. Yet we must not fail to comprehend its grave implications. Our toil, resources and livelihood are all involved; so is the very structure of our society. In the councils of government, we must guard against the acquisition of unwarranted influence, whether sought or unsought, by the **military-industrial complex.** *The potential for the disastrous rise of misplaced power exists and will persist. We must never let the weight of this* *combination endanger our liberties or democratic processes. We should take nothing for granted. Only an alert and knowledgeable citizenry can compel the proper meshing of the huge industrial and military machinery of defense with our peaceful methods and goals so that security and liberty may prosper together.* We did not guard.

(Those were different times, and our leaders were what could be called "men of state" and not the arrogant puppets which seem to multiply everywhere in the present.) Since then, the US's standing in the world has steadily declined; you do not make many friends if you are an arrogant and deadly bully. I am well aware that during Eisenhower's presidency, the US was involved in several actions that resulted in "regime change" (Iran and Guatemala) and the failed "Bay of Pigs" attack to topple Fidel Castro, inherited by President John F. Kennedy. It only shows the continuity of US international policies and the ambiguity of the thinking and actions of some protagonists. But that was then and what matters is now. The US security strategy document of 2002 mentioned above also indicates: *The presence of American forces overseas is one of the most profound symbols of the U.S. commitments to allies and friends. Through our willingness to use force in our own defense and in defense of others, the United States demonstrates its resolve to maintain a balance of power that favors freedom. To contend with uncertainty and to meet the many security challenges we face, the United States will require bases and stations within and beyond Western Europe and Northeast Asia, as well as temporary access arrangements for the long-distance deployment of U.S. forces.*

Yes, it is a "profound symbol" of imperialism.

The word "defense" is used when all it does is attack (and it is not true that the best defense is always to attack). The narrative is that we shall fight the enemy on their territory, but who exactly is the enemy? Before World War II, the US had a *war department* (war department — army and air force), and the department of the navy merged into the presently misnamed department of defense (doublespeak).

We live under the FIRM complex, which operates globally with over half a million troops, spies, technicians, trainers, and contractors in hundreds of military bases abroad, distributed in some 170 nations.[64] Thule

Air Base in Greenland is the largest one. It is by far the country with most military bases outside its territory. Imagine that tomorrow Spain or Saudi Arabia wanted to install a military base in Texas. Does it seem ridiculous? But what is the difference?

Add to that a dozen "task forces" of aircraft carriers and submarines in all oceans and seas and PMC's that perform the dirtiest tasks necessary to maintain peace and freedom.

Globalization is limited to imperial interests (and the interest of certain viruses); welfare, human rights, and democracy are not globalized despite the bombs. And under the pretext of freedom, justice is put to one side. It is crudely about "making America great," without analyzing what "great" might mean in this context, beyond the disrespect towards the rest of the planet's inhabitants. It harkens back to the 1930's "America First" movement supported among many by renowned aviator and racist Charles Lindbergh. Don't forget or deny that the US became wealthy on the shoulders of slaves. The true Americans (in all of America) inhabited the continent before the arrival of the white immigrants who slaughtered them. And yes, in the context of those times, it was what it was, but we should remember and respect that history and not hide it.

The military bases abroad maintain the empire, but they also increase the resentment of many towards it, dispersing hazardous materials in the environment without any control (also within the US) and fomenting an infamous industry that exploits the poor population that often surrounds these bases. Therefore, if the US wishes to live in a peaceful world, it must start by calling its soldiers home and reimage itself, leading by example.

Perhaps there is some hope now that Trump is gone (but he might return). Still, history tells us that we must be careful of what is to come since no matter how much better it feels to have a decent person in the White House, the general feeling of being exceptional is deep. Thus, phrases such as the "greatest democracy the world has ever known" are hubris (as is "America is back" — and many wish it weren't so).

Time will tell. It is distressing that a person like Trump could ever become president. (Perhaps following Orwell's maxim: "War is Peace/

Freedom is Slavery/Ignorance is Strength".) The former CIA chief, John Brennan, publicly said[e]: that Trump was *unstable, inept, inexperienced, and also unethical.* Former FBI director James Comey says that he is *morally unfit to be president.* And I read in a recent book[65]: *he has long been in deep with mobsters, domestic and foreign, along with corrupt union bosses and assorted swindlers. Trump even spent years deeply entangled with a major international drug trafficker who, like many of the others, enjoyed their mutually lucrative arrangements.* Is this what you want for President?

Perhaps the lesson to be learned is that in a democracy, not everyone is fit to become president and that a mechanism must be found to weed out those who clearly are inept or worse. Putin is not inept; it is worse. Democracies are delicate beasts.

It is irrelevant what party you prefer; rotten apples must be purged before more harm is done. But as already said by German poet Friedrich Schiller (1759–1805): *Against stupidity the very gods themselves contend in vain.* The idiotization of the public is part of the catastrophe, and I repeat what Eisenhower said: *Only an alert and knowledgeable citizenry can compel the proper meshing of the huge industrial and military machinery of defense with our peaceful methods and goals so that security and liberty may prosper together.*

Defense Secretary James Mattis's scathing letter of resignation to Mr. Trump said in part: *My views on treating allies with respect and also being clear-eyed about both malign actors and strategic competitors are strongly held and informed by over four decades of immersion in these issues.* He continued: *Because you have the right to a Secretary of Defense whose views are better aligned with yours on these and other subjects, I believe it is right for me to step down from my position.* In The Atlantic for December 20, 2018, I read: *Mattis is the last brake on a president that makes major life-and-death decisions by whim without reading, deliberation, or any thought as to consequences and risks,* said a senior US national-security official on Thursday, who spoke on the condition of anonymity in order to talk

---

[e] In an interview with MSNBC's. (2018). Deadline: White House. http://www.msnbc.com/deadline-white-house.

freely. *The saving grace is that this president has not been tested by a major national-security crisis. But it will come, and when it does, we are fucked.*

But as I mentioned before, individuals in government rise and fall; what matters is who we collectively think should lead. I tell you all this because the political future in the US does not look good, and those now on the sidelines will do everything possible (including rigging elections) to get back in power and continue with the Pax Americana. We should all worry.

12.2 War is a Racket

War kills more civilians than soldiers. In fact, the army is usually the safest place to be during a war. Soldiers are protected by thousands of armed men, and they get the first choice of food and medical care. Meanwhile, even if civilians are not systematically massacred, they are usually robbed, evicted, or left to starve; however, their stories are usually left untold. Most military histories skim lightly over the massive suffering of the ordinary, unarmed civilians caught in the middle, even though theirs is the most common experience of war.

— Matthew White[66]

I do not know with what weapons World War III will be fought, but World War IV will be fought with sticks and stones.

— Albert Einstein[67]

With new and terrifying weapons, the growing importance of artificial intelligence, automated killing machines, and cyber war, we face the prospect of the end of humanity itself. It is not the time to avert our eyes from something we may find abhorrent. We must, more than ever think about war.

— Margaret McMillan[68]

War is like fire; those who will not put aside weapons are consumed by them.

— Sun Tzu disciple (ca 500 BCE)[69]

The history of humanity has been one of violence from one group to another in search of political, geographical, or economic advantage. It is a history of widespread xenophobia in which the foreigner is depicted as inferior, dehumanized for easier killing. *Not a single thing that we commonly believe about wars that helps keep them around is true. Wars cannot be good or glorious. Nor can they be justified as a means of achieving peace or anything else of value. The reasons given for wars before, during, and after them (often three very different sets of reasons for the same war) are false. It is common to imagine that, because we'd never go to war without a good reason, having gone to war, we simply must have a good reason. This needs to be reversed. Because there can be no good reason for war, having gone to war, we are participating in a lie.*

So begins a book by David Swanson[70] that thoroughly debunks the idea of a "good war." Even the US entry unto WWII was not motivated by lofty ideals as it has been narrated, as you can read in Michael Zezima's book.[71] If you disagree, think of 20 years in Afghanistan, You can read in "The Afghanistan Papers,"[72] a book written by Washington Post reporter Craig Whitlock that:

…many senior US officials privately viewed the war as an unmitigated disaster, contradicting a chorus of rosy public statements from officials at the White House, the Pentagon and the State Department, who assured Americans year after year that they were making progress in Afghanistan.

War is a scourge that has been a constant of civilization (not very civilized). Although some nations conquered others and gained something by doing so, the world has changed. The globalized economy means that should you attack one of the major players in this irrational game, you would just be attacking yourself, so all this talk about the military menace of other countries is outdated demagoguery, despite what warmongering "experts" say. War has become obsolete but is still waged for the benefit of those who profit from it. The US-led global war on terror has killed nearly 1 million people globally and cost more than US$8 trillion since it began two decades ago. These staggering figures come from a landmark report issued by Brown University's Costs of War Project,[73] an ongoing research effort to document the economic and human impact of post-9/11 military operations. (Where you can learn a few other interesting facts). This without counting all those who have been hurt for the rest of their lives. But when it comes to investing trillions in rebuilding the US (creating jobs), some politicians become worried.

We hear the term used as if it could solve some problem: war on terror, war on poverty, war on drugs, war on crime. But war is not the solution to any problem. It will only leave losers[74]: *wars and militarism make us less safe rather than protect us, that they kill, injure and traumatize adults, children, and infants, severely damage the natural environment, erode civil liberties, and drain our economies, siphoning resources from life-affirming activities.*

US economist and historian of economic thought Robert Heilbroner[75] (1919–2005) pointed out that political dynamics in times of generalized anxiety caused by wars, financial problems, insecurity (fear of terrorism or immigrants), or civil instability push in the direction of authority and Fascism. Fear increases the tendency to conform. It is used by devious fearmongering politicians to manipulate the masses and seek scapegoats to blame for their ills, which is happening in many nations.

Bertrand Russell, one who you can quote about anything you might need, said[76]: *Collective fear stimulates herd instinct, and tends to produce ferocity toward those who are not regarded as members of the herd [...] neither a man nor a crowd nor a nation can be trusted to act humanely or to think sanely under the influence of a great fear.*

This was already known to the infamous German Marshal Göring, who testified at the 1946 trial of the Nazi criminals in Nuremberg[77]: *Naturally, the common people don't want war. But after all, it is the leaders of a country who determine the policy, and it's always a simple matter to drag people along whether it is a democracy or a fascist dictatorship, or a parliament, or a communist dictatorship. Voice or no voice, the people can always be brought to the bidding of the leaders. This is easy. All you have to do is tell them they are being attacked and denounce the pacifists for lack of patriotism and for exposing the country to danger. It works the same in every country.*

As explained by distinguished political theorist Sheldon Wolin,[78] the benchmarks characterizing a totalitarian system such as the German Third Reich were: *a system of power that was invasive abroad, justified preemptive war as a matter of official doctrine, and repressed all opposition at home = a system that was cruel and racist in principle and practice, deeply ideological and openly bent on world domination.*

Not much has changed. Already in 1951, Hannah Arendt, in her seminal study, said[79]: *Totalitarian solutions may well survive the fall of totalitarian regimes in the form of strong temptations which will come up whenever it seems impossible to alleviate political, social, or economic misery in a manner worthy of man.*[f] And we are on that path.

The first casualty of war is truth, and the rest are mostly civilians, as is well documented. Winston Churchill famously said: *In wartime, truth is so precious that she should always be attended by a bodyguard of lies.* If you are in a war, the safest place is being in the military. We have built many things but have also forged our obsolescence, adhering to dated ideas of solving problems by war. In the shadow of Hiroshima and Auschwitz, we cannot afford to continue with this obfuscation proper of dark minds. Propagating tales of horror and hatred, perpetuating myths of the past, inventing new threats, and avoiding reason is convenient to justify violence.

To start a war, you must identify the enemy. Easy to do if they belong to another tribe or to another nation, easier even if they are different in appearance, of another ethnic group that we can consider inferior (they are never considered superior). Sometimes very subtle differences as between the Hutu and Tutsi of Rwanda, who hacked each other to death with machetes, to give an example that many of us know about at least from a movie.[80] And if you want more details of the gruesome killing of a million, read the book written by Philip Gourevitch.[81] I think that if the population of Rwanda had been white, the story would have been different.

Xenophobia comes naturally. It is part of our mental software, originated when a human from another tribe was suspect and engendered fear. Unfortunately, we have been unable to eliminate this bias, as with other phobias for which people need psychological treatment. It can quickly turn into something much worse than a phobia by adding a narrative provoked by demagogues and ignoramuses, transforming it into hate, leading to mayhem, suffering, and murder. You need only review the story of the Shoah to understand what can happen, even to a well-educated citizenry. You start by looking the other way when a few demented fanatics do their

---

[f] Hotel Ruanda (2004).

thing, and off you go on a perilously slippery slope. I will tell you about the Shoah as a reminder.

The other way of grouping the enemy is according to what they think, and if they do not think, they can be grouped according to their beliefs. War and religion have never been far apart (President George Bush claimed he was told by God to invade Iraq[82]).

Religion allows separation, the first step on the road to genocide. Separate Christians from Jews to justify the killing of those who "killed the son of God." Split Catholics from Protestants, so they could go at each other starting one of the worst episodes in European history, the "30-year war" (1618–1648), which cost millions of lives, or more recently, another 30-year conflict in Northern Ireland where thousands lost their lives ("The Troubles" between 1968–1998). Separate the Shiites from the Sunnis to blow themselves up in the Middle East and turn an entire country into a pile of rubble (with foreign help). I could go on and on.

Already in the Bible, we read: *And Moses spake unto the people, saying, Arm some of yourselves unto the war, and let them go against the Midianites, and avenge the Lord of Midian. And they warred against the Midianites, as the Lord commanded Moses; and they slew all the males. And the children of Israel took all the women of Midian captives, and their little ones, and took the spoil of all their cattle, and all their flocks, and all their goods. And they burnt all their cities wherein they dwelt, and all their goodly castles, with fire. And Moses was wroth with the officers of the host, with the captains over thousands, and captains over hundreds, which came from the battle. And Moses said unto them, have ye saved all the women alive? Now therefore kill every male among the little ones and kill every woman that hath known man by lying with him. But all the women children, that have not known a man by lying with him, keep alive for yourselves* (Numbers 31, 3.7.9-10.14-15.17-18 KJB). Nice huh?

Those who clutch it under the arm can do all the mental contortions they want to extract themselves from what is obvious, but at least for the one who has not lost his brain, this is described plainly and simply as genocide. No way around it.

Above, I put as the first milestone of the new era the tragic events of Hiroshima and Nagasaki, not only because they were dreadful. It was also for fear that if we forget or keep denying that it was an act of genocide (however you might wish to justify it), it could happen again, with weapons much more devastating than the rudimentary weapons used then.

Until that historic moment, thousands or millions died or were injured in fierce battles in which lead and fire destroyed everything that stood in their way. In the last century's two great wars, an estimated 65 million people died between soldiers and civilians (15 million in the first from 1914 to 1918, and 50 million in the second from 1939 to 1945). This includes the millions systematically murdered by the fanatical followers of Hitler and Stalin. Many other conflicts — civil wars in Russia, China, and several African nations, the wars in Korea, Vietnam, Cambodia, Afghanistan, Iraq, Libya, Syria, Yemen, and others — amount to an estimated 200 million violent deaths. And countless humans who survived with physical or mental injuries live the rest of their lives disabled with pain, sadness, and bitterness, seek refuge, and encounter a wall.

During the grim calamity of the Vietnam war, there were marches in protest, prompted by the nightly news of body bag counts. Today not even that, and others killed by US forces do not count. Is that civilized? Is that how you get peace? Is that what you want? According to David Swanson, author, journalist, and executive director of *World beyond war*[83]: *Since World War II, during a supposed golden age of peace, the United States military has killed or helped kill some 20 million people, overthrown at least 36 governments, interfered in at least 85 foreign elections, attempted to assassinate over 50 foreign leaders, and dropped bombs on people in over 30 countries. The United States is responsible for the deaths of 5 million people in Vietnam, Laos, and Cambodia, and over 1 million just since 2003 in Iraq.*

But it is not only a matter of numbers, of who killed more or less, tortured with extra zeal and creativity, or was more brutal. Not everything that can be counted counts, and not everything that counts can be counted, as Einstein did not say. Ethical values cannot be quantified. In terms of what we aspire to be as humans, each violent death is a collective calamity, each wounded human, each tortured one is a tribulation, and they do not add up, even as they multiply.

The habits of our unfortunate heritage are difficult, although not impossible, to control. Our history of violence and warfare results from decisions made by mostly *men*. Would a world ruled by women be different?

US military expenditures are enormous and have increased dramatically in recent years. They are greater than the sum of the military expenditures of the seven nations that follow (Figure 12.1).[84] They are a high fraction of the total budget to the detriment of its own population; they bleed the republic and cheat the public. *We need global disarmament.* It is pigheaded to think that spending more money on new military hardware will help national security. One more expensive aircraft carrier, a nuclear submarine, or a new fighter jet increases costs and provides diminishing returns (except for the armaments industry).

An additional intangible cost is caused by the employment of thousands of professionals who dedicate their talent to the military effort (scientists, engineers, economists, psychologists, and doctors, among others). These are thus withdrawn from addressing the serious problems we face, which will not be solved with military hardware.

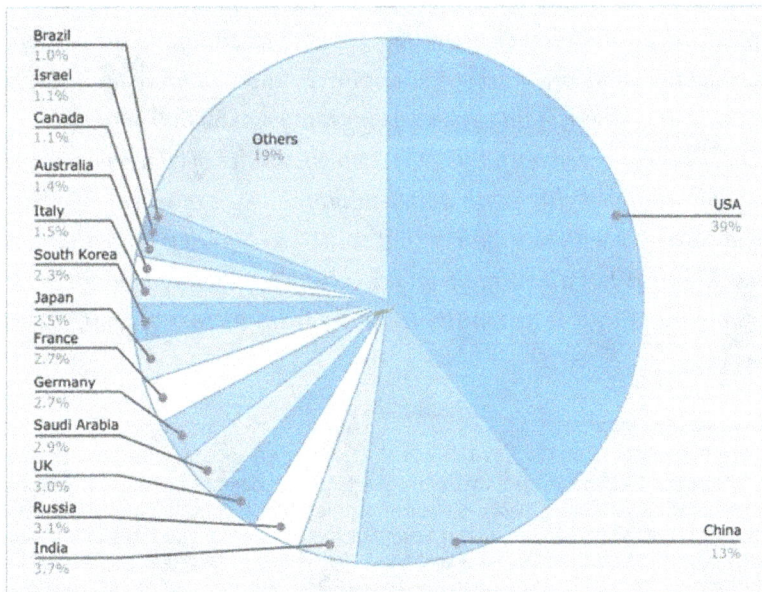

Figure 12.1. Military expenditures (SIPRI).

Add to that the cost of all those veterans who cannot cope with the things they witnessed or did during these wars and need medical help for the rest of their sorry lives.

It is estimated that US military activities account for about 70% of the US production of greenhouse gases. (In 2017, it spent $7.9 billion on 270,000 barrels of oil *per day* and emitted more than 25,000 kilotons of CO_2, being the largest institutional consumer of oil in the world.) And to what effect? This is what Tom Engelhardt has to say[85]: *It's a strange fact of this century that the US military has been deployed across vast swaths of the planet and somehow, again and again, has found itself overmatched by underwhelming enemy forces and incapable of producing any results other than destruction and further fragmentation. And all of this occurred at the moment when the planet most needed a new kind of knitting together, at the moment when humanity's future was at stake in ways previously unimaginable, thanks to its still increasing use of fossil fuels.*

In 1953 Eisenhower said[g]: *Every gun that is made, every warship launched, every rocket fired signifies, in the final sense, a theft from those who hunger and are not fed, those who are cold and are not clothed. This world in arms is not spending money alone. It is spending the sweat of its laborers, the genius of its scientists, the hopes of its children. The cost of one modern heavy bomber is this: a modern brick school in more than 30 cities. It is two electric power plants, each serving a town of 60,000 population. It is two fine, fully equipped hospitals. It is some fifty miles of concrete pavement. We pay for a single fighter with a half-million bushels of wheat. We pay for a single destroyer with new homes that could have housed more than 8,000 people. ... This is not a way of life at all, in any true sense. Under the cloud of threatening war,* **it is humanity hanging from a cross of iron.**[h] (Please, reread it!)

---

[g] Eisenhower on April 16 of 1953, in a speech to the American Society of Newspaper Editors.

[h] Cross of Iron connects to the "Cross of Gold" speech given by democratic presidential candidate William Jennings Bryan (1860–1925) at the democratic national convention of 1896. Arguing against the gold standard he said: *Having behind us the commercial interests and the laboring interests and all the toiling masses, we shall answer their demands for a gold standard by saying to them, you shall not press down upon the brow of labor this crown of thorns. You shall not crucify mankind upon a cross of gold.* Bryan lost the election to William McKinnley (1843–1901) who was assassinated 6 months into his second term.

Sixty-five years later, we can read something that says the same but from a completely different side,[86] US journalist and critic Chris Hedges: *...if we do not get control of the military spending, we're finished. We're being hollowed out from the inside like every other empire. We have expanded beyond our capacity to sustain ourselves. Our infrastructure, our public educational system, our social services — everything is crumbling for a reason; we don't have any money for it. It is consumed by the war machine.*

Eisenhower warned about the military-industrial complex (which I expanded to the FIRM complex). However, it enjoys excellent health, and every armed conflict is good news for many corporations. According to the Stockholm International Peace Research Institute (SIPRI), the US is the largest exporter of arms (and Saudi Arabia the largest client), accounting for about a third of the total, followed by Russia with 20% then China, Germany, and France. (Take notice that these numbers are very uncertain due to secrecy.) (Raytheon's Tomahawk missile and Lockheed Martin's Joint Air-to-Surface Standoff Missiles Extended Range, or JASSM-ER, cost about 1.8 million each.)

In 1935, long-forgotten United States Marine Corps Major General and two-time Medal of Honor recipient Smedley D. Butler (1881–1940) wrote[87]: *War is a racket. It always has been. It is possibly the oldest, easily the most profitable, surely the most vicious. It is the only one international in scope. It is the only one in which the profits are reckoned in dollars and the losses in lives.*

The empire invades small islands and poor lands that cannot defend themselves. Smedley explained this a long time ago when he wrote: *I spent thirty-three years and four months in active military service... And during that period, I spent most of my time being a high-class muscleman for Big Business, for Wall Street, and for the Bankers.* That was then; Butler could not have imagined what is happening today. In her study of war, historian Margaret MacMillan writes[88]: *The impacts of climate change such as the struggle for scarce resources and a large-scale movement of peoples, the growing polarization within and among societies, the rise of intolerant nationalist populisms and the willingness of messianic and charismatic leaders to exploit these will provide, as they have in the past, the fuel for conflict.*

Space is a new arena to weaponize, and the US would like to send a courtesy bomb to any address in the world. It is enough to read the astonishing and colorful document of the United States Space Command, "Vision 2020."[89] It seems to have been written by a bunch of nuts: *US Space Command — dominating the space dimension of military operations to protect US interests and investment. Integrating space forces into warfighting capabilities across the full spectrum of conflict.*

I also read as a reason: *The globalization of the world economy will also continue, with a widening between "haves" and "have-nots." Accelerating rates of technological development will be increasingly driven by the commercial sector — not the military. Increased weapons lethality and precision will lead to new operational doctrine. Information-intensive military force structures will lead to a highly dynamic operations tempo.* And then I read: *Global Engagement combines global surveillance with the potential for a space-based global precision strike capability. Need I say more? Democracy and human rights are not mentioned (I searched). And all this is paid for by "we the people".*

Even what used to be considered the domain of crackpots is becoming part of the official narrative with the government releasing footage of UFOs or, as they like to call them UAP's (for unidentified aerial phenomena). Even President Obama has fanned the flames. Some blurry images and we are all in need of a space force. The aliens are coming.

Although Russia and China are building up their capabilities to militarize space, building satellites that can attack other satellites (and making the potential debris in orbit a threat to all spacecraft), do we need to be equally stupid and escalate to nowhere? For sure, the US should be vigilant, build resilience and resistance to their new satellites (such as the all-important GPS and surveillance satellites) but weaponizing space and spending more billions is not the way to go.

Meanwhile, during the dismal Trump administration, the government could not send a monthly check to all those tethering at the brink because it mismanaged the Covid 19 epidemic. They abandoned "we the people." This is what Eisenhower meant by: *a theft from those who hunger and are not fed, those who are cold and are not clothed.*

The highest spheres of the US government are infiltrated by past executives of corporations that profit from war (a substantial number of members of congress are shareholders of defense contractors). Mixed with conservative ideologues, they promote armed conflicts and the development of new and improved weapons, which are excellent opportunities to make loads of money. The weapons factories supply whoever has the money to acquire them without much concern for their subsequent use. **Weapon producers would go broke without wars**.

Caitlyn Johnstone summarizes: *Military members who support imperialism get promoted. Those who get to the top go on to work for war profiteers. The war profiteers fund think tanks which promote more wars. The mass media report "news" stories citing those think tanks. These stories manufacture consent for more wars. The war industry reinforces itself. Those who get to the top of the war machine move on to the private sector and spend their time lobbying for more wars which create more eventual Pentagon officials who go on to lobby for more wars. Peace should be easy. This is why it's not.*

Although difficult to determine, the global cost of military matters is huge but is estimated at about $2,000 billion per year.[90] It is estimated that the cost of building a modern well equipped 500-bed hospital is about one billion, so that the money spent on killing could be used to build 1,800 hospitals *every year*. In addition, the jobless military personnel could be retrained to staff them. If the billions spent on war were invested in education, infrastructure, health, new energy sources, and to help the less fortunate in the US and on this planet, a much smarter way to fight terrorism, avoid refugees and promote peace. Think about that!

Bacevich[91] gives us a succinct picture when he writes: *We live in a country where if you want to go bomb somebody, there's remarkably little discussion about how much it might cost, even though the costs almost inevitably end up being orders of magnitude larger than anybody projected at the outcome. But when you have a discussion about whether or not we can assist people who are suffering, then suddenly we come very, you know, cost-conscious.* He also tells us: *Firing a $70,000 missile from a $28,000,000 drone flying at a cost of $3,624 per hour to kill people in the Middle East living on less than $1 per day.* Good use of your tax dollars. Insane.

For most decent persons, killing someone else is traumatic, no matter the reason, even if it were self-defense. The same thing happens with the conflicts in the Middle East and Africa. The only thing they will leave is losers planting flags on a mountain of rubble. It is the billions that count. Here is Engelhardt again[92]: *Looking back on almost fifteen years in which the United States has been engaged in something like permanent war in the Greater Middle East and parts of Africa, one thing couldn't be clearer: the planet's sole superpower, with a military funded and armed like none other and a "defense" budget larger than the next seven countries combined (three times as large as the number two spender, China), has managed to accomplish absolutely nothing. Unless you consider an expanding series of failed states, spreading terror movements, wrecked cities, countries hemorrhaging refugees, and the like as accomplishments.*

We often read or hear that China is doing this, and Russia that, but let us remember that this always refers to a small ruling class who clearly lives in a fantasy world. *The Chinese people or the Russian people or we the people do not want war.* We weed the garden so that "good plants" may thrive. What about weeding humanity, remove all those who only harm the garden?

As pointed out by historian Graham Allison,[93] the US might not be able to escape "Thucydides' trap." The Athenian historian, Thucydides (460–395 BCE), wrote the "History of the Peloponnesian War," which recounts the 5th-century BCE war between Sparta and Athens (it is recognized as the first history book). In it, we read: *What made war inevitable was the rise of Athens and the fear that this instilled in Sparta.* The trap: *When a rising power threatens to displace a ruling one, the likely outcome is war.* According to Allison, these conditions have happened sixteen times over the past 500 years. Twelve ended in war, and in the present time, we face the same conditions, China rising, the US declining, and in fear. The four that did not end in war were: Spain and Portugal in the late 15th century, United Kingdom and the United States in the early 20th century, United States, and the Soviet Union in the long "cold war" (they came close with the Cuban missile crisis), and more recently (1990 to present), United Kingdom together with France and Germany over power in Europe.

I end with the words at the end of Major General's Smedley Butler's short book: *To Hell with war!*

12.3 Inequality and Peace

We pray that peoples of all faiths, all races, all nations, may have their great human needs satisfied; that those now denied opportunity shall come to enjoy it to the full; that all who yearn for freedom may experience its spiritual blessings; that those who have freedom will understand, also, its heavy responsibilities; that all who are insensitive to the needs of others will learn charity; that the scourges of poverty, disease and ignorance will be made to disappear from the earth, and that, in the goodness of time, all peoples will come to live together in a peace guaranteed by the binding force of mutual respect and love.

— Dwight D. Eisenhower[i]

The fates of human beings are not equal. Men differ in their states of health and wealth or social status or what not. Simple observation shows that in every such situation he who is more favored feels the never ceasing need to look upon his position as in some way "legitimate" upon his advantage being "deserved" and the other's disadvantage being brought about by the latter's "fault." That the purely accidental causes of the differences may be ever so obvious makes no difference.

— Max Weber[94]

The contrast between the words of Eisenhower and what we hear today is astounding, and what German sociologist Max Weber (1864–1920) wrote could be written today. But, unfortunately, Weber died from the Spanish flu (which did not come from Spain), and his wife Marianne finished his work for publication.

Consider the enormous and growing disparity between those few who have the resources for a whole and dignified life and the vast majority of humanity, an economic proletariat to which we now add an *intellectual proletariat*, created by the avalanche of information that paradoxically contributes to lack of knowledge. This majority, immersed in ancient superstitions and unfounded beliefs (for which they are willing to kill or die), barely survives, ignorant of their slavery. At the same time, a small gang that knows, pulls strings, and enjoys and takes advantage of that knowledge. This is not new; already in

---

[i] Eisenhower's Farewell Address to the Nation, January 17, 1961.

a famous book written almost 60 years ago, Martinique-born Frantz Fanon wrote[95]: *The basic confrontation which seemed to be colonialism versus anti-colonialism, indeed capitalism versus socialism, is already losing its importance. What matters today, the issue which blocks the horizon, is the need for a redistribution of wealth. Humanity will have to address this question, no matter how devastating the consequences may be.* Unfortunately, it has only gotten worse.

At this moment, a man somewhere screams in pain because a small mass of lead, fired from an imported gun, tears up his insides, but you do not hear him. A sad and emaciated child gives up struggling to survive because hunger has left it without strength and closes his big eyes for good, but you do not see it. A woman dies after years of suffering because she was told (by a Pope, no less) that it was a sin to use a condom that would have protected her from contracting a disease that spreads silently through the nights, but you do not know her. Somewhere, in a well-guarded building, somebody gets electric shocks administered to his genitals or her vagina after first being beaten mercilessly; all they can do is groan and die. This is the daily life of the wretched of the Earth.

Meanwhile, there are fashion shows in a large city, paradigms of frivolity, while millions endure the cold to survive as they lack adequate clothes. In another, an elegant banquet is held to celebrate the sales success of some product of dubious benefit, while millions of children die for lack of medicines. Sports events, beauty pageants, and Oscar awards are followed with great interest by millions. The media is filled with stories of these stars who are the object of envy and emulation. Young people dream of 1 day triumphing just like them, and for many, their biggest worry is if their phone's battery will run out.

The peace symbol is confused with that of Mercedes Benz (It was designed in 1958 by British artist Gerald Haltom and is the superposition of the marine semaphore letters "N" and "D" for Nuclear Disarmament). People become perfect slaves of the system, working at a paltry minimum wage to buy industrial products. When young people do not protest how things are, when they become mere spectators of a screen that pollutes their minds, something is very wrong. Of course, there are exceptions, but too few. Unfortunately, it also seems that this anger venting does not achieve much: BLM is no longer news.

Brainwashing includes commercial propaganda, prepared with the advice of experts in the field, which bombards us without mercy — consume and consume this which is "new and improved," the best — in an endless spiral that stimulates the desire for material goods. It sticks to our minds like flies to dung on a humid summer day. Everything is carefully studied: the product's packaging, shape, color, and symbols that connect with the consumer's subconscious to make the product attractive, even if it is a piece of shit. The packaging is more important than the content (think of all those stupid "unboxing" videos), an idea that can be generalized to people with their clothes, jewelry, and make-up.

It is evident that faced with societies in which one group violently attacks another (recently encouraged by official discourse) where human rights are fading, we fluctuate between the emotion of wanting to intervene and the reasoning of not making things worse with more violence. The tension between the idea of countries or independent societies, on the one hand, and human rights that we consider apply everywhere, leaves us paralyzed. Karl Popper[96] alerted us about what he called the "paradox of tolerance": *Unlimited tolerance must lead to the disappearance of tolerance. If we extend unlimited tolerance even to those who are intolerant, if we are not prepared to defend a tolerant society against the onslaught of the intolerant, then the tolerant will be destroyed, and tolerance with them.* Or put another way: those who think that those with different opinions must be eliminated, must be eliminated.

The visible violence of war is not the only violence globally, nor the most significant one, even if it has the greatest mediatic impact. The daily non-violent violence of hunger, misery, and illness is much more common but less sensational, just because it is a constant of the world that affects a high portion of humanity.

Hunger, the cousin of poverty, is more than a writhing of the stomach or dizziness due to lack of food. Hunger is the sadness of mothers who see their children die malnourished, it is the anguish of parents to get food for tomorrow, it is the humiliation to see those who have much more than they need but despise them just for being poor. There's even a word "aporophobia" that describes this tendency to reject and ignore the poor simply because they are poor (and presumably have little to offer). Hunger is a consequence of impotence. The impotence of those most in need to participate effectively (not just putting a cross on some electoral ballot) in the social, political arena, and economic decisions of the nation they inhabit and the place where they work.

Of the nearly five billion people living in developing countries (development to where?), three-fifths lack essential sanitation services, a third lack access to safe water, a quarter lack adequate housing, and a fifth lacks access to modern health services. Also, a fifth of children does not reach the fifth grade. As a result, every day, 18,000 children die of hunger, *every day!* This is the bleak reality of injustice and misery. And the thing does not change if we change the numbers by some percent or change definitions to get better results as if this were a solution to the problem. Some in the US also live on the edge.

The US has the most unequal income distribution of all the industrialized countries, compounded by evident institutionalized racism, which for minorities becomes an "American nightmare." If the US boasts of being the wealthiest country, ask yourself how it is that millions could not put food on their table just 2 months into the pandemic. But, again, this is nothing new. Renowned essayist James Baldwin in a short "Letter to my Nephew," of 1962, writes[97]: *I know what the world has done to my brother and how narrowly he has survived it, and I know, which is much worse, and this is the crime of which I accuse my country and my countrymen and for which neither I nor time nor history will ever forgive them, that they have destroyed and are destroying hundreds of thousands of lives and do not know it and do not want to know it. One can be — indeed, one must strive to become — tough and philosophical concerning destruction and death, for this is what most of mankind has been best at since we have heard of war; remember, I said most of mankind, but it is not permissible that the authors of devastation should also be innocent. It is the innocence which constitutes the crime.*

The proof that industrial capitalism and the neoliberal socioeconomic system is a failure is in the data, in the evidence. If we compare the state of the planet 50 years ago with the present, we see minimal improvement, unless you are one of those who measure improvement by the number of cell phones or cars. Income inequality has increased. Twenty percent of the richest in the world own 80% of the world's wealth, possibly worse since the rich hide their wealth in tax havens. The Gross Domestic Product (GDP) of the 48 poorest nations (one-quarter of the world's countries) is less than the combined wealth of the three wealthiest persons in the world.

The famous "wine glass" graphic illustrates this global situation (Figure 12.2). The price of inequality is very high, fueling violence and ideas of tyrannical utopias. When over four billion people live on less than $5 per day, and the top 1% have a total wealth twice that of the bottom 50%, instead of justice[98] and democracy, you have a seething reservoir of hatred and anger.

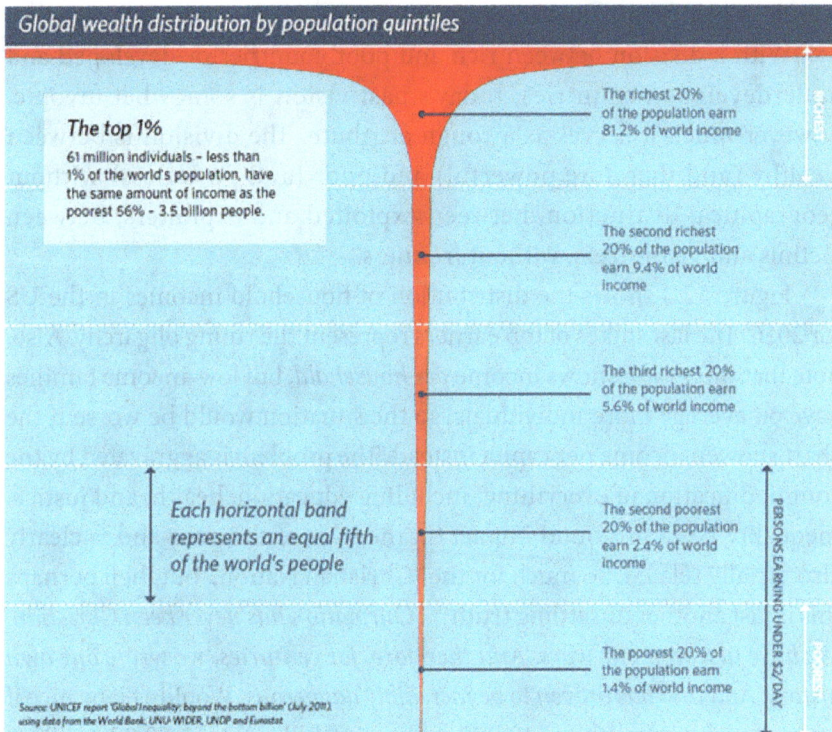

Global wealth distribution by population quintiles

The top 1%

61 million individuals – less than 1% of the world's population, have the same amount of income as the poorest 56% - 3.5 billion people.

The richest 20% of the population earn 81.2% of world income

The second richest 20% of the population earn 9.4% of world income

The third richest 20% of the population earn 5.6% of world income

Each horizontal band represents an equal fifth of the world's people

The second poorest 20% of the population earn 2.4% of world income

The poorest 20% of the population earn 1.4% of world income

Source: UNICEF report 'Global Inequality: beyond the bottom billion' (July 2011), using data from the World Bank, UNU-WIDER, UNDP and Eurostat

RICHEST

POOREST

PERSONS EARNING UNDER $2/DAY

Figure 12.2. The Wine Glass. There cannot be peace in such a world!

And this also applies to societies that think they are not capitalist, like "communist" China, which together with India composes over one-third of the world's population. (The following most populous country is the US.) They also aspire to improve their "standard of living" and follow the so-called development model of the "rich" countries (where there are lots of poor people). They are oblivious that the cost of that is the accelerated exhaustion of our resources and the possible extinction of our species.

The Nobel Prize in Economics, Joseph Stiglitz says[99]: *The top 1 percent have the best houses, the best educations, the best doctors, and the best lifestyles, but there is one thing that money doesn't seem to have bought: an understanding that their fate is bound up with how the other 99 percent live. Throughout history, this has been something that the top 1 percent eventually do learn. Often, however, they learn it too late.*

In the words of Matt Taibbi,[100] the low end is populated by *barefoot scrapers of the bottom of the international capitalist barrel.* Is there any reason to justify this situation?

With a division between rich and poor countries or developed and underdeveloped countries, today's past vision is somewhat myopic, however much it serves as a rough attribute. The division is between wealthy (and therefore powerful) and poor (and powerless) without geographical distinction, between exploited and exploiters, between victims and victimizers, without frontiers.

Figure 12.3 shows the distribution of household incomes in the US for 2016. The last spikes of top earners represent the ruling oligarchy. Also, note that this graph shows income *per household*, but low-income families have on average more individuals, so the situation would be worse if the chart showed income per capita instead. The problem is aggravated by the commodification of everything, including education, health, and justice. Inequality has a profound impact on the lives of the poor and is clearly also racially related. So much for the "Christian Nation," but then perhaps this is just another unsettling truth[101]: *Our nation has never been Christian. We have just won our wars. And therefore, for centuries, we wrote our own history. And that has proven to be incredibly dangerous.* Wouldn't it be nice if someone with a wealth of $10,000 million (10 billion) took $9,000 million and just build 90 million homes for those in need? She would still have

Distribution of Annual Household Income in the United States (2016)

Bottom 20% less than $24,002
Median $59,039
Top 20% at least $121,018

Top 5% at least $225,251

Source: U.S. Census Bureau, Current Population Survey, 2017 Annual Social and Economic Supplement

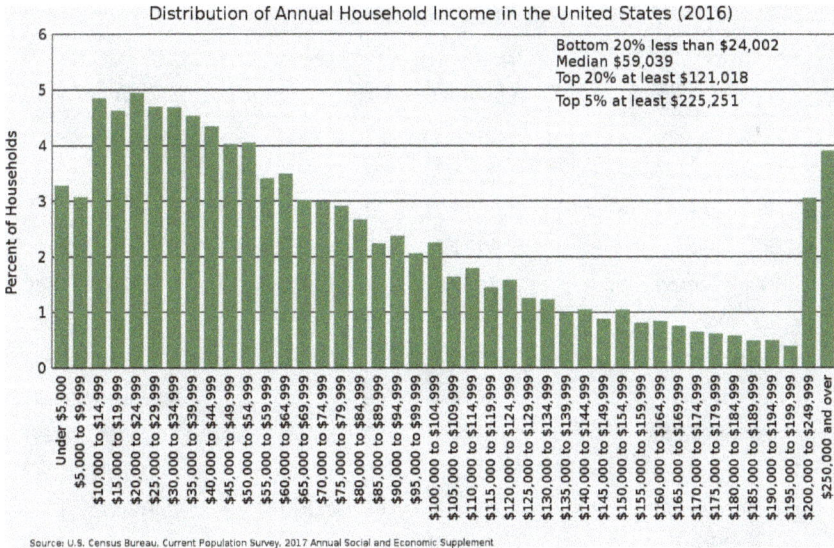

Figure 12.3.   US income distribution *per household* (2016).

a billion to live a good life. A dream, right? (Some are indeed using their fortunes to help humanitarian causes). Table 12.2 shows just the twenty wealthiest as of mid-2021.[102]

National borders have disappeared because control has been exercised for quite some time by powerful multinational industrial and financial empires supported by institutions such as the World Bank, the World Trade Organization (WTO), the European Central Bank, and the International Monetary Fund (IMF).

Most international institutions have been quite ineffective at achieving goals for a better world. As explained by Louis Menand[103] this is so because: *Establishing courts of international justice, or outlawing wars of aggression, or making pacts of collective security are simply attempts, dressed up in the language of human rights and self-determination, by the stronger powers to lock in the status quo. They instantiate winner's normality.*

According to José María Tortosa[104]: *The problem of so-called underdevelopment originates in the so-called developed countries and becomes more acute thanks to the latter with the visible collaboration of the elites of the poor countries, and it will only be solved when the so-called*

Table 12.2. Twenty billionaires.

	Name	Net worth billion $	Age	Source	Country
1	Jeff Bezos	$185.8	57	Amazon	United States
2	Elon Musk	$178.7	50	Tesla, SpaceX	United States
3	Bernard Arnault & family	$174.7	72	LVMH	France
4	Bill Gates	$131.2	65	Microsoft	United States
5	Mark Zuckerberg	$129.3	37	Facebook	United States
6	Larry Page	$118.1	48	Google	United States
7	Larry Ellison	$115.4	77	software	United States
8	Sergey Brin	$114.1	48	Google	United States
9	Warren Buffett	$103.7	90	Berkshire Hathaway	United States
10	Steve Ballmer	$97.3	65	Microsoft	United States
11	Francoise Bettencourt Meyers & family	$90.3	68	L'Oréal	France
12	Mukesh Ambani	$81.7	64	diversified	India
13	Amancio Ortega	$80.0	85	Zara	Spain
14	Carlos Slim Helu & family	$76.4	81	telecom	Mexico
15	Jim Walton	$69.6	73	Walmart	United States
16	Alice Walton	$68.7	71	Walmart	United States
17	Rob Walton	$68.4	76	Walmart	United States
18	Zhong Shanshan	$66.2	66	beverages, pharma	China
19	Phil Knight & family	$60.7	83	Nike	United States
20	Michael Bloomberg	$59.0	79	Bloomberg LP	United States

developed countries change their policy towards the underdeveloped ones, and the very rich elites of poor countries abandon their equally predatory attitude.

Perhaps facing our inner enemy is even more complex: the millenarian psychological inheritance that does not seem to allow us to change our way of thinking and our beliefs and act to effectively face threats to humanity. So instead, we shove them under the rug.

The situation is critical, but few are alarmed by it, and fewer are willing or able to do what is necessary. (How difficult is it to wear a mask to protect your fellow humans?) The reasons for this state of denial are several, but there are two that are dominant. It is partly a problem of perception and partly of deception. There is talk of a great global crisis, but we do not perceive it directly since our immediate environment is relatively small. The eyes that see the crisis are mounted on remote satellites that sense the global changes, but what they see is imperceptible for most people (the same goes for viruses).

Moreover, it is our nature to be sensitive to crises with high emotional impact, earthquakes, hurricanes, famines, and tsunamis. We are the subject of live reports (with appropriate background music to enhance drama) via satellite. It is difficult to convince people to push and accept drastic measures before the total crisis and chaos take over. And when that happens, extreme measures tend to make the situation worse.

This perverse deception results from a well-financed publicity campaign by those who would lose in the short term if the preventive measures were taken that must be taken urgently. Corporations, dependent on new markets for their profits and the tiny percentage of the planet's population that controls most of the wealth, are not willing to lose even 10% for the sake of a better future. That is why it is better to invent a bogus threat or exaggerate an existing one and thus divert attention from an "inconvenient truth."[j]

Two currents block the emergence of new ideas necessary for a world of peace: the growing totalitarianism of the state and of the spirit.

The state responds to the FIRM complex and the growing and stealthy surveyance of its intelligence services fed by our increasingly detectable footprints (or perhaps better fingerprints and voiceprints) that deal with any dissident as an "enemy." It invents a reality to suit it with "alternative facts." It stands guard over those who may pose a "threat," prepared to disappear them if necessary, as happened often during restless nights in Latin America or Africa. At a minimum, imprison you by an automated system that can profile you and arrest you for months, *accused not of crimes they have committed, but of crimes they will commit* in the words of Philip K. Dick.[105]

On the other hand, the churches interfere even in the most intimate affairs of the individual, maintaining with fear what reason cannot, and increasingly enters the political arena, making a mockery of the idea of church-state separation. Fear is a potent emotion used by those who

---

manipulate our minds leading to strong autocratic extreme leaders. Their rhetoric makes people feel safe from any invented "enemy of the state," be they immigrants, terrorists, journalists, Jews, or entire "rogue" nations. But it is mostly hubris.

The fascist motto: *Credere, obbedire, combattere*, of Benito Mussolini (shot and then hung upside down from the roof of a gasoline station in Milan — 2 days later Hitler committed suicide) serves both.

Meanwhile, the invisible monster that will cause great future violence, powerful hurricanes, disastrous droughts, devouring fires, drowning floods, famines, and endless refugee avalanches, slowly approaches.

Chapter 13

Post-truth, Conspiracies, and Denialism

There exists a subterranean world where pathological fantasies disguised as ideas are churned out by crooks and half-educated fanatics for the benefit of the ignorant and superstitious. There are times when this underworld emerges from the depths and suddenly fascinates, captures, and dominates multitudes of usually sane and responsible people, who thereupon take leave of sanity and responsibility. And it occasionally happens that this underworld becomes a political power and changes the course of history.

— Norman Cohn[1]

I think that, first, one should never, under any circumstances, fear the manipulation by power and its culture. We must behave as if this dangerous eventuality did not exist. What matters most is the sincerity and necessity of what must be said. We must not betray it in any way, much less by diplomatically keeping silent about one's principles.

— Pier Paolo Pasolini[2]

I enter a topic where the scientific temper has gone AWOL, representing a threat just as dangerous as the physical ones. We find cases ranging from innocuous to atrocious within the cacophony of opinions, news, and endorsements that cause more confusion every day to an audience that has not cultivated the scientific temper. It does not really matter if you believe that we never went to the Moon (who cares?). But if you believe that there is no such thing as anthropogenic climate change or that COVID-19 is a hoax, and you are in a position of authority but do nothing about it, then it borders on the criminal. And so, it is for many other issues. In a democracy, this also affects how people vote.

The systematic search for truth guides scientific research and should direct all our actions wherever it leads. The idea of "post-truth" is only an excuse for lying, leading to the political subordination of reality. These days one can detect something in public discourse that could be called *post-logic* to keep in the same vein. Only in this way can one understand how the same mind can accept and act following the predictions obtained by atmospheric scientific models about the trajectory of a hurricane and simultaneously deny the predictions of atmospheric scientific models about climate change and its consequences.

The widespread belief in sinister conspiracies (on the part of cults and secret societies and governments) and the denial of certain uncomfortable facts, for whatever reason, is common. It is also the subject of films, such as "Men in Black" or the canonical "Conspiracy Theory" of 1997. It is assumed that some group, entire governments, or a secret agency or order, conspire to hide certain things that those groups do not want to be known by the public, aimed at destroying our way of life and controlling the masses. (And it does happen, as recent US history shows.)

A conspiracy theory (here, the term "theory" is not used in the scientific sense), inspired by some perceived pattern (real or imaginary), seeks to explain an uncertain and inexplicable event. It has a psychological role parallel to that of superstition, responding to a need for control. It employs a reversal of the argumentative direction: starting with a conclusion and then selecting what "confirms" the theory, a process of *rationalization*, instead of reasoning.

In addition to blatant lies and the distortion of facts, the fallacy of moving the goalposts is often used. The proponents of conspiracy theories change the goalposts every time they come across evidence that refutes the conspiracy. Those who claim that we never went to the Moon find all sorts of anomalies in the original images taken by the Apollo astronauts (easily explained). Recently, high-resolution cameras photographed the sites of human activity on the Moon, which should be accepted as evidence that we *were* there. It was alleged that they were photoshops prepared by NASA; the goalposts were moved.

The conspiratorial hypothesis reduces complex social phenomena to the machinations of powerful individuals or organizations (the deep state, the "plandemic." the Illuminati) that explains more than the official version. Bad things happen because there are bad guys, and not because: "Shit happens." Both options are possible; it is difficult to distinguish between them, and most people prefer the former explanation because the latter does not provide a means of control.

Conspiracies exist at all levels, but what is essential is positive proof of such a thing, and when that is obtained, people land in jail or worse. (Think Jeffrey Epstein and Ghislaine Maxwell, and all those in the little book.) But even when false, conspiratorial thinking is harmful since it spreads suspicions and fear, delegitimizing democratic institutions and processes.[3] Fear leads to violence and, in extreme cases, to genocide, as was the case in Nazi Germany. For example, Trump claimed a conspiracy (by democrats, by manufacturers of voting machines, by republican traitors) to claim that the recent US elections were "rigged," with the result we all witnessed as those true believers assaulted the US capitol. But the evidence of fraud never surfaced, but for many, that was irrelevant.

Conspiracy theories start with the question: *cui bono*? In the face of a suspicious or unusual event, the question is who benefits, and then it is assumed that the group that benefits caused the event. But given any event. (such as an election), there are winners and losers, and those who win are not necessarily conspirators.

Indeed, there have been real conspiracies, from one group to affect another, from one nation to affect another (secret services are dedicated to this). Still, it is necessary to prove the hypothesis beyond a reasonable doubt. *The difficulty lies in separating the real conspiracies from the fictitious ones.*[4] The least plausible conspiracy theories are those that propose a vast network, effective, powerful, and secret whose purpose is to control and manipulate the rest of humanity for their nefarious goals. I will discuss these unjustified theories of a *grand conspiracy* since the others are unfortunately part of our daily lives.

The theories of grand conspiracy identify three groups:

1. The conspirators, a group of people or institutions of great power and malign intentions who want to control the population for their benefit.

(Conspiracies are not proposed for benign purposes.) They pursue their activities in absolute secrecy (smoke-filled backrooms or midnight meetings in cemeteries) and are responsible for significant events like wars, pandemics, and other calamities.

2. Most of the public: the victims of the conspiracy.
3. A handful of self-defined heroes: they have discovered the conspiracy (which was not *that* secret) and are running the barricades despite being misunderstood or ridiculed.

Certain characteristics are common to a long list of grand conspiracy theories: They explain a story or process of great emotional, political, or historical impact and therefore of great public interest, rejecting the "official" or accepted version.

They express deep distrust of social institutions, government, and the established order (in an increasing number of cases, well justified). It is alleged that these institutions frequently conspire to hide certain facts that are unfavorable to them (which is often true),

For example, the High-Frequency Active Auroral Research Program (HARP) is often associated with a secret US government installation to guide hurricanes, produce earthquakes, mind control, and weather modification. The power radiated by HARP is significant but minuscule compared with the power of an earthquake or hurricane, so that is physically impossible to achieve these things, just like a fly cannot push a car. Be that as it may, if we could guide Hurricanes, perhaps we could have avoided a few recent, very disastrous ones.

Responsibility for the burden of proof is ignored and passed on to those who disagree. Inconsistencies in the "official" version are noted — anomalous data not explained in the official version are accepted uncritically as proof that it is false. (Note that this is analogous to what happens when a certain scientific theory begins to lose credibility due to anomalous data.) The difference is that nature is neutral and does not actively "conspire" to confuse us. Without considerations of the *quality* of the evidence, that in favor (often misrepresented or taken out of context) is accepted without criticism (confirmation bias raises its ugly head), the evidence against is ignored or discarded as part of the conspiracy.

It is defended with *ad hominem* attacks (you do not know the truth, you are naive, or, worse still, are part of the conspiracy) and is formulated in an unfalsifiable way (contrary evidence is assumed to have been planted by the conspirators to confuse the issue).

Hypotheses that US government agencies conspired to assassinate President Kennedy are popular, and why not believe it if they did conspire to assassinate leaders of other countries. Also popular is the idea that the death of Princess Diana was the result of a conspiracy by MI-6 (Military Intelligence Section 6), the British secret intelligence service. There is always a group that knows that viruses are deliberately let loose for some evil goal, a "plandemic" (and let us hope they are wrong). Bioweapons are not fiction. It is a good material for books and movies and feeds the already mentioned paranoid style.

Many believe that the military hides what happened in Roswell with an alien ship despite historical documents that prove otherwise. The argument is as follows: If the government conceals that we are being visited by aliens, it would deny these visits. The government denies these visits. Therefore: the government is covering up the facts. (There was at that time a secret to be hidden, project (Mogul) by the US Army Air Forces using high flying balloons to detect Soviet nuclear tests.)

The US public health service began a terrible experiment in 1932 that only ended in 1972 (when the press exposed the study). A population of about 600 poor African American residents of Tuskegee, Alabama, was infected with syphilis under the pretext of "blood studies," and those who already had the disease were not treated. It is, for many, reason enough to think that anything is possible and lingers on in the collective memory of black citizens who are suspicious of current efforts to vaccinate against COVID-19.

It is also alleged that there was "something suspicious" in the tragic events of 9-11 (A government conspiracy to justify the invasion of Iraq and Afghanistan.) We can read5: *The 9/11 "new Pearl Harbor" was planned in astonishing detail and carried out through the efforts of a sophisticated and large network of operatives. It was more complex and far more successful than the Allende assassination, the US bombing of our own ship the "Maine" that began the Spanish-American war (and brought us Guam, Puerto Rico, Cuba, and the Philippines), the Reichstag fire that was used to justify the suspension of most civil liberties in Germany in the 1930's, and even Operation Himmler, which was used by Germany to justify the invasion of Poland, which started World War II. Whoever is responsible for bringing to grisly fruition this new false-flag operation, which has been used to justify the wars in Afghanistan and Iraq as well as unprecedented assaults on research, education, and civil liberties, must be perversely proud of their efficient handiwork. Certainly, 19 young Arab men and a man in cave 7,000 miles away, no matter the level of their anger, could not have masterminded and carried out 9/11: the most effective television commercial in the history of Western civilization.*

This was expressed by Lynn Margulis (1938–2011), distinguished professor in the department of geological sciences at the University of Massachusetts and a member of the US National Academy of Sciences, honored with several awards (including the national medal of Science), and author of several high-quality scientific books. (Margulis was Carl Sagan's first wife.) Her statement followed her reading of a book written by David Ray Griffin questioning the events.[6] Although many things can be questioned about the events surrounding this tragedy and the government's response, it is difficult to accept that a *large network of operatives* would lend themselves to an action that killed thousands. That no one with direct knowledge of this conspiracy has blown the whistle is not credible. (Defenders might retort that they were all eliminated, and then those who eliminated them were in turn eliminated.)

Serious investigators by the government or the press have not come up with what would be a mind-bending story, the way it was in other instances such as Watergate (a true but small conspiracy). I hope that I am not mistaken and that Griffin is wrong.

When asked why no one has reported the conspiracy (journalists, the opposition party, the affected group), it is explained that these groups somehow also belong to the conspiracy, which grows over time. When, as director of the Arecibo Observatory, I denied that we had made contact with aliens or that we used the "radar" to control the minds of the population, all I could do was confirm the suspicions of the questioner since my denial was expected as part of the conspiracy. On the other hand, if I invited the believers to visit the Observatory and talk freely with those they chose, it was understood that things had been prepared for the event.

Conspiracy theories have the character of a closed belief system such that all evidence against the belief becomes proof in favor. Even if there is no evidence of a conspiracy, it can be taken as support since it is supposed to be a well-kept secret. Eventually, the need to include an increasing number of people and institutions in the conspiracy reaches a point where the whole thing collapses.

Recently, a modified version can be identified, not even proposing a theory or searching for evidence. Instead, just by insinuations and repeated formulations, the idea of a conspiracy is presented, such as "Obama born in Africa," "rigged elections," or "Antifa actions." No evidence is needed. (It strikes me as curious that fascists blame Antifa for actions.) As noted by Russell Muirhead and Nancy Rosenblum[7]: *The new conspiracism sets a low bar: if one cannot be certain that a belief is entirely false, with the emphasis on "entirely" then it might be true — and that's true enough. This is the logic behind "even if it's not totally true, there's something there".*

Furthermore, we all have well-studied cognitive biases hard-wired in our brains, which makes us predisposed to conspiratorial thinking. The already mentioned confirmation bias is potent. Then there is the search for *intentionality* in events, even if they are a conjunction of natural or social random events ("it can't be a coincidence"). And there is *proportionality,* our intuitive idea that events are proportional to their causes so that something with great impact must have a big cause (without thinking that a microscopic virus can kill you). So how could just a lone gunman have assassinated President John F. Kennedy?

A current example is "QAnon." supposedly somehow connected to the "deep state" and throwing around hints of impending doom or a new awakening. The editor of The Atlantic, Adrienne LaFrance, writes[8]: *QAnon is emblematic of modern America's susceptibility to conspiracy theories and its enthusiasm for them. But it is also already much more than a loose collection of conspiracy-minded chatroom inhabitants. It is a movement united in mass rejection of reason, objectivity, and other Enlightenment values. And we are likely closer to the beginning of its story than the end. The group harness paranoia to fervent hope in a deep sense of belonging. The way it breathes life into an ancient preoccupation with end-times is also radically new. To look at QAnon is to see not just a conspiracy theory but the birth of a new religion.*

Brian Keeley[9] summarizes: *The rejection of conspiratorial thinking is not simply based on the belief that conspiracy theories are false as a matter of fact. The source of the problem goes much deeper. The world as we understand it today is made up of an extremely large number of interacting agents, each with its own imperfect view of the world and its own set of goals. Such a system cannot be controlled because there are simply too many agents to be handled by any small controlling group. There are too many independent degrees of freedom. This is true of the economy, of the political electorate, and of the very social, fact-gathering institutions upon which conspiracy theorists cast doubt. [...] Governmental agencies, even those as regulated and controlled as the military and intelligence agencies, are plagued with leaks and rumors. To propose that an explosive secret could be closeted for any length of time simply reveals a lack of understanding of the nature of modern bureaucracies. Like the world itself, they are made up of too many people with too many different agendas to be easily controlled.*

Denialism, a variation on the theme of conspiracy, challenges documented historical or current events. It is a toxic brew. Denial of the Shoah, the Armenian genocide, the American genocide of the original Americans or denial of established science such as the HIV-AIDS nexus, biological evolution, climate change, or even a spherical Earth in favor of a fictional story promulgated by a group for whom the facts are displeasing; for whatever reasons.

Sometimes it is funny, as a recent case of one who took a level in an airplane and seeing the bubble always at the center (I assume he did not look during takeoff and landing), "demonstrated" that the Earth is flat. Tragic is the story of the US rocket enthusiast "Mad" Mike Hughes, who attempted to check for himself whether the Earth is flat or a globe; his quest ended when his rocket failed, and he lost his life in February 2020.

Other times, it is not funny. The denial of the relationship between HIV and AIDS had serious consequences (in South Africa).[10] It was suggested that the pharmaceutical industry conspired with thousands of doctors and scientists worldwide to invent the HIV-AIDS connection. There is also the story that HIV was created in secret US or Russian military laboratories (there are) to infect populations (in Africa, Brazil, or Haiti) with genocidal racial aims. (Renewed now with SARS-CoV-2.) The virus was supposedly injected along with vaccination programs (and nobody noticed anything). Another version suggests that it was conspiracy by the Christian right to suppress sexuality and promiscuity. Finally, there is the conspiracy for profit, the sale of poisonous and ineffective antiretroviral drugs by pharmaceutical companies (without observing that these drugs have improved the situation in the countries in which they have been used). Unfortunately, the group of deniers was quite influential and caused the former president of South Africa, Thabo Mbeki, to pay attention to them and reject the distribution of antiretrovirals. This policy cost the lives of tens of thousands of people. Denial of a fact does not make it go away. Unfortunately, pharmaceutical companies often put money above health, and some who manufacture homeopathic medicines could be considered a conspiracy.

Those who proclaim that the Shoah never happened and that it is a fabrication maintained by Jews and their friends have to include as part of the conspiracy a large number of scholars of different persuasions in different countries that have documented and analyzed the tragedy. They also have to assume that all the captured German documents of the Third Reich are forgeries. Finally, they have to include the allied forces that photographed and liberated the extermination camps, the Nazis who confessed, and ultimately so many, that the denial falls under its own weight. But it is difficult to get someone out of a denial bubble.

Denialism often resorts to rhetorical tactics to suggest the existence of a legitimate debate among experts when there is none to cast doubt and to *ad hominem* attack if needed. Denialism contrasts with skepticism. The first takes an *a priori* motivated dogmatic stance, based on some ideological or religious position regardless of the facts and seeks to confirm their positions, while the skeptic scrutinizes arguments and data to reach a conclusion later, only motivated by a search for truth.

It is surprising to read the following: *From my readings, discussions with knowledgeable scientists close to the story, I simply conclude, as does Karry Mullis, the Nobel Laureate who wrote a foreword to Duesberg's classical work that there is no evidence that "HIV causes AIDS." I have no special expertise. I simply seek the evidence for scientific claims, especially when they have dire consequences for the science itself and the treatment.*

It is surprising because they are the words of Lynn Margulis mentioned above, who admitted a lack of expertise. Furthermore, despite his Nobel Prize, Karry Mullis has written a book[11] full of strange musings (to be polite) about aliens, astrology, and HIV not causing AIDS, and although entertaining it contains a lot of fancy conjectures.

For those who (for understandable reasons) do not trust politicians, corporate CEOs, and scientists paid for by "big interests," the conspiratorial hypotheses are plausible and supported by the many historical corruption cases. But that does not mean that everyone is corrupt and that the conditions for a broad conspiracy always exist. Of greater importance, it does not mean that there is no objective way to decide when there are good ways to check the truth of the matter.

Capitalism, for example, is not a conspiracy of the rich to dominate the world and exploit the majority, but once established as an economic system, does precisely that, if allowed to run its course without limits. But no group conspired.

It is worth asking if it is not a criminal act to deny that the cause of AIDS is HIV despite all the evidence or to state without proof that a vaccine is hazardous or to deny that smoking causes cancer. Many who believe it suffer and die. Who are the intellectual authors of these deaths?

As for Peter Duesberg, the hero of HIV-AIDS denialist, Seth Kalichman writes[12]: *Peter Duesberg's legacies will be that he both discovered the first cancer causing gene and that he brought a sort of legitimacy to a band of sad denialists and wacky pseudoscientists. How one man could be the source of so many lives saved and so many lives lost is the greatest paradox and human tragedy in this whole contorted affair.* Think back about the responsibility of scientists.

Those who deny global warming or its causes are associated and supported by the powerful energy industry to sow doubt about the scientific consensus. Here is an example from a leaked memo (written in 2002) from Frank Lunz, a Republican political consultant[13]: *The scientific debate remains open. Voters believe that there is no consensus about global warming within the scientific community. Should the public come to believe that the scientific issues are settled, their views about global warming will change accordingly. Therefore, you need to continue to make the lack of scientific certainty a primary issue in the debate... The scientific debate is closing [against us] but not yet closed. There is still a window of opportunity to challenge the science.* How wicked can you be?

For a long time, the tobacco industry denied evidence of a causal link between smoking and cancer.[14] Again, the same tactic was used as spelled out in an infamous memo from Brown & Williamson, a then-subsidiary of British American Tobacco: *Doubt is our product since it is the best means of competing with the "body of fact" that exists in the mind of the public. It is also the means of establishing a controversy.* The memo was written in 1969 and is an example of an effective weapon often used against the truth. If direct denial is not practical, then sowing doubt is a choice, to question the validity of the science, insist that "more studies are needed," so that a few more billions can be obtained.

Industry denied that Chlorofluorocarbons (CFC) caused the destruction of stratospheric ozone. Religious fundamentalist groups deny the validity of evolutionary biology. Some in government deny that we cause climate change, putting us in harm's way. People confuse familiarity with the truth, as established by research in cognitive science. This means that an oft-repeated lie mutates into perceived truth. This is put to good use by demagogues, publicists, and politicians.

Conspiracy theories and denialism are supported by *data mining* (cherry-picking) in search of those that are not congruent with the accepted view or are sources of controversy (which always can be found). Thus, a false dichotomy is established (if the accepted theory cannot explain something, it is wrong, and my alternative is correct). It also alters the burden of proof, demanding proof that what they hold is not valid. I will take a short detour to illuminate one of the most dreadful conspiracy ideas, one to which some still hold on.

13.1 The Elders of Zion

The worst illiterate is the political illiterate. He doesn't hear, doesn't speak, nor participates in political events. He doesn't know that the cost of living, the price of the bean, fish, flour, rent, shoes, and medicine all depend on political decisions. The political illiterate is so stupid that he is proud and swells his chest, saying he hates politics. The imbecile doesn't know that, from his political ignorance is born the prostitute, the abandoned child, and the worst thieves of all, the bad politician, corrupted and flunky of the national and multinational companies.

— Berthold Brecht[a]

Today I want to be a prophet once more: Should the international Jewry of finance (Finanzjudentum) succeed, both within and beyond Europe, in plunging mankind into yet another world war, then the result will not be Bolshevization of the earth and the victory of Jewry, but the annihilation (Vernichtung) of the Jewish race in Europe.

— Adolf Hitler[15]

By all these means we shall wear down the goyim they will be compelled to offer us international power of a nature that by its position will enable us without any violence gradually to absorb all the state forces of the world and to form a Super-Government. In place of the rulers of today we shall set up a bogey which will be called the Super-Government administration.

— Protocol 5E[16]

I will digress to show how believing in some somber conspiracy can lead to tragedy under certain circumstances, but also because much of this is slowly evaporating from collective memory. Consider the conspiracy by the "Elders of Zion," laid out in a pamphlet: *The Protocols of the Learned Elders of Zion*, first published in St. Petersburgh, Russia in 1903, and later circulated throughout Europe, becoming an influential antisemitic "document." (Fake news writ large.) It will show you why I stated above that you do *not* have a right to believe whatever you wish.

After assuming it on January 30, 1933 (in a proper democratic process), Adolf Hitler quickly consolidated power, eliminating any opposition, replacing those who disagreed with him, and having others murdered. His regime began with the systematic and increasing persecution of German Jews, who represented less than 1% of the population, along with the persecution of all "enemies of Nazism." The infamous Nuremberg laws of 1935, *Gesetz zum Schutze des deutschen*

---

[a] Attributed to Berthold Brecht but disputed. But whoever said this was on the money.

Blutes und der deutschen Ehre (for the protection of German blood and honor), removed the Jew's "citizens" rights, and a growing wave of verbal and physical violence culminated in the "Kristallnacht" (the night of broken glass) on November 9, 1938.

It marked the beginning of Germany's descend along an ugly slippery slope. During that orgy of anti-Semitic violence, more than a thousand synagogues burned down, thousands of businesses were destroyed, hundreds of Jews were murdered, and others were taken to concentration camps (which were *not yet* extermination camps), with few raising a voice of protest. I hope (and if you must pray, do so) that the current US madness can be contained, but there are no warranties, and I am not exaggerating.

Gerald Holton (born 1922) physicist and historian of science, comments about the Nazi regime: *The readiness with which large numbers of physicians, jurists, scientists, and other academics lent themselves to the abominations committed under the last of these shows that scientific literacy by itself provides no immunization; it also attests to the pliability of even so-called intellectuals when there is a cultural upheaval in which politics and parascience join.*

Related thoughts come from journalist Adam Hochschild[17]: *A more somber lesson offered by the events of 1917–1920 is that when powerful social tensions roil the country and hysteria fills the air, rights and values we take for granted can easily erode: the freedom to publish and speak, protection from vigilante justice, even confidence that election results will be honored.* Sounds familiar?

Saul Friedlander[18] recounts the following episode repeated in many towns during Kristallnacht: *In Wittlich, a small town in the Moselle Valley in the western part of Germany, as in most places, the synagogue was destroyed first: the intricate lead crystal window above the door crashed into the street and pieces of furniture came flying through doors and windows. A shouting SA man climbed to the roof, waving the rolls of the Torah: "wipe your asses with it Jews," he screamed while he hurled them like bands of confetti on Karnival. Jewish businesses were vandalized, Jewish men beaten up and taken away: Herr Marks, who owned the butcher shop down the street, was one of the half dozen Jewish men already on the truck... The SA men were laughing at Frau Marks, who stood in front of her smashed plate glass window with both hands raised in bewildered despair. "Why are you people doing this to us? She wailed at the circle of silent faces in the windows, her lifelong neighbors. What have we ever done to you?"*

The reason why they did this is complex and multifaceted. Still, it is the outcome of the long and sad European history of anti-Semitism promoted by attacks from the pulpit and religious literature that perpetuated the fiction of the Jew as malefic, inferior, dirty, and guilty (per *secula seculorum*) of the condemnation and death of Jesus. As the much-cited Matthew 27:25 says about the condemnation of Jesus: *And all the people answered: Let their blood fall on us and on our children.* (But those who crucified were Romans). It is *one of those phrases which have been responsible for oceans of human blood and a ceaseless stream of misery and desolation,* writes John Crossan.19 He states: *although this myth is understandable within the context of a (Jewish) sect struggling to survive, its repetition has become "the longest lie" and should be accepted for the integrity of Christianity itself.*

St. John Chrysostom (~349=407 EC), "mouth of gold," was one of the important fathers of the Church (and from his writings a vicious malcontent), bishop of Constantinople, and "Doctor of the Church," a rare title given to "Christian theologians of outstanding merit and recognized holiness" and fanatical anti-Semite. He had this to say in his Homilies Against the Jews (*Adversus Judaeos.* delivered in Antioch in 387): *How dare Christians have the slightest doings with Jews, those most miserable of all men! They are lustful, rapacious, greedy, perfidious bandits, pests of the universe. Indeed, an entire day would not suffice to tell of all their rapine, their avarice, their deception of the poor, their thievery, and their huckstering. Are they not inveterate murderers, destroyers, men possessed by the devil? Jews are impure and impious, and their synagogue is a house of prostitution, a lair of beasts, a place of shame and ridicule, the domicile of the devil, as is also the soul of the Jew. As a matter of fact, Jews worship the devil: their rites are criminal and unchaste; their religion a disease; their synagogue an assembly of crooks, a den of thieves, a cavern of devils, an abyss of perdition! Why are the Jews degenerate? Because of their hateful assassination of Christ. This supreme crime lies at the root of their degradation and woes. The rejection and the dispersion of the Jews was the work of God and because of His absolute abandonment of the Jews. Thus, the Jew will live under the yoke of slavery without end. God hates the Jews, and on Judgement Day He will say to those who sympathize with them: "Depart from me, for you have had doings with My murderers!" Flee, then, from their assemblies, fly from their houses, and, far from venerating the synagogue, hold it in hatred and aversion.*

As the Crusades, organized to recover the Holy Land from Islamic rule, transited through Jewish communities in Europe, they killed with enthusiasm (many Jews committed suicide to avoid the torturous agony). Jews were expelled from one place and then another, and finally from Spain and Portugal at the end of the 15th century. They emigrated to new lands, but they were always marked, rejected, and ended up in ghettos, the first in Venice at the beginning of the 16th century.

The gray eminence of the Shoah was anti-Semitism, molded into a biological version — it was in the blood as is clear from the Nuremberg laws — so that the solution to the "problem" was obvious. The millenarian Christian anti-Semitism, an integral part of the doctrine and liturgy of the Church, was the incentive to extermination.

Racial anti-Semitism was not a Nazi invention. Already in Spain, by the middle of the 15th century, "pure blood" statutes were established in Toledo and other jurisdictions to discriminate against "converts." Baptism was not enough to wash away sins. It was required (of the applicant who wished to enter the institutions that followed the statute) to demonstrate descent from parents who could also prove to be offspring of an "old Christian" (Christian who had no known Jewish or Muslim ancestry), a procedure like the requirements to determine the Jewish past of a person in the Third Reich.

Among this torrent of anti-Semitic literature, I highlight Pope Pius IX's encyclical of 1873 "Etsi Multa" where you can read among other things[20]: ... *When he compares them with the nature, purpose, and amplitude of the conflict waged nearly everywhere against the Church, he cannot doubt but that the present calamity must be attributed to their deceits and machinations for the most part. For from these the synagogue of Satan is formed which draws up its forces, advances its standards, and joins battle against the Church of Christ.*

Our Predecessors, as watchers in Israel, denounced these forces from the very beginnings to rulers and nations. Against them they have struck out again and again with their condemnations. We Ourselves have not been deficient in Our duty. Would that the Pastors of the Church had more loyalty from those who could have averted such a pernicious plague! But, creeping through sinuous openings, never stinting in toil, deceiving many by clever fraud, it has reached such an outcome that it has burst forth from its hiding places and boasts itself lord and master.

The Protestants were even more toxic. The essay by Martin Luther (1483–1546): *The Jews and Their Lies* written in 1543 is a virulent attack of which a few excerpts will suffice to set the tone: *These Jews are very desperate, evil, poisonous and diabolical beings to the core, and in these fourteen hundred years have been our misfortune, plague and misfortune, and remain so ... They are poisonous, harsh, vengeful, treacherous snakes, murderers and sons of the daemon, they bite and poison secretly, not being able to do this openly [...] burn their synagogues or schools and bury and cover with earth everything that cannot be burned ... I advise that their houses be razed and destroyed ... I advise that all their prayer books ... in which such idolatries, lies, curses, and blasphemies are taught, be taken away from them, ... and that your rabbis are forbidden to teach, under penalty of loss of life or limbs ... that the safe-conducts on the roads be completely abolished for the Jews ... and that all their treasures of silver and gold be taken from them .....*

Towards the end of 1938, a brochure was published by Martin Sasse (1934–1942), *Landesbischof der Thüringischen Evangelisch-Lutherischen Landeskirche* (Regional Bishop of the Thuringian Evangelical Lutheran Church) in which he pointed at Martin Luther as the intellectual author of the *Reichskristallnacht* of 1938, rejoicing that the burning of synagogues occurred on Luther's birthday, on November 10. He also wrote the following on November 24, 1938: *No German in the Christian faith can, without ignoring the good and clean cause of the freedom struggle of the German nation against the Jewish anti-Christian world Bolshevism, lament the state measures against the Jews in the Reich, especially the confiscation of Jewish assets. And to the representatives of the Church and Christianity abroad, we must point out the serious consideration of the fact that the path to Jewish world domination always leads over horrible corpse fields.*[21] It is revolting.

Hitler and his followers — Göring (1893–1946 suicide), Goebbels (1897–1945 suicide), Heydrich (1904–1942 killed), Himmler (1900–1945 suicide), Daluege (1897–1946 executed), Kaltenbrunner (1903–1946 executed), Rosenberg (1893–1946 executed), Jodl (1890–1946 executed) and the "Volk" — were good disciples. We would like to forget them, but we must remember them. We should declare January 30 a World Day of Remembrance for the 60 million deaths that resulted from the seizure of power by Hitler. We could also declare April 9 a day of remembrance, corresponding to King Leopold II of Belgium's birthday, and let us not forget December 18, Joseph Stalin's birthday. We can add Japan's emperor Hirohito, Uganda's Idi Amin, China's Mao Zedong, and …. Perhaps it will be difficult. There are too many candidates and only so many days, but perhaps an international jury could select the dirtiest hundred and build a gloomy wall with their names right at the National Mall in DC. ("The wall of shame.") I imagine they will argue for years and then give up.

Attacks on Jews multiplied on the secular plane in an endless series of pamphlets, magazines, and newspapers. Among them is the publication of the Protocols,[b,22] whose objective was to justify the pogroms suffered by the Jews and represents one of the most disreputable anti-Semitic documents of the time. The protocols were supposedly a transcript of meetings of the "Wise Men of Zion," which detail the plans of a Jewish conspiracy, aiming at world power, a recurring theme in Nazi oratory, still believed by some. The protocols were translated into many languages and widely published. (Henry Ford paid for the printing of half a million copies.) Many, beginning with Hitler, believed this fiction, as some in the present believe in other nefarious conspiracies.

The protocols were part of the foundation of the Shoah, beginning its criminal trajectory with the assassination of the illustrious German Foreign Affairs Minister Walther Rathenau (1867–1922), supposedly one

---

[b] It has been shown that they were a forgery produced in Russia by the Okhrana (the secret police of the Czar). Parts were plagiarized from a novel ("Biarritz", 1868) by the German Hermann Goedsche (1815–1878) (who recounts a Jewish conspiracy to dominate the world). In turn, the novel is plagiarized from the "Dialogue in Hell between Machiavelli and Montesquieu" (1864) by the French satirist Maurice Joly (1829–1878).

Figure 13.1.   Cover of a French edition of the Protocols circa 1934.

of the 300 wise men of Zion. The Nazi newspaper, *Völkischer Beobachter*, had written about Rathenau[23]: *How long before we have a Walter I of the dynasty of Abraham, Joseph, Rathenau? The day is coming when the wheel of world history will be put in reverse, to roll over many a corpse, of the great financier and his accomplices.* And so it happened.

In his summary at the trial conducted of those who participated in the murder, the Judge, in words that few understood in their immensity, expressed[24]: *But behind the murderers and assassins rises as the major culprit the irresponsible fanatical anti-Semitism with its hate-distorted face, which with all means of harassing and slander, of that filthy diatribe we have talked about here "The secrets of the wise men of Zion" as an example, to revile the Jews as such without regard of the person and so sows in blurred and immature heads instincts of murder. May the sacrificial death of Rathenau, who was well aware of the dangers he was exposed to when he accepted his post, may the enlightenment which these proceedings have exposed about the consequences of unscrupulous incitement, may lastly, the letter from the venerable mother[c] of the noble victim which moves all hearts that have not turned to stone, serve to clean the tainted air in Germany and that this barbarization of the mores in this critically ill Germany lead to its recovery.*

The hearts turned to stone.

Unlike the historical atrocities that we could describe as "artisan," the nuclear age, and advanced killing machines, offers indiscriminate and impersonal death, as do drones and B52 bombers. This distinguishes it from the Shoah, But in this case, although industrialized as a factory of death, it was not impersonal. Thousands of Germans and others of similar ilk, who were neither Germans nor Nazis, participated with enthusiasm[25] in acts that cannot be described for lack of adequate words to do so. (Notwithstanding the many who later claimed: *davon haben wir nichts gewusst!* — we knew nothing about that![26])

It was not just the work of a few deranged psychopaths.[27] The propagandists believed their own propaganda, the revolting Heinrich Himmler would say[d]: *We have the moral right, we had the duty to do this for our people, to kill these people who wanted to kill us.*

As Hannah Arendt observes concerning the actors of evil[28]: *The trouble with Eichmann was precisely that so many were like him, and that the many were neither perverted nor sadistic, that they were, and still are, terribly and terrifyingly normal. From the viewpoint of our legal institutions and of our moral standards of judgment, this normality was much more terrifying than all the atrocities put together.*

---

[c] Letter of Walther's mother Mathilde Rathenau written on July 2, 1922, to Gertrud Techow, mother of two conspirators Ernst Werner Techow and Hans Gerd Techow who were sentenced to long prison terms: *In unnamed pain, I reach out to you, the poorest of all women. Tell your son that in the name and spirit of the murdered man I forgive him as God might forgive him if he makes a full open confession before earthly justice and repents before the divine. If he had known my son, the noblest man to walk this earth, he would have directed the weapon of murder at himself rather than at him. May these words give peace to your soul.* In: Karl Bramer (1922). Das Politische Ergebnis des Rathenauprozesses. Verlag für Sozialwissenschaft (Berlin).

[d] Himmler, secret speech to SS officers on October 4, 1943 in Poznan, Poland. http://www. holocaust-history.org/himmler-poznan/speech-text.shtml.

The German Lutheran pastor Martin Niemöller (1892–1984) initially supported the Nazis but then realized what it meant. His poem of 1946 is memorable (there exist various versions):

First, they came for the Communists, and I did not speak out — Because I was not a Communist.

Then they came for the Trade Unionists, and I did not speak out — Because I was not a Trade Unionist.

Then they came for the Jews, and I did not speak out — Because I was not a Jew.

Then they came for me — and there was no one left to speak for me.

As Raul Hilberg[29] points out, the European Jewish community did not have a central organization; it did not have a tradition of armed struggle or resistance (against all the myths of wanting to dominate the world that was adjudicated to them). As a result, they were unprepared to face the oppressors. He notes: *In marked contrast to German propaganda, the documentary evidence of Jewish resistance, overt or submerged, is very slight. On a European wide scale the Jews had no resistance organization, no blueprint for armed action, no plan even for psychological warfare. They were completely unprepared.*

There were a few cases of resistance, all useless. They tried to appeal to reason, to argue against deportations and lost arguments because the other side did not operate with reason, or perhaps better said, acted with an unhinged reason without the most elementary ethical elements.

The Jewish Councils (Judenrat) and the rabbis were not prepared for the situation, and in many cases, with the idea of appeasing to avoid a greater evil, they abided. The head of the Jewish council of Warsaw, Adam Czerniaków (1880–1942), interceded all the time with the Nazi authorities, trying to appease the devil, a lost cause. In the end, the 450,000 Jews of Warsaw were transported to Auschwitz and killed. On July 23, 1942, Czerniaków swallowed a cyanide capsule. He left a note that said[30]: *I am powerless, my heart trembles in sorrow and compassion. I can no longer bear all this. My act will show everyone the right thing to do.*

It is a fact that Pope Pius XII[31] (Eugenio Pacelli 1876–1958) did not say anything during the entire years of the Shoah and can be cataloged as a collaborator, like the French Marshal Henri Pétain (1856–1951) or Vidkun Quisling (1887–1945 — shot) in Norway. But it is difficult to know what was going through the mind of this "tremendous and pathetic figure."[32] Research of the newly opened (and until recent secret) Vatican archives show that Pius knew as early as September of 1942 about the mass murder of Jews from the Warsaw Ghetto.[33] It is also a fact that the Roman Curia, along with the Argentine president Juan Domingo Perón (1895–1974) and Red Cross workers, helped Nazi criminals escape in the so-called rat lines through which transited figures as perverse and sordid as Eichmann and Mengele.[34] The papacy has not shown itself to be at the moral height that it pretends for others, a height that could be approached by admitting and clarifying its role.

We have heard neo-Nazis marching with a Nazi flag (concealed behind the alt-right name) demonstrate their ignorance recently in a march in Charlottesville, Virginia, shouting "Jews out.". They will never learn Pete.[e] During the protest on August 12, 2017. Heather Hyer, a 32-year-old civil rights activist, was killed, and several others were injured. POTUS 45th tweeted: *Condolences to the young woman killed today, and best regards to all of those injured in Charlottesville, Virginia. So sad!* No interest in finding out her name? "Best regards"? Indeed: So sad! and so glad this idiot is gone.

Instead of condemning these acts, which remind us of early Nazi Germany, Trump responded with the unforgettable *fine people on both sides*, Fine neo-Nazi is an oxymoron. Trump might be gone, but they are still there. (And let us not forget that although he lost, still 70 million voted for him.) In a recent book, Wall Street Journal reporter Michael Bender tells us that Trump said to his chief of staff John Kelly[35]: *Well, Hitler did a lot of good things*. Need more?

I wonder how this is possible. Bronner[36] points out that it is the product of ignorance, which feeds all prejudices. You do not need to know anything to have them. *therein lies the broader existential attraction of a prejudice like antisemitism: it may sometimes appeal to intellectuals, but it always indulges the idiot.* Remember that I wrote "Idiotization" as

---

[e] In reference to the famous song: "Where Have All the Flowers Gone?", written by Pete Seeger in 1955.

part of the problems we face. And when such idiots reach high levels of government, shit hits the fan.

In the years after the end of the Second World War, a group of German Protestant and Catholic Christians undertook the task of preventing those convicted of war crimes from serving their sentences (many to death). In letters, pamphlets, and public speeches, they argued that many defendants had their confessions extracted under intense interrogation (!) and that the prisoners suffered from incarceration (in the Landsberg prison).

The allies were accused of being unfair to prisoners who had only obeyed orders and had not been as bad as claimed. The German princess Helene of Isenburg (1900–1974), first chairperson of the *Stille Hilfe*,[37] organization founded in 1951 (*Die Stille Hilfe für Kriegsgefangene und Internierte*; "Silent assistance for prisoners of war and interned persons"), wrote a letter to Pope Pius XII: *To take men to the gallows whose sentences were largely wrong sentences, five years after the end of the war, is against all rights. We cannot understand that a victorious people ignore the command of humanity, to spread hatred and revenge.* (An example of a *circumstantial ad hominem* argument.) One wonders where she lived during the war, but then, she was a "princess." (A similar letter is unknown protesting the horrible killings perpetrated by the accused.) We are, as remarked by Schumacher, a cesspool of irrationality.

Finally, let me answer a question that might have gone through your mind as you have read me hold forth about the Shoah. Isn't it time we turn the page? This happened 80 years ago in a different world; why persist? Well, because we do not seem to have learned anything from this monstrous event, because some think in the same way ("neo-Nazis") and could lead us to another disaster. It is astonishing and troublesome that this happens even in present day Germany.

Long forgotten French sociologist Maurice Halbwachs (1877–1945; at Buchenwald concentration camp[38]) wrote in his groundbreaking "The Social Frameworks of Memory"[39]: *How can currents of collective thought whose impetus lies in the past be recreated, when we can grasp only the present?* It is an excellent reason to keep at it and see if we can understand and heal what is wrong with us.

Chapter 14

The Tragedy of Our Time

There is a cult of ignorance in the United States, and there always has been. The strain of anti-intellectualism has been a constant thread winding its way through our political and cultural life, nurtured by the false notion that democracy means that "my ignorance is just as good as your knowledge.

— Isaac Asimov[1]

I do admit to worrying late at night about that matter of time: obviously we still have to get rid of modern warfare and quickly or else we will end up, with luck, throwing spears and stones at each other. We could, without luck, run out of time in what is left of this century and then by mistake, finish the whole game off by upheaving the table ending life for everything, except the bacteria maybe, — with enough radiation even them.

— Lewis Thomas[2]

You already know enough. So, do I. It is not knowledge we lack what is missing is the courage to understand what we know and to draw conclusions.

— Sven Lindqvist[3]

You might think that you already know what I mean by the tragedy of our time: what we see every day on TV, live and with ever better resolution, or read in newspaper reports, and increasingly in cyberspace. And it is true, we seem deranged, the world is full of tragedies, big and small, personal, and communal, public and private, rivers of fresh blood on scorched earth.

Some are caused by nature, and there is little we can do to avoid an earthquake, a hurricane, a tornado, let alone an improbable but not impossible impact by an asteroid or the eruption of a supervolcano.

Others are increasingly caused by our actions, whether by religious zealots, greedy corporate CEO's, ambitious politicians, fanatics of whatever, or by a group that has control of some government and decides to kill others to keep it, always well justified with evidence that turns out to be, in many cases, a lie, or as has become fashionable: "alternative facts" or if needed a "false flag operation."

It is an old trick. For example, Hitler simulated a Polish attack on a German radio station (with German prisoners attired in polish uniforms who were then killed and photographed) to justify his invasion of Poland on September 1, 1939, which unleashed the carnage of the Second War. Likewise, the US used a confusing incident between North Vietnamese torpedo boats (probably non-existent) and the destroyer USS Maddox in the Gulf of Tonkin to get into the tragedy of Vietnam.[4] (Remember the Main?).

And do not dismiss the great potential tragedy caused by mistake. Accidents happen and are not predictable; that is why they are called "accidents." The thousands of nuclear weapons on active alert are prepared to repel any attack, real or perceived. The president of the US (Oh God!) has 12 minutes to press the button after a missile attack alert, and there are cases of false alarms that luckily did not lead to more. Behind this is the false narrative that some countries might want to attack, but not even the ex-KGB agent or the chunky guy from North Korea would want this, since there is nothing to gain and a lot to lose as already discussed.

You might have heard of the accident of a B52 bomber in Goldsboro, North Carolina on January 24, 1961. It carried two 24 megaton nuclear bombs (about a thousand times more powerful than Hiroshima and Nagasaki). It was reported that four of the five safety devices had been activated for one of the bombs. The bomb did not explode because the fifth switch in the aircraft cockpit was not activated. Dozens of accidents involving nuclear weapons are known,[5] yet we hear vehement opposition to nuclear energy but none against these weapons, which could be nearer to where you live than what you imagine.

The case of Stanislav Yevgrafovich Petrov (1939–2017), retired Lieutenant-Colonel of the Soviet Air Defense Forces, commander of a radar monitoring station near Moscow, is not so well known. It happened on the night of September 26, 1983, 3 weeks after the Soviet military had shot down Korean Air Lines Flight 007. In response to an alarm indicating an apparent attack by a US missile (and then further alarms indicating four additional missiles detected by the radars, he nevertheless decided that he was not going to give the alert to initiate a possible counterattack. He later told what he thought: *People do not start a nuclear war with only five missiles.* He is considered by some as the man who saved the world. But as the saying goes: *errare humanum est*, and we are human. The saying attributed to Seneca continues with: *sed perseverare diabolicum.*

That is how we are, psychologists tell us, it's part of human nature that makes us very little human in the sense that the term is usually used, as when we say that torture is inhuman when we humans do it. The urge to kill the insignificant other is much deeper than our skin. Of course, we are the others for the others. It is the basis of an ethic of exclusion when we need one of inclusion. Robin Fox says[6]: *You are constructed to survive in a world without philosophers, popes, or policemen, but with an often hostile environment, including groups like yours but hostile to it, and your own kin as your only resources.* That is the origin of our tendency to nepotism and xenophobia driven by fear.

But no, I do not mean the above, quite tragic indeed, and I also do not mean the 2016 presidential election in the US (and in an increasing number of nations), which many consider a tragedy of our time.

The tragedy of our time is that we do not have enough. Enough to rid ourselves of the slow-moving monster we have created, enough to undo all the harm we have done. Bill McKibben has this to say[7]: *By 2075 the world will be powered by solar panels and windmills — free energy is a hard business proposition to beat. But on current trajectories, they'll light up a busted planet. The decisions we make in 2075 won't matter; indeed, the decisions we make in 2025 will matter much less than the ones we make in the next few years. The leverage is now.*

On a personal level, time is scarce enough to live in this complex modern connected world. There is so much to know, so much to see and appreciate, that one life is not enough, however long it may be. Think of all the good books you will never read, all the beautiful music you will never hear, all the great movies you will never see, all the amazing landscapes you will never experience. Sure, if you belong to that minority who crave a cocktail party with the rich and famous, then you are already dead, do not worry about it.

Most cannot afford this even if they had the time, most are too tired after being exploited to even try, most will sit there and amuse themselves to death, which is one good book you should read.[8] When Neil Postman wrote this famous book, he was referring to television. With the internet, the idiotization of people is even more powerful and dangerous, and the border between truth and falsehood evaporates in their minds. One who lived in a Middle Ages village did not have this problem, perhaps a book (if he could read), a horseback trip to the nearest town, and maybe live music from a passing band or in the local church. Blissful ignorance, few options, ample time even if life was shorter.

We need, and many want, knowledge, and that is not the same as information. But, unfortunately, most of the human-animal does not even know what it can learn, happy in their ignorance, while others *long for immortality who don't know what to do with themselves on a rainy Sunday afternoon.* as Susan Ertz points out.[9]

The substitution of knowledge by "pseudo-knowledge" is alarming, as if it were a virus transmitted at the speed of light through the internet and spread by institutions and people of dubious reputation and against which we have not found an effective vaccine. They seek prominence and profit and confuse an unsuspecting and ill-prepared public. We see them every day on screens: from astrologers and seers to the "gurus" of alternative therapies, to "experts" who too often do not know what they are talking about and babble inanities. This pseudo-knowledge is supported by various means of mass communication that disseminate light and frivolous culture that has contaminated people's minds.

Recently, two modalities have emerged (or resurfaced) that are endangering the idea of an open society: "alternative facts" and "fake news," which are now official tools. The mixture of reality and fiction is an old and proven tactic to persuade the public. With meritorious exceptions, the media have lost the distinction of being the "fourth estate," journalists becoming mere "stenographers for the powerful," in the words of Chris Hedges,[10] and instruments of manipulation, regardless of the truth. Ideologies penetrate the media and paint everything in one color. For a citizen of that color, that is the truth. To be fashionable, we could say that we are in the post-knowledge era, which is nothing other than ignorance, which some say is bliss, but it is death.

We must resist the old and the new superstitions, often presented (to top it off) as if they were science. These groundless inventions cause damage, especially in public health, with alternative therapies and false medicines. There is no such thing as "alternative medicine." Something is effective and therefore is medicine, or it is not; there is no alternative.

The problem arises from the information revolution, the full consequences we do not know, which possibly represents a fundamental epistemological change. We are immersed in a network full of valuable information and practically infinite bullshit (I think the latter is winning), and it is easy to get lost.

It was easy in the past. To learn, a book was sought, a good book written by a recognized expert, a scholar on the subject, and the chances of acquiring reliable information and knowledge on the subject in question were high.

Nowadays, many do not read books; they are too long, there is no time, and the alternative is fast, less expensive, and easy, just ask Google. It has become fashionable to tell potential readers how long it will take to read something; a 10-minute read might be acceptable for now; soon, we will just tweet. We do not have time because if we live a 100 years (let's be optimistic), we have a total of 36,500 days that correspond to 876,000 hours, and if we discount a third for sleep, we have 584,000 hours left for everything else. And if we remove another third since most of us work about 8 hours or more per day (or spend hours looking for work), then we have 292,000 hours left for everything else. And if you use 10% of that time to read books, assuming you can read one in 48 hours, you have time to read about 600, which would be great if you choose carefully from the millions that exist, and I hope you don't regret this one. Google can find lots of data (also about you), and Google tells me that: *44% of Americans with a high school degree or less surveyed in early 2019 hadn't read a book in the last 12 months.*

But for each question, there are thousands of sites, and most contain less than the truth or more than the truth. Of course, some sites are more reliable than others; the problem is how to distinguish.

Public discourse is not only characterized by what it includes but even more so by what it excludes. With laudable exceptions, not much is said about disarmament, global injustice, let alone the role of religions, among other things that affect humanity's future and are infinitely more important than the lives of the rich and famous as you have been reading. When discussing a social problem, representatives of different religions participate in contributing their hackneyed dogmatic positions quoted in some obsolete ancient text. These gentlemen do not contribute to the solution of problems, in many cases not because they do not have good intentions, but because our *real* problems, (I say this because there are problems that are not real such as for example, "gay" marriages), are not solved with ready-made phrases and God help us. If it were so easy, we would no longer have them (the problems). Questioning these questionable religious premises is considered disrespectful, although a position based on questionable premises does not deserve respect, per *saecula saeculorum.* The concept of "politically correct" discourse has become popular in many spheres. It is nothing more than instituting hypocrisy, diverting attention from the most critical issues, however uncomfortable they may be. We question aliens, miraculous healings (but not miracles), and indigenous cults. We question the psychics who communicate with the deceased, but not the religious who communicate with God. But what is the difference? As H.L. Mencken put it[11]: *...in the fields of morals and government: their discussion is often so contaminated by pseudo-religious considerations that a rational and realistic dealing with them becomes impossible. It accounts unquestionably for the general feeling that religion itself is a highly complicated and enigmatical thing, with functions so diverse and sinister that plain men had better avoid thinking of them, as they avoid thinking of the Queen's legs and the Kings death.*

Nevertheless, it is quite simple at bottom. There is nothing secret or complex about it, no matter what its professors may allege to the contrary.

Whether it happens to show itself in the artless mumbo-jumbo of a Winnebago Indian or in the elaborately refined and metaphysical rites of a Christian archbishop, its single function is to give man access to the powers which seem to control his destiny, and its single purpose is to induce those powers to be friendly to him. That function and that purpose are common to all religions, ancient or modern, savage or civilized, and they are the only common characters that all of them show. Nothing else is essential.

To know something, we need to justify it, and it is *not* enough to find it through the internet without further groundwork. I also use Wikipedia, but only as the starting point of an inquiry and not as the final point. *Information* can be summarized in two or three pages, but *knowledge* requires much more study, and the misfortune is that many no longer know that distinction nor have the time to find out. It gets even worse since knowledge, important as it is, is not equivalent to wisdom, something we dearly need. University intellectuals, a species which is also undergoing extinction with the increasing mercantilist view of universities, are the few who are paid to do this, and everyone needs to defend scholarship, and it is important for intellectuals to "go public," especially scientists which for reasons I do not understand are often not seen as intellectuals. Some have done so and have been attacked by idiots.

People have been brainwashed since childhood all the way to university (if they get there). Citizens become mere voters choosing by emotion and not by reason, and since emotions are easy to manipulate, we obtain a double whammy for democracy. Or, as the Argentine singer, songwriter, and philosopher Facundo Cabral (1937–2011; murdered in Guatemala)[a] liked to say: *You have to fear the buttheads, who are dangerous because as they are the majority, they even elect presidents.* And you see the results, and not only in the US. Some propose a system with weighted votes, such that the best-educated carry more weight, but, although it sounds elitist, in principle, it is not a bad idea. It would also not be wrong to demand from those who govern a minimum of relevant education, over and above, knowing how to read and write, be a citizen and be over a certain age. But then it never was as they say, "one man, one vote" (which changed after Susan B. Anthony (1820–1906)

---

[a] Cabral was assassinated during a tour in Guatemala City.

(and the nineteenth amendment to the US constitution). "One dollar one vote" would be closer to the truth, with all those PAC's pouring money for a candidate. Nevertheless, democracy is the best way we know in theory to organize a society, as remarked a long time ago by Winston Churchill[12]: *Indeed it has been said that democracy is the worst form of Government except for all those other forms that have been tried from time to time.*

Ignorance of science in a scientific world leads to damaging postures.

It is becoming increasingly urgent to educate so that the individual knows how to navigate this complex world and separate the grain from the chaff. We need a program of studies for future public servants[13] without which they should not be allowed to aspire. After all, you cannot get a driver's license without first passing an examination. ("examination" is a bit of an overstatement.) Information is within a finger's reach; knowing how to handle it, evaluate it, and connect it, is the difficult part, and this you do not learn by yourself.

14.1 Cybernetia

In a world populated by people who believe that through more and more information, paradise is attainable, the computer scientist is king. But I maintain that all of this is a monumental and dangerous waste of human talent and energy. Imagine what might be accomplished if this talent and energy were turned to philosophy, to theology, to the arts, to imaginative literature or to education? Who knows what we could learn from such people — perhaps why there are wars, and hunger, and homelessness and mental illness and anger.

— Neil Postman[14]

You can't pursue any kind of inquiry without a relatively clear framework that's directing your search and helping you choose what's significant and what isn't... If you don't have some sort of a framework for what matters — always, of course, with the proviso that you're willing to question it if it seems to be going in the wrong direction — if you don't have that, exploring the Internet is just picking out the random factoids that don't mean anything... You have to know how to evaluate, interpret, and understand... The person who wins the Nobel Prize is not the person who read the most journal articles and took the most notes on them. It's the person who knew what to look for. And cultivating that capacity to seek what's significant, always willing to question whether you're on the right track — that's what education is going to be about, whether it's using computers and the Internet, or pencil and paper, or books.

— Noam Chomsky[15]

These are dangerous times. Never have so many people had so much access to so much knowledge and yet have been so resistant to learning anything. In the United States and other developed nations, otherwise intelligent people denigrate intellectual achievement and reject the advice of experts. Not only do increasing numbers of lay people lack basic knowledge, they reject fundamental rules of evidence and refuse to learn how to make a logical argument.

— Tom Nichols[16]

Whatever we think about the future of mankind and about the threats to our collective survival depends on what we believe is true. An important protagonist in this is the public discourse: that which is heard, seen, and read by the public. Mainstream media (MSM) is increasingly concentrated in fewer hands, becoming a system of conditioning more effective than that imagined by Aldous Huxley[17] and more insidious than the control imposed by a totalitarian regime because it is not done explicitly.

Caitlin Johnstone summarizes[18]: *Western mass media outlets are propaganda. They are owned and controlled by wealthy people in coordination with the secretive government agencies tasked with preserving the world order upon which the media-owning plutocrats have built their kingdoms, and their purpose is to manipulate the way the mainstream public thinks, acts and votes into alignment with the agendas of the ruling class.* (Eastern media is the same.)

Editorial control will decide what the public learns about the world and can readily manipulate public opinion. This constitutes an inversion of Greek thought which placed taking care of your mind in the first place. But then, there has been "progress," right?

Some understand it, but few take note, although it is not something new. In 1591, Giordano Bruno in De Immenso wrote[19]: *Wisdom and justice began to abandon the world when the learned, organized into sects, started to use their doctrine to make money ... Both religion and philosophy are vanquished by such attitudes; states, kingdoms, and empires are overturned, ruined, and outlawed together with sages, princes, and peoples.*

The World Wide Web is a revolution in the making and a marvelous technology, but like all of them, it has a double-edged sword, and it is too soon to say where it will cut deepest. For those of us who know how to take advantage of this resource, it is excellent. However, criminals and swindlers also use the Internet and have taken more than one to an unhappy end. It can function as a modern incarnation of the great Library at Alexandria or as a cesspool filled with misinformation and outright lies often backed by powerful interests, including politicians, with the public unable to separate truth from fiction. Again, lies and disinformation are of infinite scope, whereas the truth is one, so it is easily overwhelmed. It can also imprison you and transform you into a useful idiot or kill you. This reminds me of science fiction writer Theodore Sturgeon, who in the '50s declared what has come to be known as Sturgeon's law: *90% of everything is crap.* (For the www, I would guess it is 97%.)

It was described as the *Information Superhighway*, but it really is a whole country: "Cybernetia," with highways, cities, towns, roads, and trails, some so muddy that you can easily get stuck. There are no maps or tour guides, and you can get lost on the wrong path without noticing. It is a country of unstoppable growth, of marvels and humbugs, in which there are no immigrants or refugees. Anyone can enter through its gate without the need for a passport or a visa and no TSA lines. I would put a clearly visible sign on this gate: *Protect your mind and privacy before entering.* People spend hours exploring Cybernetia, but they can only scratch the surface, just like those tourists who visit eight countries in 2 weeks, with no time or means to dig deeper. They can also make "friends" or find themselves in a cul-de-sac.

We live increasingly connected, plugged in to music, exploring Cybernetia, glued to cell phones, and absorbing what is fed to us by whatever screen we look at or on the radio as if it were the sun's light that tans us. For the Sun, we resort to lotions that protect us from the dangerous part of its radiation (UV — recall that we almost destroyed that protection provided by ozone). Still, there is no such filter for the information tsunami, and we absorb facts, lies, and bullshit without knowing which is which. We protect our skin but damage our minds, and it is addictive. Just imagine living without your phone for a month. About TV (and more generally anything similar such as YouTube), Pierre Bourdieu[20] wrote: *The political dangers inherent in the ordinary use of television have to do with the fact that images have the peculiar capacity to produce what literary critics call a reality effect. They show things and make people believe in what they show. This power to show is also a power to mobilize. It can give a life to ideas or images, but also to groups. The news, the incidents and accidents of everyday life, can be loaded with political or ethnic significance liable to unleash strong, often negative feelings, such as racism, chauvinism, the fear — hatred of the foreigner or, xenophobia.*

The WWW (one acronym that takes longer to say than World Wide Web) potentially democratizes the globalized world. I say potentially because it may democratize, or it may enslave, transforming the individuals into automatons that see a screen, with content filtered by some corporation, government agency, or special interest groups. When the Internet gets highjacked by the FIRM complex, it will be worse than Goebbels's propaganda machine.

It constitutes a powerful communication tool and an unparalleled means of convocation, allowing voices to be heard and perhaps politicians to take notice, a forum in which all kinds of issues can be presented and debated. It is a place of expression for those who did not have one, a new and dynamic component of the public discourse, with Facebook, YouTube, Twitter and all the rest. It also supports a gigantic and growing market, exemplified by Amazon and eBay. It can convene a march for BLM or an assault on the US capitol. Facebook's inventor and billionaire Mark Zuckerberg wrote[21] *One of the most painful lessons I've learned is that when you connect two billion people, you will see all the beauty and ugliness of humanity.* What is frightening is that such persons have colossal power overall; what they decide can change the world, for good or bad, and determine our behavior. Two edges.

Many convey their anger marching in the virtual streets of Cybernetia, but the effect on those in power is mostly just as virtual. We vent our frustrations on Facebook or Twitter and feel pleased about the "likes," but it is fiction, a generally inconsequential noise, very different in its effect from the great manifestations of the past when marches and strikes (sometimes violent) could result in changes. So, for example, when the economy depended greatly on coal and manual labor, a coal miners' strike could result in changes (but were often violently suppressed). That is behind Karl Marx's and Friedrich Engels's famous 1848 "Workers of the world unite." With modern sources of liquid energy (oil and gas), which flow through computer-managed pipes, things have changed, and coal is no longer what it was 100 years ago. (And we should stop burning it even if some idiot says it is beautiful.)

Do you think that those who are in power care about what you write or what I am writing right now? They are too busy tweeting their own views and scheming their next moves. We could write with white ink on white paper, and it would be the same. Roy Scranton writes[22]: *No matter how many people take to the streets in massive marches or in targeted direct actions, they cannot put their hands on the real flows of power, because they do not help produce it. They only consume.*

We shall see the result of what is currently happening in the US thanks to brutal police behavior towards minorities, particularly African Americans. Protests primarily by young people who soon will replace the old warmongering farts will have some effect and hopefully improve things locally, but this is not enough because of what I have already discussed.

Ominously, Cybernetia is also a threat to our privacy, to our democracy, to our liberty, and enables the establishment of unprecedented state control, "Big Brother" with his artificial intelligence (AI) knows your every move even if your GPS is off. He also knows your tastes and habits (or do you think it is a coincidence that windows are opened that contain ads for something you searched for a few days ago? Indeed, AI is one of those incipient technologies which will have two very sharp edges. One difficulty with AI is that it looks for patterns for correlations in big data, but correlation (as we saw) does not entail causation, and this is where things can go very wrong.

Everything you do on the internet, every place you visit, can be known, so that Big Brother, fed by the likes of Google, Amazon, Facebook, Microsoft, Apple, ATT, and others, can control, and classify you and sell you what you do not need. But more to the point, what is being sold to the advertisers is you: If you are not paying for the product, you are the product.[b]

Shoshana Zuboff[23] calls it surveillance capitalism: *Surveyance capitalism literally claims human experience as free raw material for translation into behavioral data. Although some of these data are applied to product or service improvement, the rest are declared as a proprietary behavioral surplus. Fed into advanced manufacturing processes known as "machine intelligence" and fabricated into prediction products that anticipate what you will do now, soon, and later. Finally, these prediction products are traded in a new kind of marketplace for behavioral predictions that I call behavioral futures markets. Surveillance capitalists have grown immensely wealthy from these trading operations, for many companies are eager to lay bets on our future behavior.*

AI can also decide what it lets you see, what it lets you read. My spam e-mail is full of not-spam. But no matter how intelligent, it will make mistakes and put you in a box where you do not belong, from which it will be difficult to extricate yourself. You will find yourself in deep doo-doo, or even worse, in a dark site being tortured and questioned for something you know nothing about. There is no need to get paranoid, but it can also take over your computer or smartphone, including your camera and microphone. (You can buy a small switchable cover for your device's camera.[c]) They do it in China. Do you want this for our democracy?

Big Brother has his biggest mansion in the US (but also resides in Russia, China, and other countries) for the simple reason that most communication infrastructure (software and hardware) resides in the US or US-controlled territory. We contribute unwittingly to this by providing all sorts of personal information for free, and without our knowledge, our

---

[b] I recommend Netflix's documentary: The Social Dilemma.
[c] Example: https://www.amazon.com/EYSOFT-Webcam-Cover-0-7MM-Thin/dp/B075FCNF4B/ref=sr_1_2?dchild=1&keywords=lens+cover+for+laptop&qid=1595453627&sr=8-2.

computers, our internet-connected security cameras (LOL), and telephones supply a stream of "metadata," that is, data about our communications, allowing government agencies, corporations, and anyone who pays, to classify and characterize you. Zuboff writes: *We are the objects from which raw materials are extracted and expropriated for Google's prediction factories. Predictions about our behavior are Google's products, and they are sold to its actual customers but not to us. We are the means to others' ends.*

Furthermore: *Surveyance capitalism operates through unprecedented asymmetries in knowledge. and the power that accrues to knowledge. Surveillance capitalists know everything **about us** whereas their operations are designed to be unknown **to us**. They accumulate vast domains of new knowledge **from us** but not **for us**. They predict our futures for the sake of others' gain not ours. As long as surveillance capitalism and it's behavioral futures markets are allowed to thrive, ownership of the new means of behavioral modification eclipses ownership of the means of production as the fountainhead of capitalist wealth and power in the twenty first century.*

On rare occasions, there is a warning, someone who reveals what is happening, as Edward Snowden and a few others have, but in a curious reversal of logic, those in power resort to obliterating or discrediting the messenger. Snowden[24] tells us: *Metadata can tell your surveillant the address you slept at last night and what time you got up this morning. It reveals every place you visited during your day and how long you spent there. It shows who you were in touch with and who was in touch with you.* Authoritarian and totalitarian leaders love it. The World Wide Web has lost its innocence.

It is unconstitutional to gather private information about all citizens without a warrant and a good reason. The fourth amendment to the US Constitution states: *The right of the people to be secure in their persons, houses, papers, and effects, against unreasonable searches and seizures, shall not be violated, and no warrants shall issue, but upon probable cause, supported by oath or affirmation, and particularly describing the place to be searched, and the persons or things to be seized.* Furthermore, in the already mentioned 1948 UN Universal Declaration of Human Rights, Article 12 states: *No one shall be subjected to arbitrary interference with his privacy, family, home or correspondence,* (there were no emails then) *nor to attacks upon his and reputation. Everyone has the right to the protection of the law against such interference and attacks.* You do not have to be a lawyer to understand.

Australian journalist Julian Assange, the founder of Wikileaks, has been indicted by the US government (recently exposed to have used false testimony,[25] but it was not news for MSM). Charged with violating the Espionage Act of 1917 (not to be taken lightly), he has been deprived of freedom for years. But the Espionage Act is meant to be used against spies, not against journalists. It is difficult to follow all the legal arguments in these cases (again, ethics and common decency does not count, and Trump had no qualms in pardoning convicted criminals). Difficult to judge what harm was done (really none to "national security" in this case). It is a move to threaten investigative journalism, which relies on leaked information. Misdeeds are classified as secret, and then whoever blows the whistle is accused of breaching security and being a traitor; but what is wrong with us knowing the government's wrongdoings? Now we learn that under the directorship of Mike Pompeo the CIA had plans to kidnap or murder him. The 2021, Nobel Peace Prize was awarded to two worthy journalists: Russian journalist Dmitry Andreyevich Muratov and Filipino-American Maria Angelita Ressa. Assange is suspiciously absent. I wonder. But then, they also gave the prize to Kissinger.

Pentagon Papers whistleblower Daniel Ellsberg wrote[26]: *the current state of whistleblowing prosecutions under the Espionage Act makes a truly fair trial wholly unavailable to an American who has exposed classified wrongdoing.* (In other countries, journalists are routinely assassinated, the worst one is Mexico.)

The "Pentagon Papers," a long-secret document that revealed all the sordid events related to the Vietnam War, were made public by Daniel Ellsberg (Born 1931) and Anthony Russo (1936–2008) in 1971, when Ellsberg became convinced of that war's injustice. They did this, risking careers and freedom. In 1973, they were charged under the Espionage Act of 1917, but soon the charges were dismissed as revelations about government misconduct against Ellsberg surfaced. The Nixon administration even had Ellsberg's psychiatrist's office broken into in search of damaging information. In his memoirs, Ellsberg writes[27]: *Lying to the public, about anything, but above all on the issues of life and death, war and peace, was a serious matter; it wasn't something you could shift responsibility for. I wasn't going to do it anymore. It came to me that the same thing applied to violence. No one else was going to tell me over again that I (or anyone else) had to kill someone, that I had no choice, that I had a right or a duty to do it that someone else had decided for me.*

The powerful tool of state intelligence constitutes a secret means to influence collective behavior, as seems to have happened in the last US presidential election. The US is all upset about foreign interference in its politics, but before being so offended it should stop doing the same or worse to other countries. Practice what you preach. Journalist, author, and lawyer Glenn Greenwald tells us[28]: *Taken in its entirety, the Snowden archive led to an ultimately simple conclusion: the US government had built a system that has as its goal the complete elimination of electronic privacy worldwide. Far from hyperbole, that is the literal, explicitly stated aim of the surveillance state: to collect, store, monitor, and analyze all electronic communication by all people around the globe.* German chancellor Angela Merkel's telephone, was intervened by US intelligence in cahoots with German intelligence (BND). Amazing.

In the past, this dirty work was done by the secret services of many countries through informants and spies. Today cybernetic espionage is stealthy, automatic, and continuous, with the excuse of keeping us safe (from what?). Instead of freedom, we give our lives away to some "cloud" that then shunts this information to increasingly more powerful AI (don't think there is a big room with thousands of people reading billions of emails or tweets) that analyze our activities, and this has perhaps helped thwart some ugly events. However, it also takes us into the realm of "psychopolitics"[29] without knowing for what purpose that information might be used, going from vigilance to active control. But who watches Big Brother? *Understanding the mechanism by which people may have been surveyed, let alone gathering reliable evidence, is a task for experts, well beyond the abilities of average citizens. The issue is also complicated by the proportional nature of the right to privacy: most people agree that states ought to have some surveillance powers, but these must be used properly and against the right people. How this should happen is the million-dollar question.*, writes human rights investigator Clara Usiskin.[30]

And equally important, who owns your data? Your data is you, your tastes, thoughts, friends, and other things about you that you might wish to keep private. Your right to privacy, your right to be let alone, is nothing new, although it is increasingly violated. In an important publication by Louis Brandeis (I went to the University that bears his name) and lawyer Samuel Warren published in 1890 Harvard Law Review, they write[31]: *The principle which protects personal writings and all other personal productions, not against theft and physical appropriation, but against publication in any form, is in reality not the principle of private property, but that of an inviolate personality.*

You might think, as I do, that you have nothing to hide, you have done nothing wrong, which is what Jews believed in Nazi Germany.

Snowden again: *At any point, for all perpetuity, any new administration — any future rogue head of the NSA — could just show up to work and, as easily as flicking a switch, instantly track anybody with a phone or a computer, know who they were, where they were, what they were doing with whom, and what they had ever done in the past.*

Few people know what goes on and what the millions who work for the national security apparatus do since they operate on a "need to know" basis, and their job is to lie, hide, and deceive. It has become an independent beast, and it is not clear who is in control, if anyone or if it is an organization running amok doing what it does because that is what it does.

And if someone gets out of hand and really bothers them, their hounds will find them and come in the early morning hours to get rid of the problem, or perhaps just shoot outright or put some plutonium in tea. Even entering your countries embassy can be deadly, as we have recently witnessed to the dismay of many. As collateral damage, the notion that you live in a surveillance state fosters a submissive population, fearing that any challenge to orthodoxy might entail retribution. Still, we need rebels to guide society into new avenues which might be much better for our journey. The US could start with full pardons for Assange, Snowden, and Manning. (And lesser-known cases.) Hello Biden.

But this is not the only worry about the other sharp side of this technological sword. Cybernetia is also full of "Merchants of Doubt."[32] promoted by corporations or political action groups who use any means (even paying corrupt scientists) to misinform and sow doubt which propagates like the plague. With advancing technologies, you will see and hear experts and politicians saying things they never said, and it will become increasingly harder to distinguish the true from the false.[33] It is a real danger. Action groups (foreign and domestic) can inundate media with false allegations before an election, changing the outcome. It matters little if later it turns out to have been a pack of lies. Malicious groups use social media to spread misinformation, which can lead to wrong decisions by the public, can create economic havoc, jeopardize public health, and incite violence. Epidemiologist David Michaels, who served for 8 years as Assistant Secretary of Labor for the Occupational Safety and Health Administration (OSHA) writes[34]: *Following the US election of Donald Trump, the fundamentals of evidence-based policymaking came under unprecedented attack. Just as unwelcome news became fake news, unwelcome science became fake science. Incredibly, the federal government elevated studies conjured by product defense specialist over the studies done by independent academic scientist. Worse, perhaps, the scientist whose careers have been defined by their science-for-sales studies exonerating toxic chemicals were brought inside, running, or advising the very agencies that regulate those chemicals.* A lot of harm was done.

Spies can hack into anything of interest, political, military, or industrial, to get information. This is part of the "Shadow War" currently underway, as described by Jim Sciutto.[35] Well-meant cybersecurity is of utmost importance. And I say "well-meant" since it can easily be transformed into censorship, as it is in China and other countries, being a great temptation for all those in power, a dangerous path.

It is curious that in 2001, Saint Isidore of Seville (560–636) was proposed as the Internet's patron. He was the first of the medieval compilers and published in 20 volumes and 448 chapters *Originum sive etymologiarum libri viginti* a work that summarized the knowledge of the time. The "Etymologies" contained many items, with translated versions of the great Greek thinkers (whose originals were not available) in a blend of valuable things and rubbish.

The overwhelming tsunami of stuff that shows up on our screens every day generates a virtual world that gradually becomes real. A person no longer finds some truth to avoid sinking into the depths of madness or getting lost in the desert of alienation. A few insiders warn us, but few will heed or can kick the habit. Jaron Lanier, computer scientist and a founder of the field of virtual reality, suggests addiction is established through our neurotransmitters (dopamine) that provide a brief pleasure stimulus with every "like." Lanier writes[36]: *the core process that allows social media to make money and also does the damage to society is* **behavior modification**. *Behavior modification entails methodical techniques that change behavioral patterns in animals and people. It can be used to treat addiction, but it can also be used to create them. The damage to society comes because addiction makes people crazy. The addict gradually loses touch with the real world and real people. When many people are addicted to manipulative schemes, the world gets dark and crazy.*

Research suggests that information technologies will also impact our cognitive abilities. Our ability to read and write honed over centuries will diminish since we were not born with it. This will modify the way we learn, affect our critical thinking skills, and lead future generations to a shallow zombie-like existence. As stated by Maryanne Wolf[37]: *The quality of our reading is not only an index of the quality of our thought, it is our best-known path to developing whole new pathways in the cerebral evolution of our species. There is much at stake in the development of the reading brain and in the quickening changes that now characterize is current, evolving iterations.*

Perhaps it will not be long before writing and reading are forgotten skills no longer needed. Children will not have to spend years learning the three R's, and books will gradually become obsolete. We will enter a new oral-visual age and just talk and listen, automatically translated by AI into any language, those that will survive current language extinction, which will impoverish world culture. Books can be burned for fuel (more CO_2!), and we will have again lobotomized civilization.

14.2 Bubbles

… the world in which we live is very nearly incomprehensible to most of us. There is almost no fact — whether actual or imagined — that will surprise us for very long, since we have no comprehensive and consistent picture of the world which would make the fact appear as an unacceptable contradiction. We believe because there is no reason not to believe.

— Neil Postman[38]

What do you think happens to you after hours of watching YouTube, surfing the WWW, and watching TV? Given the number and variety of issues that might interest you, depending on your prejudices or tastes, it will be easy to find hundreds if not thousands of channels and websites that support your point of view, no matter how extraordinary or peculiar. In any case, there is no time to look at everything, and less time for a background study or fact-check, so you go with what appeals to you. If you believe that vaccines cause autism you can easily confirm it, or that global warming is a hoax, search and you shall find, or that biological evolution is not for humans, no problem, and you can also confirm that the Shoah never happened and that we never went to the Moon. You can confirm your beliefs no matter how wrong they may be, especially if you have no reason not to believe. I will not mention rigged elections (I just did!).

As cognitive psychologists teach us, we are wired to seek confirmation of our previous beliefs — *confirmation bias* — that will do its dirty work without any effort. We are also wired to enjoy the company of like-minded people who think and are like us and are wired to follow the crowd since this had survival value. If you are walking on a street and see an oncoming group of people apparently fleeing from something, you will turn around and join, or you might not and leave the gene pool. The already mentioned Dunning Kruger bias also contributes to people accepting fake as true: *I believe this is true because I am brilliant and would know if something is fake.*

This enables the formation of global virtual *bubbles* like those we already have locally with our churches, mosques, temples, and various social clubs and political parties, which can manufacture any worldview they wish. Caitlin Johnstone writes[39]: *any attempt to understand the world which fails to take into account the fact that extremely powerful people are pouring massive amounts of money and resources into manipulating your understanding of the world will necessarily result in a distorted worldview.*

And if you are alienated enough, just become a fanatic football fan (any ball shape). The Internet makes all this more accessible and global.

One crucial thing about data, things you might believe are true, is that to gain knowledge you must look at *all* of it, not just "cherry-pick" data that support your beliefs. Just one well-documented fact that disagrees with what you believe is more important than one hundred that support it.

This was already well expressed by Francis Bacon (1561–1626), who could be considered as the forerunner of modern cognitive science in his *Novum Organum*[40]: *The human understanding, when any proposition has been once laid down (either from general admission and belief, or from the pleasure it affords), forces everything else to add fresh support and confirmation; and although most cogent and abundant instances may exist to the contrary, yet either does not observe or despises them, or gets rid of and rejects them by some distinction, with violent and injurious prejudice, rather than sacrifice the authority of its first conclusions. It was well answered by him who was shown in a temple the votive tablets suspended by such as had escaped the peril of shipwreck, and was pressed as to whether he would then recognize the power of the gods, by an inquiry, but where are the portraits of those who have perished despite their vows? All superstition is much the same, whether it be that of astrology, dreams, omens, retributive judgment, or the like, in all of which the deluded believers observe events which are fulfilled, but neglect and pass over their failure, though it be much more common. But this evil insinuates itself still more craftily in philosophy and the sciences, in which a settled maxim vitiates and governs every other circumstance, though the latter be much worthier of confidence. Besides, even in the absence of that eagerness and want of thought (which we have mentioned), it is the peculiar and perpetual error of the human understanding to be more moved and excited by affirmatives than negatives, whereas it ought duly and regularly to be impartial; nay, in establishing any true axiom the negative instance is the most powerful.*

Hundreds of sites will tell you about the merits of homeopathic remedies and other strange therapies, and thousands will document any fashionable madness. Repeatedly hear that there is no such thing as climate change (or at least that it is not caused by us), and it will become true. Any "fact" spreading through social networks and multiplying virally becomes "fact" solely based on constant repetition.

You have a collection of very persuasive anecdotes, experiences of things that have happened to someone, and others expand with "me too" or "like." But all the anecdotes together are not data with which you can proceed to conclude. They are just that, anecdotes, stories that might be or not true, and are not a good reason to generalize (which is a fallacy we all commit). We love stories (much more than statistics), and that is the trap. "Experts" who are not, have a way to reach millions at the click of a button and mislead you to death. Edward Louis Bernays (1891–1995), recognized as the founder of this field of public relations and author of the classic book *Propaganda* expressed[41]: *The conscious and intelligent manipulation of the organized habits and opinions of the masses is an important element in democratic society. Those who manipulate this unseen mechanism of society constitute an invisible government which is the true ruling power of our country. We are governed, our minds are molded, our tastes formed, our ideas suggested, largely by men we have never heard of. This is a logical result of the way in which our democratic society is organized. Vast numbers of human beings must cooperate in this manner if they are to live together as a smoothly functioning society.*

Scary, right?

In the past, if someone expressed what was clearly nonsense (in the context of that time), other people in his village might let him know, he would be ignored or, still worse, banished. Today the global village is so extensive that you will always find other fellow travelers who share the nonsense. Together, they can meet and think about what they want, feel supported and stay there, comfortably traveling in a cyber bubble. Instead of being alone in his madness, a fanatic or extremist now finds a support network, so devotees (and terrorists) are recruited. The bubbles grow, they separate, and polarization and radicalization increase. Bubbles for peace lovers, bubbles for warmongers, and for conspiracies. William Gibson's 1984 novel "Neuromancer" mentions "a consensual hallucination." Johnathan Haidt[42] writes: *Each matrix* (bubbles) *provides a complete, unified, and emotionally compelling worldview, easily justified by observable evidence and nearly impregnable to attack by arguments from outsiders.*

The bubbles are narrow in time, history loses importance, and the present is what matters, which is forgotten tomorrow. We fall into *presenteeism* where all that matters is this moment.

In today's increasingly interconnected world (with about 5 billion internet users), we are overinformed. However, we can only absorb a tiny fraction of all the information zooming through the planet's nervous system. The hardest problem is to gain knowledge from this, very different from being informed, as already mentioned.

Thomas Jefferson wrote[d]: *Whenever the people are well-informed, they can be trusted with their own government.* Today I would say people with knowledge instead of well informed. Public ignorance is nothing new, and every year, because of the advance of knowledge, every one of us is a bit more ignorant. What is alarming is the increasing rejection of expertise, the idea that one's opinion about a subject is as valid as expert knowledge, which endangers democracy.

Even for those who understand and seek expertise, it is quite difficult to determine who is expert, but a few points should be considered. The first thing is any vested interests that might affect their impartiality and independence. In other words, if in a situation the expert has something to gain or lose depending on his or her (I will for brevity use him henceforward) advice, he may lose credibility if the advice supports action that would benefit him, and conversely. For this reason, he who criticizes from within a group (a whistleblower) is given more weight

---

[d] Thomas, Jefferson to Dr. Richard Price, 1789. http://www.let.rug.nl/usa/presidents/thomas-jefferson/letters-of-thomas-jefferson/jefl73.php [Accessed February 11 2018].

Table 14.1. Criteria that serve as a guide for evaluating an expert.

Conflict of Interest	Does the organization employing the expert (or the expert himself have something to gain with the expert's statements? For example, a neumologist who works for a cigarette factory is suspect. If so, find other experts.
Credentials	Years of study, research, and a degree in the area of competence conferred by a recognized school. Active work in an institution in the area of competence.
Affiliations	Is the expert affiliated with a recognized institution (university, research institute, government agency, industry, hospital)? If not, you must ask how he makes a living.
Peers	Has the expert's work been published in recognized and specialized journals whose articles are peer-reviewed? Has he published well-reviewed books with reputable publishers? Is he respected by his peers and recommended by those you trust?
Expertise in relevant field	Find out if the expert has done work in the specific field in question. For example, a tennis expert is not a soccer expert. Often, well-known but not expert people (politicians, celebrities) are invited to comment on things.

than an outsider and will be generally persecuted with more vigor since he is seen as a "traitor."

Next, we look at the credentials. These are important, although they are not always enough, and sometimes, they are unnecessary, since you can be an expert without a degree, perhaps a self-taught person with much experience. For example, a surgeon with many years of experience from a lesser-known school might do better than a recent graduate from a prestigious school.

Table 14.1 presents some criteria that serve as a guide for evaluating an expert.

If you visit a doctor and on the wall, you see a framed diploma that says: *Issued by the Royal Academy of Sciences of Paraguay* and then (to obtain a second opinion) visit another whose diploma says: *Issued by the Johns Hopkins School of Medicine*, whose opinion will you give more weight? This presupposes that you know the difference between the Royal Academy of Sciences of Paraguay and Johns Hopkins, starting with the fact that the former does not exist. Nor is it significant that someone was "nominated for the Nobel Prize," as I have heard several times — my colleague and friend can write a letter to the Swedish Academy and nominate me. Note also that a Nobel Prize in Chemistry does not make you a medicine expert.

As journalist and novelist Upton Sinclair (1878–1968) famously wrote: *It is difficult to get a man to understand something, when his salary depends on his not understanding it.* We have recently witnessed over 70 million people who voted for a president who showed ignorance and malice every time he opened his mouth or tweeted. Johann Wolfgang von Goethe's (1749–1832)[43]: *There's nothing more horrible than ignorance in action* comes to mind. Perhaps this dangerous episode that almost ended with US democracy (with all its flaws) should inspire us to be more careful when choosing a candidate for this job.

Neil Postman wrote[44]: *What Orwell feared were those who would ban books. What Huxley feared was that there would be no reason to ban a book, for there would be no one who wanted to read one. Orwell feared those who would deprive us of information. Huxley feared those who would give us so much that we would be reduced to passivity and egotism.*

Orwell feared that the truth would be concealed from us. Huxley feared the truth would be drowned in a sea of irrelevance. Orwell feared we would become a captive culture. Huxley feared we would become a trivial culture, preoccupied with some equivalent of the feelies, the orgy porgy, and the centrifugal bumble puppy. As Huxley remarked in Brave New World Revisited, the civil libertarians and rationalists, who are ever on the alert to oppose tyranny, "failed to take into account man's almost infinite appetite for distractions." In 1984, people are controlled by inflicting pain. In Brave New World, they are controlled by inflicting pleasure. In short, Orwell feared that what we fear will ruin us. Huxley feared that our desire will ruin us. I think I side with Huxley.

Chapter 15

Ethics and Economics

For years the nation has been losing the taxpayer supported public goods that are available to all. In their place has come a vast outcropping of private goods available mainly to the well off. At the same time America's rich have been paying less in taxes to support the common good. And more government expenditures have been finding their way into bailouts, subsidies, and government contracts, going to favored industries like coal, oil, Big agriculture, Wall Street, and industries specializing in production for the military. There is something dreadfully wrong with this picture.

— Robert B. Reich[1]

The ideology of industrial society, driven by notions about economic growth, ever-rising standards of living, and faith in the technological fix, is in the long run unworkable. In changing our ideas, we have to look forward towards the eventual target of a human society in which population, use of resources, disposal of waste, and environment are generally in healthy balance. Above all we have to look at life with respect and wonder. We need an ethical system in which the natural world has value not just for human welfare but for and in itself. The Universe is something internal as well as externa.

— Crispin Tickell[2]

15.1 Obsolete Ethics

The most important human endeavor is the striving for morality in our actions. Our inner balance and even our very existence depend on it. Only morality in our actions can give beauty and dignity to life. To make this a living force and bring it to clear consciousness is perhaps the foremost task of education. The foundation of morality should not be made dependent on myth nor tied to any authority lest doubt about the myth or about the legitimacy of the authority imperil the foundation of sound judgment and action.

— Albert Einstein[3]

We may spend most of our waking hours advancing our interests, but we all have the capacity to transcend send self-interest and become simply a part of a whole. It's not just a capacity; it's the portal to many of life's most cherished experiences.

— Jonathan Haidt[4]

We have become obsolete. It has to do with the exponential increase of our power that has surpassed our ethical principles and the ideas that we maintain about the norms that should govern that power. We can annihilate a million people in an instant. Still, we cannot imagine the meaning of this possibility. We kill persons from a comfortable air-conditioned room, where a screen with crosshairs shows us what a distant drone sees. To an indoctrinated soldier, it looks like a game. Soon we will not need the soldier. Do you really think that this will solve anything?

It has led to a worldwide pandemic of corruption and violence, driven by an international class of oligarchs who use the power of money extracted from workers, hidden in tax-havens, and used to drive favorable government actions to push what they think are their interests, forget the rest.

The massive leak (October 2021) of secret financial records called "Pandora Papers" contains documents showing how over 100 billionaires, hundreds of politicians from across the world, and many celebrities, hide their fortunes in so-called offshore tax heavens, not only to avoid paying their share but also to hide how they cheat and steal. (Of course, when asked, they all deny any wrongdoing or claim it is all legal, but it is not about legality but about morality, an often-repeated confusion.)

Oligarchy destroys democracy. Realize that the wealthy do not work for money; money works for them. Most have little to no social conscience, do not care about the dead and injured because of their action or inaction, about those that are hungry or sick. Webster defines a psychopath as: *a person having an egocentric and antisocial personality marked by a lack of remorse for one's actions, an absence of empathy for others, and often criminal tendencies*, and that is what many are. They think they are superior and that they deserve what they got (often inherited). They only care about the Dow Jones average, which for most citizens is irrelevant.

We need a renewed secular ethic updated under the fading shadow of Hiroshima and Auschwitz and the many other tragedies that result from our arrogance and false beliefs. We need an ethic that rejects the biblical dominion and submission to the non-existent supernatural, and advocates the fraternity between humans and the need for harmony and balance with nature. An ethic that places being over possessing, quite difficult within the structure of voracious industrial consumerism without borders, and adherence to religions that claim to be an ethical gateway but do not measure up to this assertion. We need to place the *relationships* between humans over the values of freedom and individualism especially in the US where citizens, with a distorted concept of freedom and liberty, refuse to wear a mask to protect society (relationships) from the spread of Covid19 appealing to their misunderstood freedoms.

They will, however, dutifully use their seat belts, but when it comes to abortion, the same people will not appeal to free choice. It is astonishing and stupid. "Liberty" and "freedom" have become convenient words to use when appropriate. Chris Hedges writes: *You cannot use the word "liberty" when your government, as ours does, watches you 24 hours a day and stores all your personal information in government computers in perpetuity. You cannot use the word "liberty" when you are the most photographed and monitored population in human history. You cannot use the word "liberty" when it is impossible to vote against the interests of Goldman Sachs or General Dynamics. You cannot use the word "liberty" when the state empowers militarized police to use indiscriminate lethal force against unarmed citizens in the streets of American cities. You cannot use the word "liberty" when 2.3 million citizens, mostly poor people of color, are held in the largest prison system on earth.*

I visited the Soviet Union (when it still was) and found the people friendly and eager to share. I was invited to a party, and many wanted to drink vodka with the "Amerikanskiy" until I almost passed out. They wanted peace as much as we did and loathed (once convinced you were not a government agent) their oligarchs as much as we do but were afraid of US intentions, as are many nations in the present and for good reasons.

Some think that the current pandemic might change all this for the better, as we understand the interconnectedness of everything and everyone. Still, although I wish this were true, most people want to go back to "normal," even those for which normal was miserable. Just witness the resistance to change modest behavioral matters, and here I am advocating huge changes by comparison. No chance. "Freedom" eclipses collective wellbeing.

We should not confuse obsolete with age, nor think that old is obsolete (although some view old people as such and store them in "homes" waiting for the grim reaper and the testament). Some old ethical teachings are not obsolete. Among them, I propose that Confucius's (in Analects 5:12) golden rule of 2,500 years ago has supremacy: *What I do not wish others to do to me, I do not wish to do to others.* (But we bomb other countries and kill and injure innocent persons and then are enraged if somebody strikes back). Or the later and somewhat different version in Luke 6:31: *Do unto others as you would have them do unto you?*

It is difficult to understand how someone can go to church, praise the Lord and repeat one of the great commandments: *You shall love your neighbor as yourself* (Mark 12:31), and also the sixth of the Decalogue: *Thou shalt not kill*, which was downloaded from the cloud on Moses's tablet, and then on Monday go back to a job in the armament industry, or go and mistreat immigrants, or citizens of different ethnic backgrounds who do no harm. If the sixth were obeyed, without any fine print, we could live in peace. But that requires a new ethic, a new religion, if necessary, a conviction that the road to the future cannot be covered with twisted corpses if we want to get there with some dignity. But the forces that oppose this path are powerful and seem to be growing and, *everyone* must accept this new ethic, and this does not even happen among Christians who are supposed to obey them.

The epic poem "The People Yes" written by Carl Sandburg (1878–1967) in 1936 has this line[5]: *Sometime they'll give a war, and nobody will come.*

Wouldn't it be nice? (As was sung by the Beachboys) In this context, "friendly fire" is one of the best oxymorons I know ("corporate ethics" is another one). Corporate social responsibility is just a phrase — public relations. Of course, citizens who need a salary are often unable to consider this moral dilemma or might rationalize their actions by thinking, "if I don't do it, others will," and they are probably right.

Furthermore, they probably have imbibed the narrative that it is for a good cause, and so toiling at $7.25 an hour do not question why we need to spend a huge amount of tax dollars for war to "defend" the nation. (Again: defend from whom?) Unions (also an endangered species) representing the workers are despised by those in power.

We need a new global ethic that forms the basis on which the future citizen is educated, filling the lack of scruples of the dark mind. Reason devoid of ethical considerations produces *Homo demens* — leading to disaster as our grandparents witnessed during World War II. We are on our way. Only a new meta-education of global solidarity, evoked by the image of Apollo 8, can extract us from barbarism, which, as French philosopher and sociologist Edgar Morín points out, accompanies civilizations.[6]

Although we nurture our past, remembering supposedly heroic deeds (of the victors), cultivating traditions and customs, erecting the wrong statues, and looking for roots, we do not seem to care about the future, beyond ours. We do not consider those who will think of us as the past, and the painful question which we will not hear: Is this what you left us? Or in the words of young Greta Thunberg: *We will never forgive you.*

The necessary ethical transformation presupposes two premises that seem not to be part of the present *zeitgeist*: That we care about the well-being of those we do not know, and almost as a corollary, that we care about the well-being of those we do not know because they have not yet been born — the generations of the future. Or maybe we will not do anything for future generations since they have not done anything for us, a good market argument.

Jorge Riechmann,[7] professor of moral philosophy at the Autonomous University of Madrid, in his monumental Trilogy of Self-containment (unfortunately only available in Spanish), proposes three basic ethical assumptions:

- Moral universalism: The necessities of every human being are of equal moral value.
- Rejecting exclusionary anthropocentrism: The needs of humans are not the only ones that count morally: the principle of moral universalism extends beyond the barrier of our species.
- Special responsibility of humans: Because we are the only known moral agents within our biosphere — with respect to other living beings and the integrity of the biosphere.

And three dimensions around which the moral question is articulated in our present world: humanity/future generations, North/South, and human/other beings.

There is nothing complicated here if we aspire to live in a peaceful world. Revised ethical thinking is needed underscoring the fraternity among humans — really among all forms of life — based on human values instead of superhuman myths. We need to value the internal qualities of individuals instead of what surrounds them, including the color of their skin, which seems to annoy so many. We need solidarity that considers that the suffering of a stranger in a distant country has the same weight as that of a friend. An ethic that emphasizes that the Earth, belongs equally to all of us and to no one in particular. An ethic that implements the faded motto that adorns French walls and monuments: *Liberté, Égalité, Fraternité*. (Which is also the motto of the destitute Republic of Haiti.) An ethic that teaches us that the only thing that violence produces, both individual and collective violence, is more violence and more pain. I observe the opposite and I do not think it is my myopia. We need Enlightenment.

In his excellent book[8] that analyzes our current predicament Kevin MacKay ends with: ... *the only way humanity can meet the challenge of our current crisis is to reassert the pro-social framework of a moral community, and to expand the bounds to include all peoples, all nations, and the natural world.*

The path of the developing countries leads only to an unattainable fantasy, nothing but a fata morgana. Besides, we cannot talk about development without considering its sustainability where we follow the provisions of the third principle of the Rio Declaration of 1992 that adopts the definition given in the cardinal "Brundtland report" of the UN of 1987 (over 30 years ago !)[a]: *The right to development must be fulfilled to equitably meet developmental and environmental needs of present and future generations.*

But years go by, and nothing happens, and there is no reason to think that this will not continue as is until it is too late. Rachel Carson had this to say[9] in 1953! *The real wealth of the Nation lies in the resources of the earth — soil, water, forests, minerals, and wildlife. To utilize them for present needs while ensuring their preservation for future generations requires a delicately balanced and continuing program, based on the most extensive research. Their administration is not properly, and cannot be, a matter of politics.*

The following might be a worse simplification than that of the parallelepiped cow, but I think it serves to clarify a difficulty. The problem among humans lies in the asymmetry between the "good" and the "bad," and since we all want to become famous, I will call it Altschuler's Law: *In the short term, the bad guys have everything to prevail. In the long term we are all dead.* By any definition, good people are honest; they do not cheat, torture, kill, and they respect others. Without these or other scruples, bad guys, especially when they attain a certain critical mass and are organized, will prevail. Review the history of the German Third Reich to see this dynamic. And bad apples are everywhere, as we have been made painfully aware recently, all the way to Congress and the White House.

The lack of ethical conscience and solidarity with the less fortunate of the planet is appalling. Peter Singer[10] puts it this way: *We are responsible not only for what we do but also for what we could have prevented. We would never kill a stranger, but we may know that our intervention will save the*

---

[a] Report of the World Commission on Environment and Development. "Our Common Future". http://www.unesco.org/education/pdf/RIO_E.PDF. Gro Harlem Brundtland, was elected several times Prime Minister of Norway and between 1998 and 2003 she was Director General of the World Health Organization.

lives of many strangers in a distant country, and yet do nothing. We do not then think ourselves in any way responsible for the deaths of these strangers. This is a mistake. We should consider the consequences both of what we do and of what we decide not to do.

Contrast this with the years of the Vietnam War and the tragic events of the recent Latin American past, years of upheaval, torture, and death, of *Masters of War* by Bob Dylan, *American Tune* by Paul Simon and Art Garfunkel and *Buscando América* by Rubén Blades. It is necessary to awaken the slumbering conscience of those in charge and make them understand that the *Status quo* is not acceptable. As famously said by the late supreme court justice Ruth Bader Ginsburg (1933–2020): *Real change, enduring change, happens one step at a time.* The sad thing is that we have painted ourselves into a corner because of years of negligence, and now we need to jump and run. Unfortunately, one step at a time will not extricate us from the mess we have created.

It is in that sense that Karl Jaspers (1883–1969) (who directed Hannah Arendt's thesis) wrote: *What a fatality, when people yield in good faith renounce violence, because they believe in non-violence! They are then only more radically overcome by the violence that hides behind the veil of this fraudulent teaching.*

Chris Hedges, an essential critic of modern civilization, agrees[11]: *In the face of modern conditions, revolution is inevitable. The rampant inequality that exists between the political and corporate elites and the struggling masses; the destruction wreaked upon our environment by faceless, careless corporations; the steady stripping away of our civil liberties and the creation of a monstrous surveillance system — all of these have combined to spark a profound revolutionary moment. Corporate capitalists, dismissive of the popular will, do not see the fires they are igniting.* But then again, as he writes elsewhere[12]: *Perhaps in our lifetime we will not succeed. Perhaps things will only get worse. But this does not invalidate our efforts. Rebellion — which is different from revolution because it is perpetual alienation from power rather than the replacement of one power system with another — should be our natural state.* If you look coldly at the state of the world, you might agree; I believe he is right but also that it is too late.

15.2 Amber Alert and Red Card

> There's no time to lose, I heard her say
> Catch your dreams before they slip away
> Dying all the time
> Lose your dreams and you will lose your mind
> Ain't life unkind?
>
> — The Rolling Stones, Ruby Tuesday

An AMBER alert has been issued for her. Few remember the last time they saw her, nor where it was, the testimonies are ambiguous, and she may be dead or perhaps kidnapped. Some say they have seen her in the Amazon, among lost tribes. Others murmur that she is a refugee in a northern European country, incognito, because she is ashamed and feels abandoned. Others say that it is only a myth, that she never existed in the first place despite much talking, that it is a dream that some have had. But the thing is that there are people who have killed for her and others who have died for her.

Little is known about her family, and it is not known with certainty where she was born. Many fell in love with her as soon as they got a glimpse, even if in a dream. She must have been beautiful when she was young, although some who say they have seen her more recently say that despite her many disappointments, she continues to dazzle. It seems that she is of French or English descent, there are even some who say that she is related to famous philosophers, but they disagree if it was Diderot, or Hume, or one of the great Greeks she often quoted. No doubt she had something magical; I admired her as a young man and thought she deserved to be the queen of the planet, one we should all love and venerate as if she were a goddess, even by atheists. She had her moments, like every star, but it seems that she has been forgotten like so many others, forgotten like Brigitte Bardot or Marlon Brando. It is not that people do not wish to see her again, but how to love someone you do not even know? That is what happens to most humans. And there are some influential forces that loath her. The search continues despite the many who have lost interest or hope; although there is no specific reward offered, many have said that finding her would be enough and wonderful; it would change the world. If you see her somewhere, warn everyone, put her image on Facebook, or just tweet: "I found *Social Justice*."

I do not see nor hear but a sporadic protest from most young people about what is happening around them, which is worrisome. (Although as I write, I witness days of protest about institutional racism in the US, which is to be celebrated, instead of squashed by militarized police.) But many accept the lies, hypocrisy, and selfishness of those who govern (especially when it is about the wellbeing of others, and Trump excelled, and I am not flogging a dead horse since he or his ghost will be there for a while). Not a stone is thrown in protest, partly because when they leave school or the university where they had the time and a support group for this (I remember the seventies), they need to earn a living. Some must pay a large debt, and the reality of the world hits them in the face, and many think that if you can't beat them, join them, which just compounds the problem. I perceive a conformism and apathy that must be a consequence of the miss-education they receive and the brainwashing perpetrated by the public discourse that entertains them until death, as was pointed out some time ago by Neil Postman.[13] They walk like robots with their faces looking at a screen instead of looking at each other and the real world they will inherit.

It is also true that justice is not the same concept for everyone. But we all can identify things that are simply not just, no matter what else we believe. For example, huge wealth inequality is not just, discrimination for whatever reason is not just, killing innocent people is not just, and you can go on and on.

Some of the most pressing problems we face are imperceptible slow processes such as climate change, the growing population, antibiotic-resistant bacteria, and the poisoning of the planet. Others are potential catastrophic events of low probability but high impact such as nuclear war or a pandemic. Our brain is not "wired" for this; on the contrary, as you can check right now: you are not aware of the contact of your watch with your skin or your glasses on your nose or were, as now you are. We only perceive the short term and the local environment, which allows some to deny global warming because "it snowed yesterday." Complacency will sink us as the Titanic did in 1912 to the beat of its orchestra composed of eight musicians ("that glorious band" according to Admiral Lord Fisher) who continued playing while she went down.[14] In a 1907 interview, her captain, E. J. Smith (1850–1912) declared[15]: *In all my experience I have never been in an accident of any sort worth speaking about. [...] I will say that I cannot imagine any condition which could cause a ship to founder. I cannot conceive of any vital disaster happening to this vessel. Modern shipbuilding has gone beyond that.* There are lots of Smith's in this world.

In an excellent book[16] Naomi Klein explains how in the face of a natural or artificial crisis, unpopular changes occur after the episode, when people are still disoriented, generally for the benefit of capitalist interests, in what she calls the *shock doctrine.* She quotes Milton Friedman, *guru of the movement for unfettered capitalism,* who in an op-ed column for the Wall Street Journal wrote: *Most New Orleans schools are in ruins, as are the homes of the children who have attended them. The children are now scattered all over the country. This is a tragedy. It is also an opportunity to radically reform the educational system.* Nice Guy.

There is a war between the economic system and the planet, which is becoming a huge "Shithole" that will include the US no matter how high a wall is built.

Let us forge *Homo ecologicus* and show a RED CARD to *Homo economicus* by an education that enables students to choose with courage after deliberation, as Savater[17] says: *capable of persuading and willing to be persuaded.* But to decide, we need an ethical basis.

The situation will not improve until we stop thinking about "mother nature" in the passive sense that we associate with femininity, with an attitude of dominance that characterizes men's relationship with women. If we view nature simply as inert raw material, as a scenario in which economic activity meets our needs, an economy that harmonizes with her will not emerge.

We must fight the dark-minded enemy at all costs, but not with an AK-47, AR 15 or a sacred book, but with a renewed secular ethic and the humble force of reason instead of the arrogant reason of force.

Unfortunately, I do not see what could widen the narrow crack in the block through which it would be possible to introduce the wedge to break the monolith with new ideas so that some light might penetrate. One might think that these could arise from a truly autonomous university. Yet, as early as 1961, President Eisenhower expressed his concern for university independence[b]: *The prospect of domination of the nation's scholars by federal employment, project allocations, and the power of money is ever present and is gravely to be regarded.*

Were we to lose the independence of our universities, were they to become just factories with technicians as output, and research at the behest of corporations and the military, it would be a catastrophe, perhaps as important as the final destruction of the great library at Alexandria. And it is happening.[18] University research is being co-opted by the interests of the FIRM complex, and programs that do not lead to immediate "returns" are threatened. There have also been cases of attacks on scholars who are not in accord with some political or religious groups.

Cornel West and Jeremy Tate reacted[19] to the news that Howard University (and I fear many will follow) will dissolve its classics department: *The Western canon is an extended dialogue among the crème de la crème of our civilization about the most fundamental questions. It is about asking "What kind of creatures are we?" no matter what context we find ourselves in. It is about living more intensely, more critically, more compassionately. It is about learning to attend to the things that matter and turning our attention away from what is superficial. Howard University is not removing its classics department in isolation. This is the result of a massive failure across the nation in "schooling," which is now nothing more than the acquisition of skills, the acquisition of labels and the acquisition of jargon. Schooling is not education. Education draws out the uniqueness of people to be all that they can be in the light of their irreducible singularity. It is the maturation and cultivation of spiritually intact and morally equipped human beings.* Instead, we put more money into the military.

Abraham Flexner (1866–1959), founding Director from 1930 to 1939 of Princeton's Institute for Advanced Study, wrote in 1939[20]: *Is it not a curious fact that in a world steeped in irrational hatreds which threaten civilization itself, men and women-old and young-detach themselves wholly*

---

[b] Eisenhower's Farewell Address to the Nation, January 17, 1961.

or partly from the angry current of daily life to devote themselves to the cultivation of beauty, to the extension of knowledge, to the cure of disease, to the amelioration of suffering, just as though fanatics were not simultaneously engaged in spreading pain, ugliness, and suffering? Truly, a curious fact.

15.3 Obsolete Economics

A life wholly absorbed in need and its satisfaction, be it on the level of conspicuous consumption or of marginal survival, falls short of realizing the innermost human possibility of cherishing beauty, knowing the truth, doing the good, worshiping the holy. A techné which would set humans free from the bondage of drudgery, to be the stewards rather than the desperate despoilers of nature, should surely not be despised.

— Erazim Kohak[21]

You cannot have untold obscene wealth unless you have untold obscene poverty. That's the law of the so-called free market.

— Tweet of the DSA[c]

We live in a revolutionary moment. The disastrous economic and political experiment that attempted to organize human behavior around the dictates of the global marketplace has failed. The promised prosperity that was to have raised the living standards of workers through trickle-down economics has been exposed as a lie. A tiny global oligarchy has amassed obscene wealth, while the engines of unfettered corporate capitalism plunders resources; exploits cheap, unorganized labor; and creates pliable, corrupt governments that abandon the common good to serve corporate profit

— Chris Hedges[22]

Scientific temper and critical analysis also matter here, and evidence must serve to judge the adequacy of economic theory. Something must be wrong with a financial system where the rich get richer and the poor poorer; those are the facts. The Nobel prize for economics Wassily Leontief[23] criticized the mathematical abstraction and lack of empirical contrast of neoclassical economics many years ago. Economics cannot

---

[c] Tweet of the Democratic Socialist of America. March 22, 2017.

ignore sociology, biology, ethics, and ultimately neither can it ignore physics. As remarked by economist Michael Hudson, economic models are science fiction.

The primary driver of industrial capitalism is the selfishness and greed of an alienated (and often criminal) oligarchic ruling class. When it gets out of hand, as it has, it leads to the world we inhabit and inhibits justice. For them, solidarity is just a word, and socialism is an obscenity. Let me point out as if it were "Exhibit A," an infamous memo written by the chief economist of the world bank Lawrence Summers in 1991, later (1999) appointed as U.S. Treasury Secretary in the Clinton Administration and afterward president of Harvard University, which shows the callousness of those that rule and advise:

DATE: December 12, 1991
TO: Distribution
FR: Lawrence H. Summers
Subject: GEP

"Dirty" Industries: Just between you and me, shouldn't the World Bank be encouraging MORE migration of the dirty industries to the LDCs [Less Developed Countries]? I can think of three reasons:

1) The measurements of the costs of health impairing pollution depends on the foregone earnings from increased morbidity and mortality. From this point of view a given amount of health impairing pollution should be done in the country with the lowest cost, which will be the country with the lowest wages. I think the economic logic behind dumping a load of toxic waste in the lowest wage country is impeccable and we should face up to that.

2) The costs of pollution are likely to be non-linear as the initial increments of pollution probably have very low cost. I've always thought that under-populated countries in Africa are vastly UNDER-polluted, their air quality is probably vastly inefficiently low compared to Los Angeles or Mexico City. Only the lamentable facts that so much pollution is generated by non-tradable industries (transport, electrical generation) and that the unit transport costs of solid waste are so high prevent world welfare enhancing trade in air pollution and waste.

3) The demand for a clean environment for aesthetic and health reasons is likely to have very high income elasticity. The concern over an agent

that causes a one in a million change in the odds of prostrate [sic] cancer is obviously going to be much higher in a country where people survive to get prostrate cancer than in a country where under 5 mortality is 200 per thousand. Also, much of the concern over industrial atmosphere discharge is about visibility impairing particulates. These discharges may have very little direct health impact. Clearly trade in goods that embody aesthetic pollution concerns could be welfare enhancing. While production is mobile the consumption of pretty air is a non-tradable.

The problem with the arguments against all of these proposals for more pollution in LDCs (intrinsic rights to certain goods, moral reasons, social concerns, lack of adequate markets, etc.) could be turned around and used more or less effectively against every Bank proposal for liberalization

A good example of what I mean by "dark minds" and of how logic without ethics can become criminal. If we have people who think like this in leadership positions, not much will change. (It does not matter if, as he later argued, the memo was written by someone else, *he signed it.*) There was not much talk about CO_2 then, but the thinking could apply to that, and although China's growing economy means increasing CO_2 production, consider that a high fraction of this is for export.

The premise of unlimited growth, unlimited production, and insatiable consumption is false because the world, although large, is limited and not as robust as might be thought, and all the wishful thinking about technological "fixes" to the mounting problems this creates is bull, a fairy tale. We already saw how our "ecological" footprint overwhelms our planet.

In 1994, economist Hermann Daly and theologian and philosopher John Cobb pointed out in a profound critique of conventional economic theories and policies[24] that the primordial abstraction in contemporary economic theory — *Homo economicus* — is a creature quite different from flesh and blood *Homo sapiens. Homo economicus* is highly individualistic, greedy, and selfish. It (I say it this way since it is genderless) only needs goods and services obtained from the market without caring about the community in which it resides. It is insensitive to the needs of others, nor cares about things that are not parts of the market, such as the purity of the air or water, or the rivers and soils, never mind ethical questions. It does not care for the beauty of a landscape, for happiness, love, solidarity, and justice. (Other than when this affects it personally.) *Homo economicus* is not interested in inhabiting a planet in peace since this does not have a market value (and militarism is profitable). The above, which is important for an average *Homo sapiens*, is incorporated by economists under "externality," elements of no interest to the economic calculation because they do not have a market value.

It is difficult to assign a market value to these externalities when they are far-reaching and thus incorporate them into the economic system. However, it is feasible to provide an estimate of the economic value of losing fish from a lake because of contamination by industry, and we can imagine that this cost could be included in the price of the product, perhaps to pay for the necessary equipment and procedures to avoid or repair the damage.

But attempting an economic calculation in more far-reaching cases, such as the climate, or the future uselessness of antibiotics due to their excessive use quickly leads to absurdity since it is not just an economic problem.

If we destroy a lake, we could survive moving elsewhere, but if we destroy the planetary ecosystem, the economy stops having meaning because there will not be an economy. Thus, the so-called economic externalities should not be considered as a secondary effect that can be, at least in theory, internalized but should be the starting point of economic policy.

The cost of electricity, for example, is specified as an amount per kilowatt-hour. However, this does not include the cost of polluting the seas due to oil spills, the cost of coal-miners health, the aesthetic value of stinking landscapes of oil refineries (if you can smell it then there is something in the air, right?), nor the costs of accidents such as Chernobyl or Fukushima Daiichi. Not to mention the incalculable global cost of the effects of the increase in atmospheric CO_2 concentration (which will also affect the health of plants and animals). All these are externalities.

And if you buy an electric car and feel good about it, think again, since the electricity that feeds its batteries is most likely produced by a plant that emits CO_2. In addition, the lithium in the batteries comes from mines (Australia, Chile, and China are the biggest producers), which cause ecological damage.

Since we are on the subject, be aware that every time you send an e-mail, you add a bit of CO_2 to the atmosphere (yes, I know that individually it is insignificant, but we are many). True, there is much less ecological damage done by sending an e-mail than a physical letter, so it is an improvement if you add and subtract. That is the kind of addition and subtraction necessary for all the activities of modern society, but it is a vast spreadsheet. Take as an example the absurdity of having to travel to an office far from your home to spend the day in front of a computer, when you could do the same work from home (probably happier and with higher efficiency. (We have been forced to do much of this by a virus.) Since jobs are scarce, why not have a 4-day workweek? Where is it written that we must work 5 days a week? (I know where) but you know what I mean. Everyone would be happier and able to lead a fuller, better life. Michael Sandel[25] writes: *The reach of markets and market-oriented thinking, into aspects of life traditionally governed by nonmarket norms, is one of the most significant developments of our time.*

In a similar vein, Wendy Brown expresses[26]: *Political, personal, and social relations are rendered by neoliberal reason in market terms, from learning to eating, becomes a matter of speculative investments — ranked, rated, balanced in your portfolio. And democracy itself is devalued and transformed.* The economy measures social well-being by statistics such as the gross domestic product (GDP), ignoring the distribution of wealth (very skewed as we saw), and the value of social interactions that emerge from society. Still, they are not part of the market. What is the monetary value of having good friends, of clean air, or of studying philosophy?

Governments love to flaunt statistics about employment, crime, etc. (especially if they improved), but often job increase is in low-paying shitty jobs nobody really wants. The quality of our lives is determined by much more than by a number, without underestimating that everyone needs a minimum for a dignified life, which should be considered a right. (But then we get back to that dirty s-word.) It is determined by the environment in which we reside, our physical and mental health, the lived experience, and the satisfaction of our non-material goals. Otherwise, life is not worth it, even if you are a billionaire and own an island. Humans are not what they own, nor are they their beliefs. Both things are acquired during life, and we will leave them forever when we cease to exist. Basing our life, the reason for being on these things is regrettable, but that is what the system teaches us.

The market responds to the needs and desires of those who can pay and not to those of the needy. The notorious free market gives those with economic and political power (is it not the same?) the freedom to do as they wish, but it is not a *fair* market. Just one example: for the European Union, the Astra Zeneca Covid vaccine costs $2.75 per dose, whereas, for South Africa, it is $5.25.[27] In the words of Robert Reich[28]: *Not all wealthy people are culpable, of course… The abuse has occurred at the nexus of wealth and power, where those with great wealth use it to gain power and then utilize that power to accumulate more wealth. This is how oligarchy destroys democracy.* Because of its very nature, the market is incapable of providing health care for all, justice for all, equal education, and opportunity for all. Robert Reich writes a final word to Mr. Dimon, the billionaire chairman and CEO of J.P. Morgan Chase, the largest American bank: *But you are partly responsible for the system we have today, a system that almost everyone knows is not working, which most Americans see as rigged against them, which has siphoned of much of the wealth of the country for itself, invited demagogues to run for political office, and pushed voters to elect them, done little to protect the planet from devastating climate change, and drained much of our democracy of its vitality. And because you are responsible for the system, I believe that you at least have a moral duty **not** to use your formidable political power to stop the movement towards a more just system.* Yes, but it will not happen.

The neoliberal economic system transcends the merely economic and has become a toxic ideology. As Wendy Brown writes[29]: *More than merely saturating the meaning or content of democracy with market values, neoliberalism assaults the principles, practices, cultures, subjects, and institutions of democracy understood as a rule by the people.* Caitlin Johnstone writes[30]: *As near as I can tell the major problems with the world I am leaving to my children ultimately boil down to the fact that money tends to elevate the very worst kinds of people: those who are willing to step on anyone to get ahead, even if it means impoverishing everybody else, or starting wars, or destroying the ecosystem we all depend on for survival.*

A little-known feature of the world economic system also transcends the local economic system and traduces into "super imperialism," a significant F of the FIRM: The financial control of the world's economy. This is laid out in economist's Michael Hudson recently revised book, "Super Imperialism: The Economic Strategy of American Empire."[31] He writes: *The United States now rules not through its position as world creditor but as world debtor, making other countries lenders to itself simply by building up their own central bank reserves in the form of U.S. Treasury securities. This rigged game of dollarizing the world's central bank reserves has enabled America to flood the world with dollars without constraint as it appropriates foreign resources and companies, builds military bases and outposts, and imports foreign goods and services giving nothing in return except treasury IOU's of questionable (and shrinking) value. Rather than America's debtor position being an elemental weakness, it has become the foundation of the world's monetary and financial system. The rationale for America's ability to retain its role as world banker and key currency status no longer reflects the 1945 postwar faith in its moral leadership in the rhetoric of open markets. Its diplomats have shown a readiness to play the role of wrecker if foreign central banks stop re-lending their dollar inflows to the U.S. Treasury.* This is a complex issue, well documented in this important book.

In biology and society, new forms cannot arise if there is no variety if there are no mutations in thinking that lead to new ideas to give us a chance to survive. Richard Wolff,[32] one of the leading socialist economists in the US, summarizes as follows: *To preserve their accumulating wealth, large corporations and those they enrich wield ever more undemocratic power over the political and cultural realms of society. Their goals are self-preservation and self-aggrandizement. These features of capitalism are all social failures in terms of justice, democracy, equality, liberty, and ecological sanity. Yet mainstream media, politicians, and academics doggedly act and speak as though capitalism were the obviously "optimal" system to be continued, reinforced, and celebrated. By proceeding as though we are not in fact experiencing capitalism's systemic failures, they perform their ideological assignments.* Failure indeed.

One of the basic features of capitalist economics is that it siphons wealth from those who create it (the workers) to the upper echelons, producing an uncontrolled increase in inequality, subjecting workers to a corporate dictatorship even while they live in a democracy. This "regime" often dominates a citizen's life. This is how we end up with a situation where the CEO's wealth is hundreds or thousands of times greater than the wealth of workers. This is how we get to the absurdity of billionaires; they should be outlawed. It is outrageous that they use their wealth for a joyride into space (thinking about more money developing "space tourism") instead to try and help the less fortunate. This unregulated and rapacious capitalism is characterized by an increasing number of mergers and acquisitions. Big corporations swallow up smaller enterprises until they can control the market and the invisible hand becomes visible. As well said by Sirvent and Haiphong[33]: *Capitalism **does** indeed involve a redistribution of wealth, just not in the way a just society is supposed to.*

You can read about this in Jonathan Tepper's book[34] that tells us how: *America has gone from an open, competitive marketplace to an economy where a few very powerful companies dominate key industries that affect our daily lives. Digital monopolies like Google, Facebook, and Amazon act as gatekeepers to the digital world. Amazon is capturing almost all online shopping dollars. We have the illusion of choice, but for most critical decisions, we have only one or two companies, when it comes to high-speed Internet, health insurance, medical care, mortgage title insurance, social networks, Internet searches, or even consumer goods like toothpaste. Every day, the average American transfers a little of their paycheck to monopolists and oligopolists.*

Richard Wolff summarizes: *Over the last 30 years, the vast majority of US workers have in fact, gotten poorer, when you sum up flat real wages, reduced benefits (pensions, medical insurance, etc.), reduced public services, and raised tax burdens. In economic terms, "American exceptionalism" began to die in the 1970s.*

If you let capitalists run amok, they will seek ways to charge you for every breath of oxygen you inhale and tax you for every breath of CO_2 you exhale, and there is talk already of taxing solar energy. Increasingly the motto is "you get what you pay for," including justice, health, and education, among many other things, which should not follow that rule.

There is something else that corporations "siphon" from society for free: *knowledge*. Our educational system and a great deal of the research done at universities and government laboratories are mostly funded by tax dollars. Once employed by a corporation, the knowledge that the new employee provides to the corporation is not paid back to society. (That employee may get a fine salary, perhaps sufficient to pay back a hefty student loan). On the contrary, it is used to expand the profit of the corporation. Societies train people at their own expense, and corporations take advantage of that contribution.

The famous TINA (there is no alternative) regarding our economic order is part of the belief system of many, and any deviation is dismissed as communist or socialist or even liberal. However, Senator and presidential candidate (2020) Bernie Sanders exposed good and sensible ideas for a better inclusive future, where rights were guaranteed by the state for *everyone* (otherwise they are not *rights*), among them: health, housing, and education. I would think that few would consider this as undesirable (except for the usual assholes), and the cost of not investing in this is much higher than you imagine. An uneducated and sick society spells its own ruin.

But as soon as the senator used the scare word *socialism,* he was destined to fail, eliciting a knee-jerk reaction by those who think socialism is wicked.

The problem is distortion and ignorance of what socialism means,[35] part of the idiotization process (and there are many versions). Robert Reich tells us[36]: *Every time over the last century Americans have sought to pool their resources for the common good, the wealthy and powerful have used the boogeymen of "socialism" to try to stop them.* But then, when the government spends billions to bail out the rich, there is no problem. Matt Taibbi tells us[37]: *Far from taking care of the rest of us, the financial leaders of America and their political servants have seemingly reached the cynical conclusion that our society is not worth saving and have taken on a new mission that involves not creating wealth for all but simply absconding with whatever wealth remains in our hollowed-out economy. They don't feed us; we feed them.*

If any nation pursues an alternative economic system, the empire can resort to economic sanctions, bombs, and regime change.

Michael Hudson writes: *Excluding the USSR, Cuba and other country seeking an independent path from the U.S.-centered "Free World" had been the essence of the Cold War. Instead of trying to buy off socialist or state capitalist economies with aid, they were marginalized by trade and investment exclusions, sanctions, and, where necessary, military force or sponsorship of coups as in Guatemala, Iran, Vietnam, Indonesia, the Congo, Chile and its Latin American neighbors.*

And after the intervention, what remains is not a pretty sight. Of course, there *are* alternatives (there better be), but if you belong to the oligarchy, you probably will not favor them; you are OK. Or you might believe the myth of the infamous "trickle-down economics,," which as William Blum[38] writes: *Will this mean any better life for the multitudes than the Cold War brought? Any more regard for the common folk than there's been since they fell off the cosmic agenda centuries ago? "By all means," says Capital, offering another warmed-up version of the "trickle down" theory, the principle that the poor, who must subsist on table scraps dropped by the rich, can best be served by giving the rich bigger meals.*

The boys of Capital, they also chortle in their martinis about the death of socialism. The word has been banned from polite conversation. And they hope that no one will notice that every socialist experiment of any significance in the twentieth century — without exception — has either been crushed, overthrown, or invaded, or corrupted, perverted, subverted, or destabilized, or otherwise had life made impossible for it, by the United States.

Individual freedom does not mean much if we do not clarify freedom of what, for what, and for whom. Furthermore, how free is that large sector of citizens who must work 24/7 just to make ends meet? There are freedoms such as that of thought, expression, and movement, which should have few limits, but there are other freedoms that we need to limit since they are exclusive. For example, my freedom of expression in no way limits your freedom of expression. Still, my freedom to accumulate land limits your freedom to obtain land in a finite world, finite in all dimensions.

Freedom of expression and thought are inhibited by the unbounded freedom to accumulate wealth, for the simple reason that with wealth, one obtains political and economic power over others. That power can lead to the control of all sorts of media (as is the case) and an increasing number of corrupt politicians (as is also the case). The wealthy support politicians they favor, spend their millions to influence elections, fund think tanks that think like them, and provide funds to hundreds of universities to hire professors who further their cause, thereby corrupting education, science, and democracy. As quoted by McKibben,[39] one of those professors expressed: *Only idiocy would conclude that mankind's capacity to change the climate is more powerful than the forces of nature.* An idiot indeed.

Economist Robert Heilbroner (1919–2005)[40] pointed out that the issue is not about capitalism versus socialism or communism in its various versions, but rather the problem arises from the industrialization of the economy that imposes similar values on all these modes of socio-economic organization. They are values that subordinate the human scale to the industrial scale optimized to increase efficiency, the "conquest" of nature regardless of the consequences, and the priority of production and material achievements at the expense of other values. We must become aware of the fallacy of progress (that progress is always a good thing). It forces us to live according to the clock and the rhythm of the factory or the office in a lifestyle that contrasted with non-industrial societies, *seem superbly rich in every dimension except the cultivation of the human person.*

"Efficiency" is not always a good thing. It might be more efficient for a store (and therefore increase earnings) to have self-check-out, something you can see happening, but it comes at a cost which is the loss of jobs (admittedly not a great job, but a job nonetheless — now the customer does the job for free). The Jevons paradox (named after British economist and logician William Stanley Jevons (1835–1882) happens for certain processes when increased efficiency can lead to a reduction in the cost of the process but increases the demand for a resource, draining it at a higher rate. Take, for example, fuel efficiency (we all believe that higher miles per gallon are desirable). Different studies[41] have determined that better mileage leads to more travel (and all associated adverse effects of more cars on the road) and can result in a higher demand and use of fuel (rebound effect) estimated at around 20%.

Economist Walter Weisskopf wrote[42]: *The real "economic" problem is indeed created by scarcity: not the economic scarcity of factors of production like natural resources, labor and capital, or the goods and services produced for sale to others, but the existential scarcity of the penultimate resources of life, time, and energy, a scarcity created by our finitude and mortality [...] Economist are right when they talk about an optimum allocation problem; but the problem of optimum allocation of life-time and life-energy to the various dimensions of human existence and not only to the one economic dimension that is involved in traditional production and consumption for and trough the market* — the tragedy of our time.

In the end, whatever economic system you favor, and although there are built-in evils in all of them, what matters is the ethical attitude of the individual. If it is OK to accept billionaires (or even millionaires) while most live in poverty, then the economic system you adopt does not matter much. The much-cited Adam Smith knew this when he wrote: *No society can surely be flourishing and happy of which by far the greater part of the numbers are poor and miserable.*[43]

Chapter 16

What to Do?

There is always a well-known solution to every human problem neat, plausible, and wrong.

— H. L. Mencken[1]

The last days of any civilization, when populations are averting their eyes from the unpleasant realities before them, become carnivals of hedonism and folly. Rome went down like this. So did the Ottoman and Austro-Hungarian Empires. Men and women of stunning mediocrity and depravity assume political control. Today charlatans and hucksters hold forth on the airwaves, and intellectuals are ridiculed. Force and militarism, with their hypermasculine ethic, are celebrated. And the mania for hope requires the silencing of any truth that is not childishly optimistic.

— Chris Hedges[2]

To prevent widespread misery and catastrophic biodiversity loss, humanity must practice a more environmentally sustainable alternative to business as usual. This prescription was well articulated by the world's leading scientists 25 years ago, but in most respects, we have not heeded their warning. Soon it will be too late to shift course away from our failing trajectory, and time is running out. We must recognize, in our day-to-day lives and in our governing institutions, that Earth with all its life is our only home.

— 15,000 scientists warning to humanity[3]

Most of us occasionally suspect that the world we've created is too complex and fast-paced to understand, let alone control Most of us sometimes guess that even the "experts" don't really know what's going on; and that as individuals and as a species we've unleashed forces that we cannot manage. [...] The challenges [...] converge, intertwine, and often seem to be largely beyond our ken.

— Thomas Homer-Dixon[4]

After all this, your inevitable question will be: What to do? Let us not pretend to change things if we continue doing the same things (and I would add thinking), said the genius, Albert Einstein. We continue doing and thinking the same things, even writing books to see if we stop doing and thinking the same things and organizing wars. It seems that 2,000 years of bloody history had not taught us anything. Of course, if everything were wonderful on this planet, there would be no reason to change much, but unless you live on another, surely you agree that there are more than enough reasons. From all I have witnessed, I see a grim future.

We know what needs to change to deal with the problems we face; the evidence is compelling, but some responses seem almost impossible to implement, and if you know an easy solution, it will be, as Mencken says in the epigraph. Speaking of Mencken, here is another of his irreverent statements for you to ponder[5]: *As democracy is perfected, the office represents, more and more closely, the inner soul of the people. We move toward a lofty ideal. On some great and glorious day, the plain folks of the land will reach their heart's desire at last, and the White House will be adorned by a downright moron.*

Many years ago, we had a chance but made an unwitting choice about the road to follow, led by the blind, not understanding that it headed to hothouse earth. The other path, as suggested by Georgescu-Roegen, led to an ecological economy and a different world. But it is too late now, and it has made all the difference.

The problems we face have stealthily been transformed from relating to human or divine laws to natural laws. But we still think we shall find a solution appealing to the former ones. That is the danger of myths. Furthermore, we are not dealing with just one problem, but with an interrelated system of problems, so that unless we approach it as such instead of fixing a little here and there, it will not work. And then there is the problem of time.

One answer would be to resign oneself, go play Pickleball or visit Belgium, and ignore everything. When I finish writing this, I will consider it. Others, believers of the divine, with a cruel logic, will not do anything convinced that God will not abandon them or see it all as the prelude to the second coming. I have been studying the answers of others for years, experts in all areas. For years I have witnessed the struggles to change the world, dear friends disappeared in the night, I participated in the marches of the sixties and seventies. I breathed tear gas, and in the end, little has changed for the better, and much has inexorably worsened. Public discourse distracts us from the real problem posed by the catastrophic convergence, the public getting all riled up because of some trivial issue, and we go to battle for oil we should stop using.

Revolutions have occurred, mostly exchanging roles but not leading to a betterment of the lives of most. We need rebels who can denounce the worst and make governments understand the power of "we the people." Rebels to remind governments that their authority arises from the people and ceases with their sovereign presence.[a] But perhaps not even this will be enough, the problem is overwhelming, and the public does not seem to understand, is brainwashed, or does not care. Moreover, science education, despite all efforts, has failed. Observe how we are dealing with Covid19. And, as written by Roger Hallam, co-founder of Extinction Rebellion[6]: *Needing to keep voters, and having their hands tied by powerful vested interests, governments worldwide are seemingly unable to make the necessary changes. That is why a Rebellion is necessary.* But rebels are barely tolerated and often punished.

It would be nice to revive the Enlightenment, starting with our culture, which must defend itself, but only attack with persuasion, by example, and an outstretched hand. It will not be easy, as Todorov writes[7]: *We are all children of the Enlightenment… at the same time the ills fought by the spirit of the Enlightenment turned out to be more resistant that eighteenth-century theorists thought. They have grown even more numerous. The traditional adversaries of the Enlightenment — obscurantism, arbitrary authority, and fanaticism — are like the heads of the Hydra that keep growing back as they are cut. There is reason to fear that these attacks will never cease. It is therefore all the more necessary to keep the spirit of the Enlightenment alive. The age of maturity that past authors were hoping would come seems not*

---

[a] As declared in 1813 by the great Uruguayan leader Jose Gervasio Artigas (1764–1850).

to be the destiny of humankind: humanity is condemned to seek the truth rather than possess it.

The resurgence of religions with a pretense of political impact — the "R" in FIRM — has added an ingredient of justification to violence since it is carried out in God's name. Just listen to speeches from some government officials or from some pulpit to realize this. I do not mention those with the black flags; the color says it all. The only thing that seems reasonable is to contain them and wait for them to come out of their own Dark Age. Bombs will not do it.

But the thing is much harder still. I intend for everyone to read this (aspiration of any writer), not seeking fame but from my conviction that ignoring it leads to disaster. I guess my readers will agree with a lot of what I say (another bubble) concerning what we could do to achieve a culture of peace based on a new ethic (not easy, but possible). But social action by intellectuals (including scientists, who for some reason are not included) is necessary, however contrary to custom. It is essential to leave the academy so that everything is not merely academic. We will have to jump over university walls and then over the national ones (at a time when an intellectually challenged one wanted to build a vast wall to divide humanity). Then cross social, geographical, and cultural borders since a world of peace is only possible if everyone signs on. Faced with the reality of a world in crisis, university members must actively insert themselves into society, offering the necessary knowledge and leadership to solve the most pressing problems, *Sagesse oblige*. The main problem: Who will listen?

Carl Sagan, referring to the destruction of Alexandria, wrote[8]: *The glory of the Alexandrian Library is a dim memory. Its last remnants were destroyed soon after Hypatia's death. It was as if the entire civilization had undergone some self-inflicted brain surgery, and most of its memories, discoveries, ideas, and passions were extinguished irrevocably. The loss was incalculable.*

New ideas, new ways of ordering a society will hardly become new goals to implement since those who have the power of implementation have the power which comes with wealth, as I have indicated. I repeat it since it is a controlling factor. The politicization of critical issues has poisoned the well. And it does not matter if you live in a democracy (once elections are over) or a dictatorship. So the current pandemic might kill millions, but it will pass (until the next one), and we will go back to designing better fighter planes and nukes. That is how stupid we are.

It will not be easy to overcome the narrow and anachronistic nationalisms even though we are all on the same spaceship, more interconnected and interdependent than ever. No better lesson than that provided by an invisible virus. To rid us of the mythical punishment of Babel seems impossible. This is not a pessimistic stance; it is merely realistic.

Politicians seem to forget who pays their salaries (although they get a lot "extra" from the FIRM). It is not difficult and has been said often and by many; citing again one of my heroes, *There are certain things that the world quite obviously needs: tentativeness, as opposed to dogmatism, in our beliefs; an expectation of co-operation, rather than competition, in our social relations; a lessening of envy and collective hatred. These are things which education could produce without much difficulty.* That was Bertrand Russell 65 years ago,[9] and we are still waiting.

Without a doubt, a factor of great importance is the citizen's education, his vision of the world, including a solid updated ethical component. In that sense, I believe a change could happen, that we can have "a weak but genuine hope." Well-known British author H.G. Wells (1866–1946) wrote: *Human history becomes more and more a race between education and catastrophe.* But when people violently refuse to wear a mask to protect themselves and others, you wonder.

So the first step is for youth to understand what is happening, to realize they are slaves of the system and that if they do not demand an urgent change, what awaits them is serious.

The educational work must begin in the schools since it is too late when students enter a university (if they do). We cannot make a better investment in the future than investing in education, and if some think it is too expensive, then, as Derek Bok, a former Harvard president (from 1971 to 1991), once said: *If you think education is expensive, try ignorance.* We must educate to forge a new human being who is not merely a new resource for the FIRM complex, a well-prepared slave, but a creator of a new dawn. But time is running out. As Amanda Gorman recited at the inauguration of President Biden: *For there is always light, if only we're brave enough to see it. If only we're brave enough to be it.*

But it is already five o'clock in the morning. And unfortunately, those who control education are the ones who see no reason for change, preparing their own demise. They do not see the light.

In the words of Darrell Fasching[10]: *No civilization that has the power to destroy itself can afford the luxury of such a laissez-faire view of education. A civilization that has no wisdom to pass on to the next generation about what makes life valuable and worthwhile is a civilization preparing its own demise. After Auschwitz and Hiroshima, we can no longer afford the luxury of such a naive worldview, which tends to reduce knowledge to facts and skills and treat values as personal preferences that we are free to select from as if we were shopping in some intellectual supermarket. Professors need to be models of persons who understand and are prepared to be answerable for the social consequences of their views and require this of their students as well. We can no longer afford to live in a world where everything is "a matter of opinion" or personal preference.* That was 25 years ago.

And over 50 years ago, in 1970, Austrian philosopher Ivan Illich (1926–2002) wrote[11]: *The pupil is thereby "schooled" to confuse teaching with learning, grade advancement with education, a diploma with competence, and fluency with the ability to say something new. His imagination is "schooled" to accept service in place of value.* The situation has not changed. It would be good to teach the human animal to be more human and less animal.

Natalia Ginzburg (1916–1991) a forgotten heroine whose husband Leone Ginzburg was murdered in a German prison in Rome, provides us with a vision (written in 1960)[12]: *As far as the education of children is concerned, I think they should be not the little virtues but the great ones. Not thrift but generosity and an indifference to money; not caution but courage and a contempt for danger; not shrewdness but frankness and a love of truth;*

not tact but a love of one's neighbor and self-denial; not a desire for success but a desire to be and to know. Wouldn't it be nice?

Rob Riemen has this to say about our current predicament, in particular, the turn to authoritarianism as a response to social stress[13]: *Education is no longer intended as a process of character formation to help people live in truth and create beauty, carry out justice and convey a certain wisdom. It has degenerated into an instrument for the transfer of everything useful, knowledge that is usable for the economy and everything you need to know in order to earn money. [...] Only once we rediscover our love for life and decide to devote ourselves to what truly gives life — truth, goodness, beauty, friendship, justice, compassion, and wisdom — only then, and not before, will we become resistant to the deadly bacillus called fascism.*

And then we need the US to heed the words of Richard Falk[14]: *...for the US government to renounce aggressive war and regime-changing intervention as a means of altering the political destiny of a foreign country. Such undertakings fail miserably on the level of policy; worse, they unavoidably implicate the United States in massive human suffering and criminality that embitters any society so victimized and, in the ugly unfolding, alienates world public opinion.*

Meanwhile, we have lost precious time and nature does not play overtime. Indeed, as expressed by Thomas Friedman[15]: *We are the first generation for whom "later" will be the time when all of Mother Nature's buffers, spare tires, tricks of the trade, and tools for adapting and bouncing back will be exhausted or breached. If we don't act quickly together to mitigate these trends, we will the first generation of humans for whom later will be **too late**.*

In Table 16.1, as a summary, I present what we are facing, estimate a (subjective) probability that we shall succeed in implementing a solution (from very low to very high), the theoretical difficulty faced (something can be easy in principle — like eliminating all nuclear weapons — but difficult to achieve due to obstinacy) and a few comments.

Table 16.1. Subjective evaluation of some problems we face.

Problem	Probability of success	Theoretical difficulty	Comment
Catastrophic convergence — Climate Crisis	Very low	Most difficult	After 50 years of pompous declarations with minimal action and even opposition, the future looks ghastly, and the US is not helping (Neither does China). Every year we get closer to the point of no return.
Nuclear Weapons	Medium	Easy	It would be easy to scrap them all, but for this to happen, every country in the "nuclear club" must agree and trust each other. Then, the US could lead by example and significantly reduce its arsenal. We do not need nor can we use nuclear weapons.
Inequality	Low	Easy	We are moving in the wrong direction with unrestrained ruthless capitalism. A redistribution of wealth is unlikely because the very rich are not going to tax themselves. So it is easy in theory: change the rules.
Population growth	Little we can do	—	There is little we can do and accept that we shall reach 10 billion soon.
Other Problems — Idiotization	Medium	Easy	All you need is a large investment for excellent education, but it does not look good. And when the uneducated majority rules, we are in trouble.
War/Violence	Low	Easy	Need for the US (and a few others) to stop military interventions. To change perspective and offer a helping hand when asked instead of a closed fist.
Deforestation	Low	Difficult	Population growth and corporate interests drive deforestation.
Promote Research	Medium	Easy	Put scientific research as a high priority item.
Ice loss, Water levels	Low	Difficult	As the temperature rises, ice melts. The water will come no matter what.

(Continued)

Table 16.1. (*Continued*)

Problem	Probability of success	Theoretical difficulty	Comment
Food production	Low	Difficult	Changing weather patterns and lack of water will negatively impact food production in many regions.
Pandemic	Medium	Medium	Increasing temperatures and encroachment on wild habitats will facilitate zoonosis of new pathogens, as has the current Covid19 pandemic.[16]
Natural Disasters	Low	Low	Little we can do except to avoid dangerous places.
Chemical pollution	Low	High	As a byproduct of industrialization it is almost impossible to turn back

Some think we should escape to one of those planets we are discovering to reach a better world and abandon the burning ship (January 2020, Australia is burning, August 2020, California is burning, July 2021, Midwest heatwave is killing people). But other planets, as I have discussed, are not the solution; we must deal with this one.

But if those we "choose" to lead ignore or discredit the warnings by scientists and experts, who else will lead us away from the wrong path? It will not be the FIRM complex, entrenched in its ways. Should the blind leaders not be cursed? As said by German poet, and literary critic Heinrich Heine (1797–1856): *In dark ages people are best guided by religion, as in a pitch-black night a blind man is the best guide; he knows the roads and paths better than a man who can see. When daylight comes, however, it is foolish to use blind, old men as guides.* What Hedges says in the epigraph (might as well call it epitaph) says it all.

As for the monster we are feeding, alternate sources of energy, which do not provide a constant source of power, can delay its arrival. But it is difficult, mainly because we have little time left for the huge enterprise transforming our societies and our complex global economic systems. Of course, if we managed to transform our economy into a new system, we might not need a constant flow of energy, and we might even be happier. Is it necessary for industry to churn out junk 24/7?

We are before a canonical case of "too little, too late." "Or maybe just maybe," as Homer-Dixon writes[17]: *before we've exhausted nature and ourselves in a futile effort to produce meaning from material things, we'll reconsider our values and recognize that we can choose another path into the future. Our first step down that path must be to acknowledge that our global situation is urgent — that we're on the cusp of a planetary emergency — and then to begin a wide-ranging and vigorous conversation now about what we can and should do.* That was written in 2006, 16 years ago, and we have not moved one inch forward, and the vigorous conversation, if anything, is a shouting match! As well said by Bacevich[18]: *Simply put, the way that humankind in the 21st century aspires to live is pushing Earth to the brink of exhaustion. These issues will define the balance of the 21st century: an eastward shift of global power; technological dystopia; and potentially irreversible environmental degradation. To none of these does neoliberal globalization, the pursuit of militarized hegemony under the guise of global leadership and freedom defined as the removal of limits provide anything even approximating an adequate response. Quite the opposite.*

The monster will arrive sooner rather than later (many experts expect it even sooner). It will be preceded by the violence caused by the growing climatic upheavals that will affect several regions of the planet, causing famines, epidemics, and huge migrations that no wall will stop. Mosquitoes that are "the preeminent and globally distinguished killer of humankind"[19] (we are second) can easily fly over any border wall like flying syringes carrying deadly tropical diseases. They will thrive in the southern US as it gets warmer. Mosquitos do not like it if it gets overly hot (they like it if temperatures are 50–80-degree Fahrenheit), so the tropics might become a better place in this regard, but humans cannot survive extreme heat either. We have witnessed what happens during a heatwave, and they have been just a minor sample.

In the past, facing slow and bounded changes, we adapted, modified lifestyles, and migrated to more livable places. It is a story that will not be repeated for the simple reason that there is nowhere to go (we can see how difficult it is when only a million need to escape from where they live), that our societies are more vulnerable because they are complex, and that the changes we face are multiple and rapid. You cannot pick up and move a huge cornfield 100 miles to where the rain now falls.

Brian Fagan ends his book as follows[20]: *But if we've become a supertanker among human societies, it's an oddly inattentive one. Only a tiny fraction of the people on board are engaged with tending the engines. The rest are buying and selling goods among themselves, entertaining each other or studying the sky or the hydrodynamics of the hull. Those on the bridge have no charts or weather forecasts and cannot even agree that they are needed; indeed, the most powerful among them subscribe to a theory that says storms don't exist, or if they do, their effects are entirely benign, and the steepening swells and fleeing albatrosses can only be taken as a sign of divine favor. Few of those in command believe the gathering clouds have any relation to their fate or are concerned that there are lifeboats for only one in ten passengers. And no one dares to whisper in the helmsman's ear that he might consider turning the wheel.*

In a recent book that follows a parallel line to this one, Roy Scranton writes[21]: *Wars begin and end. Empires rise and fall. Buildings collapse, books burn, servers break down, cities sink into the sea. Humanity can survive the demise of fossil-fuel civilization and it can survive whatever despotism or barbarism will arise in its ruins. We may even be able to survive in a greenhouse world. Perhaps our descendants will build new cities on the shores of the Arctic Sea when the rest of the Earth is scorching deserts and steaming jungles. If being human is to mean anything in the Anthropocene, if we are going to refuse to let ourselves sink into the futility of life without memory, then we must not lose our few thousand years of hard-won knowledge, accumulated at great cost and against great odds. We must not abandon the memory of the dead.*

Maybe it is less pessimistic, but to get to that stage the pain will be immense. Twenty years ago, I wrote[22]: *I imagine that eons from now, the stratigraphic record will document this mass extinction. Someone, although I can't imagine who, will be looking at an ancient cliff from a distance, and note that a strange thick and twisted mile-long layer of material of different color paints its walls. Careful study by a large team of scientists will slowly reveal the fascinating origin of these deposits. These are the amalgamated and metamorphosed remains of entire cities, mostly stone and steel, with a few bits of very interesting materials of puzzling origin squashed together by tectonic processes as if they were the body of a car which has been reprocessed by one of those special machines found in junkyards.* Human fossils will be abundant below the layer; none will be found above.

Chapter 17

Two Gates

The question of questions for mankind — the problem which underlies all others and is more deeply interesting than any other — is the ascertainment of the place which Man occupies in nature and of his relations to the universe of things. Whence our race has come; what are the limits of our power over nature, and of nature's power over us; to what goal we are tending; are the problems which present themselves anew and with undiminished interest to every man born into the world.

— Thomas Henry Huxley[1]

The history of life on earth has been a history of interaction between living things and their surroundings. To a large extent the physical form and the habits of the earth's vegetation and its animal life have been molded by the environment. Considering the whole span of earthly time, the opposite effect, in which life actually modifies its surroundings has been relatively slight. Only within the moment of time represented by the present century has one species — man — acquired significant power to alter the nature of his world.

— Rachel Carson[2]

Everything you have read so far is not encouraging. We are destroying our home. It is a dystopian vision, but what do you want me to do? Cover the sky with my hand? Affirm what against all evidence is not true, just because I would like it to be true?

I think it was the Spanish writer Antonio Gala who said that a pessimist is a well-informed optimist. Renowned German philosopher Arthur Schopenhauer (1788–1860) said[3]: *For the rest, I cannot here withhold the statement that optimism, where it is not merely the thoughtless talk of those who harbor nothing but words under their shallow foreheads, seems to me to be not merely an absurd, but also a really wicked, way of thinking, a bitter mockery of the unspeakable sufferings of mankind. Let no one imagine that the Christian teaching is favorable to optimism; on the contrary, in the Gospels, world and evil are used almost as synonymous expressions.* So, yes, I am not optimistic.

What remains is a clumsy animal who believes its own myths and has deceived himself to avoid the metaphysical horror of knowing reality. Few accept that we are much less than we think we are, but if you play with fire, eventually you get burned, and this time burned means that this planet will become a world without us. An old Danish proverb says: *Predictions are difficult, especially those of the future* (Some allege it was said by the Danish physicist Niels Bohr, but it is older.) So, perhaps I am wrong; I wish I were. But, on the other hand, it is also possible to view this as an optimistic vision: the universe will have a little less pain and stupidity.

Perhaps you have heard of Hungarian biologist and Nobel Laureate for 1937 Albert Szent-Györgyi. I end quoting from the preface of a short book written 50 years ago which begins by asking: *Why does man act like a perfect idiot?* and ends it with ... *the ultimate question: Will mankind be able to survive the machinations of present-day men who often appear to act more like crazy apes then sane human beings?* By now, you know what I think.

17.1 A Dreadful Gate

It was the most terrible task and the most terrible order which could have been given to an organization: the order to solve the Jewish question. In this circle, I may say it frankly with a few sentences. It is good that we had the severity to exterminate the Jews in our domain.

— Heinrich Himmler[4]

Not long ago, I read in the press that the original gate of the Dachau concentration (Figure 17.1) camp near Munich that some idiots had stolen had been recovered after 2 years (In Norway). A replica replaced the original, but the feeling was that it was not the same. Somehow, although

Figure 17.1. The entrance gate at Dachau.

the two gates are just that, iron gates with the cynical motto: *Arbeit Macht Frei*, in the style of Orwell, the original had something that the replica did not have, a relic of superstition.

We imagine that the original was a "witness" of all those who entered through that gate to never leave. Yet, those already famished beings mistreated by prejudice and ignorance, transported in trucks or trains like cows that go to the slaughterhouse, still had something: hope.

I have always asked myself, and I have asked others, particularly Jews, where that benevolent and almighty god was in those times of torture and murder and why his representative here on Earth did not say anything. Nobody has answered with more than the usual: "the ways of the Lord are inscrutable" (or something equivalent), that more than an answer is an insult to intelligence.

Jason Stanley writes[5]: *In fascist ideology, in times of crisis in need, the state reserves support for members of the chosen nation, for "us" and not "them." The justification is invariably because "they" are lazy, lack a work ethic, and cannot be trusted with state funds and because "they" are criminal and seek only to live off state largesse. In fascist politics, "they" can be cured of laziness and thievery by hard labor. This is why the gates of Auschwitz had emblazoned on them the slogan ARBEIT MACHT FREI — work shall make you free.*

But that gate is old history, and in time, it will be forgotten, no matter how many photos you find on the internet, no matter how many movies of this genre you see, just because old news is not news since we do not even remember last years. You will know when it hits you in the face (or worse), but by that time, it will be too late. Increasingly, what matters is today, the present without past or future. What matters is to tweet "I am here now" for all the world to know. The *presenteeism*. We live from "hand to mouth," with a brain bypass.

This is an excellent place to quote historian Walter Laqueur concerning the Nazi regime[6]: *Democratic societies demonstrated on this occasion as on many others before and after, that they are incapable of understanding political regimes of a different character. Not every modern dictatorship is Hitlerian in character and engages in genocide, but everyone has the potential to do so. Democratic societies are accustomed to think in liberal, pragmatic categories; conflicts are believed to be based on misunderstandings and can be solved with a minimum of goodwill; extremism is a temporary aberration, so is irrational behavior in general, such as intolerance, cruelty, etc. The effort to overcome such basic psychological handicaps is immense. It will be undertaken only in the light of immediate (and painful) experience. Each new generation faces this challenge again for experience cannot be inherited.* And we witness this as we see morons marching with the Nazi flag.

17.2 Dante's Gate

Through me you pass into the city of woe:
Through me you pass into eternal pain:
Through me among the people lost for aye.
Justice the founder of my fabric moved:
To rear me was the task of power divine,
Supremest wisdom, and primeval love.
Before me things create were none, save things
Eternal, and eternal I shall endure.
All hope abandon, ye who enter here.

— Dante Alighieri, Divine Comedy, Canto 3

Figure 17.2.   The gate to Hell, Gustave Doré (1862).

And when we are all happily exploring Cybernetia, inside a colorful distorting bubble, a long and winding road will take us to the second gate, which will become visible as the mental fog slowly lifts. Now it is not about the millions who went through the gates of *Arbeit Macht Frei* when apparently God was out for lunch. Now everyone will have to cross the gate, the almost eight thousand million that we are, and enter the Anthropocene transition as if it were the "Total Perspective Vortex." As in all tragicomedies, those who thought they would be spared will also irremediably pass, even Zaphod Beeblebrox. (Do not worry if you do not know what I am talking about, it is only for insiders.) As said by someone you might know[a]: *We will all go together when we go.*

Arthur Schopenhauer also had something to say about this[7]: *In like manner, what was said by the Father of History has not since him been contradicted, — that no man has ever lived who has not wished more than once that he had not to live the following day. According to this, the brevity of life, which is so constantly lamented, may be the best quality it possesses.*

---

[a] Tom Lehrer

If, finally, we should bring clearly to a man's sight the terrible sufferings and miseries to which his life is constantly exposed, he would be seized with horror: and if we were to conduct the confirmed optimist through the hospitals, infirmaries, and surgical operating-rooms, through the prisons, torture chambers, and slave kennels, over battle-fields and places of execution; if we were to open to him all the dark abodes of misery, where it hides itself from the glance of cold curiosity, and finally allow him to glance into Ugolino's dungeon of starvation, — he too would understand at last the nature of this "best of possible worlds." For whence did Dante take the materials for his hell, but from this our actual world? And yet he made a very proper hell of it. And when, on the other hand, he came to the task of describing heaven and its delights, he had an insurmountable difficulty before him; for our world affords no materials at all for this. Therefore there remained nothing for him to do, but, instead of describing the joys of Paradise, to repeat to us the instruction given him there by his ancestor, by Beatrice, and by various saints.

But from this it is sufficiently clear what manner of world it is. Certainly human life, like all bad ware, is covered over with a false luster. What suffers always conceals itself. On the other hand, whatever pomp or splendor any one can get, he openly makes a show of: and the more his inner contentment deserts him, the more he desires to exist as fortunate in the opinion of others, — to such an extent does folly go; and the opinion of others is a chief aim of the efforts of every one, although the utter nothingness of it is expressed in the fact that in almost all languages vanity, vanitas, originally signifies emptiness and nothingness. But under all this false show, the miseries of life can so increase — and this happens every day — that the death which hitherto has been feared above all things is eagerly seized upon. Indeed, if fate will show its whole malice, even this refuge is denied to the sufferer; and in the hands of enraged enemies, he may remain exposed to terrible and slow tortures without remedy. In vain the sufferer then calls on his gods for help: he remains exposed to his fate without grace.

Those who believe the myth of Apocalypse and a new Kingdom refrain from celebrating. You have been deceived. Lose all hope, as Dante wrote. When we pass and glimpse back, we will see, accompanied by Slartibartfast, a post-Holocene planet drifting around the Sun. A torrid suffocating landscape perhaps burst by nuclear weapons caused by the last futile attempt to avoid the worst or possibly littered with bodies exterminated by an uncontrollable pandemic[8] brought on by the heat and our growing population encroaching on the habitats of wildlife. Perhaps I should have added this to the catastrophic convergence, but it is bad enough without it. It is a matter of time and seeing reality as it is, raw and naked, and not as some would like it to be. We made our way steadily towards that gate, led by leaders we cannot trust pushed by the FIRM complex and we can do little, "sleepwalking into the future," in the words of James Howard Kunstler.[9]

Salsa singer and composer Ruben Blades tells us: *Citizens prepare yourselves, what was given is over, take the last drink. You cannot complain; the show was good and cheap. In the face of pain, good humor is essential. Grab your partner and dance to the song about the end of the world. Do not be afraid, do not start shouting, control, no nerves, and refrain from crying. For good or bad, we sent for it, and now the bill has come, and we must pay. Say goodbye to your neighborhood and the world in general, and that on earth nobody be left without dancing the song about the end of the world.*

The tragedy arises from the paradoxical situation in which we find ourselves: After what is a cosmic second, although for us it is a long history, after remarkable material and mental effort, we find ourselves at the dawn of understanding what this is all about. We know what lies ahead, even if a few details escape us, but we see pretty clearly that the road we are traveling on leads to an abyss.

It is as if a sinister hand turned off the light switch, a moment after the mind was illuminated, allowing us only a fleeting glimpse of beautiful works of art, only with time to express admiration: Ah, that is what it was! Before we immerse ourselves in eternal darkness.

I am going to play Pickleball...

References

Chapter 1

1. Bertrand Russell (1943). *An Outline of Intellectual Rubbish*. Haldeman-Julius, Kansas.
2. Ella Wheeler Wilcox (1914). *Poems of Problems*, pp. 154–55, Biblio Bazaar (2009).
3. James Lovelock (2007). *The Revenge of Gaia: Earth's Climate Crisis* and *The Fate of Humanity*, Basic Books, New York.
4. https://harvardmagazine.com/breaking-news/james-watson-edward-o-wilson-intellectual-entente.
5. A term coined by Christian Parenti (2011). *Tropic of Chaos. Climate Change and the New Geography of Violence*. Nation Books, New York.
6. Charles S. Pierce (2009). *Idiot America. How Stupidity Became a Virtue in the Land of the Free*. Doubleday, New York.
7. Leszek Kolakowski (1988). *Metaphysical Horror*. Blackwell, Oxford.
8. Roberto Sirvent and Danny Haiphong (2019). *American Exceptionalism and American Innocence: A People's History of Fake News: From the Revolutionary War to the War on Terror*. Skyhorse, p. 37, New York.
9. Caitlin Johnstone (2021). *Notes from the Edge of the Narrative Matrix*. Lightning Source, UK.
10. Jason Stanley (2018). *How Fascism Works: The Politics of Us and Them*. Random House, New York.
11. Max Blumenthal (2019). *The Management of Savagery*. Verso, p. 42, New York.
12. William Blum (2014). *America's Deadliest Export: Democracy*. ZED Books, London.
13. Michael Parenti (2011). *Face of Imperialism*. Pergamon, p. 120, Elmsford, New York.

14. Jim Sciutto (2019). *The Shadow War: Inside Russia's and China's Secret Operations to Defeat America.* Harper, New York.
15. https://www.youtube.com/watch?v=TMrtLsQbaok.
16. Julio Cortázar (1967). *Around the Day in Eighty Worlds* (translation from Spanish, 1989). North Point Press, New York.
17. Blaise Pascal. *Pensées sur la religion et sur quelques autres sujets.* NABU Press, Charleston, SC.
18. Lloyd Dumas (2011). *The Technology Trap: Where Human Error and Malevolence Meet Powerful Technologies.* Praeger, Westport CT.
19. Chalmers Johnson (2010). *Dismantling the Empire: America's Last Best Hope.* Metropolitan Books, New York.
20. Primo Levi (2008). *Survival in Auschwitz (If This is a Man).* The Orion Press, Miami, FL.
21. https://jovempan.com.br/programas/panico/defensor-da-ditadura-jair-bolsonaro-reforca-frase-polemica-o-erro-foi-torturar-e-nao-matar.html. Retrieved November 2019.
22. Brazil president weeps as she unveils report on military dictatorship's abuses. *The Guardian*, December 10, 2014. https://www.theguardian.com/world/2014/dec/10/brazil-president-weeps-report-military-dictatorshipabuses.
23. Alberto Brandolini. http://ordrespontane.blogspot.com/2014/07/brandolinis-law.html.
24. Andrew J. Bacevich (2020). *The Age of Illusions: How America Squandered Its Cold War Victory.* Metropolitan Books, New York.
25. Congress of the United States of America (July 4, 1776). The French "Declaration of the Rights of the Man and of the Citizen of 1789" states as its first article: "Men are born and remain free and equal in rights".
26. Dwight Macdonald (1957). *The Responsibility of Peoples and Other Essays in Political Criticism.* Victor Collantz, London.
27. Eugen Drewermann (1998). *Der Sechste Tag: Die Herkunft des Menchen und die Frage nach Gott.* Walter Verlag (translation by the author). Olten, Switzerland.
28. Matthew White (2012). *The Great Book of Horrible Things.* Viking, New York.
29. Eduardo Galeano (1994). La función social, el arte de un escritor... y las palabras mejores que el silencio (Social function, the art of an author … and the words better than silence.) Interview published in the newspaper *El Mundo*, of Perú, on November 19, 1994.

30. George Santayana (1906). *The Life of Reason*. Scribner, New York.
31. David Rieff (2016). *In Praise of Forgetting: Historical Memory and Its Ironies*. Yale University Press, New Haven, CT.
32. Darrell Huff (1993). *How to Lie with Statistics*. W. W. Norton & Company, New York.
33. William Bruce Cameron (1963). *Informal Sociology, a Casual Introduction to Sociological Thinking*. Random House, p. 13, New York.
34. Garson O'Toole (2017). *Hemingway Didn't Say That: The Truth Behind Familiar Quotations*. Little A, New York.
35. Robert Graef (2019). *Ignorance. Everything You Need to Know About not Knowing*. Prometheus Books, p. 36, Amherst, New York.
36. Larry Diamond (2019). *Ill Winds, Saving Democracy from Russian Rage, Chinese Ambition, and American Complacency*. Penguin Press, London.
37. Rob Riemen (2018). *To Fight Against This Age*. Norton, New York.
38. Daniel Immerwahr (2019). *How to Hide an Empire: A History of the Greater United States*. Farrar, Straus and Giroux, p. 77, New York.
39. Richard Rorty (1999). *Achieving Our Country: Leftist Thought in Twentieth Century America*. Harvard, Cambridge, MA.
40. Sarah Churchwell (2018). *Behold, America: A History of America First and the American Dream*. Basic Books, New York.
41. https://www.reaganfoundation.org/media/128652/farewell.pdf.
42. Jeremy Scahill (2007). *Blackwater: The Rise of the Most Powerful Mercenary Army*. Nation Books, New York.
43. Jean Meslier (1664–1729). *Testament: Memoir of the Thoughts and Sentiments of Jean Meslier*. Translated by Michael Shreve. Prometheus Books (2009), p. 43, Amherst, New York.

Chapter 2

1. Bertrand Russell (1992). *The Basic Writings of Bertrand Russell*, Robert Egner and Lester Denonn (eds.). Routledge, p. 411, New York.
2. Gerald James Holton (1993). *Science and Anti-science*. Harvard University Press, Cambridge, MA.
3. Stephen Eric Bronner (2020). *The Sovereign*. Routledge, New York.
4. https://www.amnesty.org/en/what-we-do/universal-declaration-of-human-rights/

5. Isaiah Berlin (1956). *The Age of Enlightenment*. New American Library, p. 113, New York.

6. Galileo Galilei, Académico Linceo (1615). *Letter to the Grand Duchess Christina of Tuscany*.

7. Jonathan Haidt (2012). *The Righteous Mind: Why Good People Are Divided by Politics and Religion*. Vintage, New York.

8. Thomas Homer-Dixon (2000). *The Ingenuity Gap: Facing the Economic, Environmental, and Other Challenges of an Increasingly Complex and Unpredictable World*. Vintage, p. 17, Canada.

9. Daniel R. Altschuler and Fernando J. Ballesteros (2019). *The Women of the Moon: Tales of Science, Love, Sorrow, and Courage*. Oxford University Press, Oxford, England.

10. Catherine Nixey (2017). *The Darkening Age: The Christian Destruction of the Classical World*. Houghton Mifflin Harcourt, Boston, MA.

11. http://www.zeno.org/Literatur/M/Goethe,+Johann+Wolfgang/Briefe/1812.

12. Peter Gay (1964). *The Party of Humanity, Essays in the French Enlightenment*. Alfred Knopf, New York.

13. Philipp Blom (2010). *A Wicked Company. The Forgotten Radicalism of the European Enlightenment*. Basic Books, New York.

14. Isaiah Berlin (1997). "Of our century". In *The Proper Study of Mankind*, Henry Hardy (ed.). Farrar, Strauss & Giroux, p. 1, New York.

15. Gerald Holton (2000). The rise of postmodernisms and the "end of science." *Journal of the History of Ideas*, 61(2), 327–341.

16. Anthony Gottlieb (2016). *The Dream of Enlightenment*. W. W. Norton, p. 233, New York.

17. Stephen Bronner (2006). *Reclaiming the Enlightenment: Toward a Politics of Radical Engagement*. Columbia University Press, New York.

18. Tzvetan Todorov (2010). *In Defense of the Enlightenment*. Atlantic Books, p. 22, London.

19. Thomas Homer-Dixon (2000). *The Ingenuity Gap: Facing the Economic, Environmental, and Other Challenges of an Increasingly Complex and Unpredictable World*. Vintage, New York.

20. Robert Heilbroner (1991). *An Inquiry into the Human Prospect*. W. W. Norton, New York.

21. Paul and Anne Ehrlich (1996). *Betrayal of Science and Reason*. Island Press, p. 11, Washington, DC.

Chapter 3

1. Anicius Manlius Severinus Boethius (2008). *The Consolation of Philosophy.* World Publications Group, p. 56, East Bridgewater, MA.
2. Bertrand Russell (1928). *Sceptical Essays: Dreams and Facts.* Routledge Classics (2001), p. 26, Oxfordshire, England.
3. For an excellent exposition of the basic ideas of science I refer you to: Peter Atkins (2003). *Galileo's Finger.* Oxford University Press. Oxford, England.
4. Daniel R. Altschuler (2002). *Children of the Stars.* Cambridge University Press, Cambridge, England.
5. Neil deGrasse Tyson and Donald Goldsmith (2004). *Origins: Fourteen Billion Years of Cosmic Evolution.* Norton, New York.
6. Yuval Harari (2015). *Sapiens: A Brief History of Humankind.* Harper, New York.
7. Brian Fagan (2004). *The Long Summer: How Climate Changed Civilization.* Basic Books, New York.
8. Juan Grompone (2009). *La Danza de Shiva, Libro II.* La flor del Itapebí, Montevideo.

Chapter 4

1. Rachel Carlson (1962). *Silent Spring.* Houghton Mifflin, p. 277, Boston, MA.
2. Toyofumi Ogura (1948, 1997). *Letters from the End of the World.* Kodansha, Tokyo.
3. Rachel Carlson (1962). *Silent Spring.* Houghton Mifflin, p. 8 (reprinted 1994 with introduction by Al Gore.) (How sad to see that my copy has a stamp: "DISCARDED by Lake Region High School Library" and was checked out just four times.)
4. Donella Meadows, Jorgen Randers, and Dennis Meadows (2004). *Limits to Growth: The 30-Year Global Update.* Chelsea Green, Vermont.
5. Farman, J. C., Gardiner, B. G. and Shanklin, J. D (1985). Large losses of total ozone in Antarctica reveal seasonal CLO_x/NO_x interaction. *Nature*, 315, 207–10.
6. Stephen Bronner (2005). *Blood in the Sand: Imperial Fantasies, Right-Wing Ambitions, and the Erosion of American Democracy.* University Press of Kentucky, Lexington, KY.
7. Thomas Friedman (2016). *Thank You for Being Late.* Farrar, Strauss and Giroux, New York.

8. Christian Parenti (2011). *Tropic of Chaos. Climate Change and the New Geography of Violence*. Nation Books, New York.

9. Paul R. Ehrlich and Anne H. Ehrlich (2012). Our unrecognized emergency. In Watson, R. (ed.), *Environment and Development Challenges: The Imperative to Act*. Univeristy of Tokyo Press, Tokyo.

Chapter 5

1. Lewis Thomas (1995). *Late Night Thoughts on Listening to Mahler's Ninth Symphony*. Viking, p. 62, New York.

2. Fritjof Capra (1982). *The Turning Point Science, Society, and the Rising Culture*. Simon and Schuster, New York.

3. Bill McKibben (2019). *Falter, Has the Human Game Begun to Play Itself Out?* Henry Holt, p. 72, New York.

4. Günther Anders (1980). *Die Antiquiertheit des Menschen II band 2*. C.H. Beck, Munich.

5. Herman Melville (1849, 2020). *Redburn: His First Voyage*. Independently published, p. 220.

6. Emma Lazarus (1883). *The New Colossus*. Poem on a Plaque on the Pedestal of the Statue of Liberty.

7. Richard Hofstadter. *Anti-Intellectualism in American Life, The Paranoid Style in American Politics, Uncollected Essays 1956–1965*. Library of America, New York.

8. Anne Appelbaum (2018). A Warning from Europe. *The Atlantic*, October 2018, p. 53.

9. Bertram Gross (1990). *Friendly Fascism. New Face of Power in America*. Black Rose Books, Castroville, TX.

10. Bertrand Russell (1933). "*The Triumph of Stupidity*" in *Mortals and Others: Bertrand Russell's American Essays*. Routledge, pp. 1931–5, Oxfordshire, England.

11. Dean Burnett (2016). *Idiot Brain*. Norton, New York.

12. Hannah Arendt (1951). *The Origins of Totalitarianism*. Shoken (2014), p. 448, New York.

13. *The Guardian*, March 27, 2018, on the killing by police of unarmed Stephen Clark in Sacramento, California.

14. Frank Edwards, Hedwig Lee, and Michael Esposito (2019). Risk of being killed by police use of force in the United States by age, race–ethnicity, and sex. *PNAS* August 5, 2019. https://doi.org/10.1073/pnas.1821204116.

15. Timothy Snyder (2017). *On Tyranny: Twenty Lessons from the Twentieth Century*. Tim Duggan Books, p. 71, New York.
16. Noam Chomsky (February 23, 1967). The Responsibility of Intellectuals. *The New York Review of Books*.
17. Shinobu Kitayama and Cristina E. Salvador (2017). Culture embrained: Going beyond the nature nurture dichotomy. *Perspectives in Psychological Science*, 12, 841–54.
18. Shawn Lawrence Otto (2016). *The War on Science: Who's Waging It, Why It Matters, What We Can Do About It*. Milkweed Editions, Minneapolis, MN.
19. Samuel P. Huntington (1996). *The Clash of Civilizations and the Remaking of World Order*. Simon & Schuster, New York.
20. Thomas Homer Dixon (2006). *The Upside of Down: Catastrophe, Creativity, and the Renewal of Civilization*. Island Press, p. 16, Washington, DC.
21. Nicholas Georgescu-Roegen (1971). *The Entropy Law and the Economic Process*. Harvard University Press, p. 21, Cambridge, MA.
22. Thomas Homer Dixon (2006). *The Upside of Down: Catastrophe, Creativity, and the Renewal of Civilization*. Island Press, Washington, DC.
23. Kevin MacKay (2017). *Radical Transformation, Oligarchy, Collapse, and the Crisis of Civilization*. Between the Lines, p. 27, Toronto, Canada
24. Joseph E. Stiglitz (2017). *Globalization and Its Discontents Revisited: Anti-Globalization in the Era of Trump*. Norton, p. 303, New York.
25. Kishore, Nishant, *et al.* (2018). Mortality in Puerto Rico after hurricane Maria. *New England Journal of Medicine*. https://www.nejm.org/doi/full/10.1056/NEJMsa180397.
26. Seth G. Jones and Martin C. Libicki. How terrorist groups end: lessons for countering Al Qa'ida. https://www.rand.org/content/dam/rand/pubs/monographs/2008/RAND_MG741-1.pdf.
27. Noam Chomsky (2017). *Who Rules the World?* Metropolitan Books, New York.
28. Greg Bean. https://www.michaelwest.com.au/media-dead-silent-as-wikileaks-insider-explodes-the-myths-around-julian-assange/.
29. Daniel Ellsberg (2002). *Secrets: A Memoir of Vietnam and the Pentagon Papers*. Viking, p. 43, New York.
30. Ralph W. McGehee (1999). *Deadly Deceits: My 25 Years in the CIA*. Ocean Press, London.
31. Max Blumenthal (2019). *The Management of Savagery*. Verso, New York.
32. Stephen Grey (2006). *Ghost Plane: The True Story of the CIA Torture Program*. St. Martin's Press, p. 226, New York.

33. Extraordinary rendition: A backstory. *The Guardian*, August 31, 2011, Consulted May 10, 2017.
34. Clara Usiskin (2019). *America's Covert War in East Africa: Surveillance, Rendition, Assassination.* Hurst & Co., London.
35. https://www.deepstatedeclassified.com/weapons-of-the-deep-state-part-1-the-federal-reserve/.
36. John Shumacher (1990). *Wings of Illusion: The Origin, Nature, and Future of Paranormal Belief.* Prometheus, Amherst, NY.
37. Wendel, J., Showstack, R., and Kumar, M. (2017). Dan Rather's vision for scientists in an era of "fake news". *Eos*, 98, https://doi.org/10.1029/2017EO088793. Published on 12 December 2017.

Chapter 6

1. Isaac Asimov. A Cult of Ignorance. *Newsweek* (January 21 1980).
2. http://www.unesco.org/new/en/education/themes/education-building-blocks/literacy/resources/statistics/ (Consulted February 2021).
3. Science and Technology: Public Attitudes, Knowledge, and Interest. https://ncses.nsf.gov/pubs/nsb20207/public-familiarity-with-s-t-facts.
4. Harry Frankfurt (2005). *On Bullshit*. Princeton University Press, p. 63, Princeton, NJ.
5. Errol Morris in an interview in *Nautilus*, http://nautil.us/issue/63/horizons/thomas-kuhn-threw-an-ashtray-at-me.
6. Simon Singh and Edzard Ernst (2008). *Trick or Treatment: The Undeniable Facts about Alternative Medicine.* W. W. Norton, p. 118, New York.
7. Shang, A., *et al.* (2005). Are the clinical effects of homoeopathy placebo effects? Comparative study of placebo-controlled trials of homoeopathy and allopathy. *Lancet*, 366(9487), 726–32.
8. Ernst E., Pittler M.H. (1998). Efficacy of homeopathic arnica: a systematic review of placebo-controlled clinical trials. *Archives of Surgery* (Chicago, Ill.: 1960) 133(11), 1187–90.
9. Ernst Mach (1898). *Popular Scientific Lectures*. Open Court Publishing, p. 280, Chicago.
10. Richard P. Feynman. *What Do You Care What Other People Think?: Further Adventures of a Curious Character*. W. W. Norton, New York.

11. Carl Sagan (1997). *The Demon-Haunted World: Science as a Candle in the Dark*. Ballantine Books, New York

12. Bertrand Russell (1948). *Human Knowledge: Its Scope and Limits*. London: George Allen & Unwin, p. 180, Sydney, Australia.

13. Adwallader, J. V. and Cadwallader, T. C (1990). Christine Ladd-Franklin (1847–1930). In *Women in Psychology: A Bio-bibliographic Sourcebook*, O'Connell, A. N., and Russo, N. F (eds.). Greenwood Press, pp. 220–5, Westport, CT.

14. Peter Godfrey-Smith (2003). *Theory and Reality*. University of Chicago Press, p. 176, Chicago, IL.

15. Lewis Thomas (1995). *Late Night Thoughts on Listening to Mahler's Ninth Symphony*. Penguin Books, London.

16. Daniel R. Altschuler and Fernando J. Ballesteros (2019). *The Women of the Moon: Tales of Science, Love, Sorrow, and Courage*. Oxford University Press, Oxford, England.

17. Andrei Sakharov (1970). Reflections on Progress, Peaceful Coexistence, and Intellectual Freedom. *The New York Times*, 22 July 1968, also: *Progress, Coexistence, and Intellectual Freedom*. W. W. Norton, New York.

18. Roger N. Shepard (1990). *Mind Sights: Original Visual Illusions, Ambiguities, and Other Anomalies*. W.H. Freeman, New York.

19. John R. Searle (1993). Rationality and Realism, What is at Stake? *Journal of the American Academy of Arts and Science*, Fall 1993.

20. Thomas Nagel (1997). *The Last Word*. Oxford University Press, Oxford, England.

21. Julian Huxley (1923). *Essays of A Biologist*. A. Knopf, p. 12, New York.

22. Naomi Oreskes (2019). *Why Trust Science?* Princeton University Press, Princeton, NJ.

23. Lee McIntyre (2019). *The Scientific Attitude*. MIT Press, p. 91, Boston, MA.

24. Lewis Thomas (1995). Late Night Thoughts on Listening to Mahler's Ninth Symphony. Penguin Books, London.

25. Bertrand Russell (1928). *Skeptical Essays*. Routledge Classics, p. 11, Oxfordshire, England.

26. Lyall Watson (1973). *Supernature: A Natural History of the Supernatural*. Hodder & Stoughton, London.

27. Charles Sanders Peirce (1878). How to make our ideas clear. In *The Essential Peirce* (Vol. 1), Nathan Houser and Christian Kloesel (eds.). Indiana University Press, p. 138, Bloomington, IN.

28. The Bible (various authors). King James Version. Genesis 11: 5–7.

29. Shawn Lawrence Otto (2016). *The War on Science: Who's Waging It, Why It Matters, What We Can Do About It.* Milkweed Editions. Minneapolis, MN.

30. Francis Bacon (1597). "ipsa scientia potestas est" ('knowledge itself is power') occurs in Bacon's *Meditationes Sacrae* (1597).

31. Proverbs 24:5. KJV Bible.

32. Carl Sagan (1995). *The Demon-Haunted World: Science as a Candle in the Dark.* Ballantine Books, p. 25, New York.

33. Neil Postman (1985). *Amusing Ourselves to Death.* Penguin Books, London.

34. Julius Lukasiewicz (1994). *The Ignorance Explosion.* Carleton University Press, Ontario, Canada.

35. Aldous Huxley (1958). *Brave New World Revisited.* Harper Perennial Modern Classics, p. 78, New York.

36. Neil Postman (1992). *Technopoly. The Surrender of Culture to Technology.* Alfred Knopf, p. 28, New York.

37. Torres, Ricard (June 2013). QWERTY vs. Dvorak Efficiency: A Computational Approach. https://www.onacademic.com/detail/journal_1000032538870310_0797.html.

38. John Cornwell (2004). *Hitler's Scientists: Science, War, and the Devil's Pact.* Penguin, p. 34, London.

39. Christopher Simpson (1988). *Blowback: The First Full Account of America's Recruitment of Nazis and Its Disastrous Effect on Our Domestic and Foreign Policy.* Weidenfeld and Nicolson, London.

40. Clifford Conner (2020). *The Tragedy of American Science: From Truman to Trump.* Haymarket Books, Chicago, IL.

41. Lloyd Dumas (2002). *Lethal Arrogance.* St Martin's Press, p. 6, New York.

42. Harry Collins y Trevor Pinch (1998). *Golem, and The Golem at Large (2002).* Cambridge University Press (Canto), Cambridge, England.

43. Jarred Diamond (1997). *Guns, Germs, and Steel: The Fates of Human Societies.* Norton, New York.

44. Rolfe Humphries (translator, 1968). *De Rerum Natura of Titus Lucretius Carus.* Indiana University Press, p. 24, Bloomington, IN.

45. W.V. Quine and J.S. Ullian (1978). *The Web of Belief.* McGraw Hill, pp. 119–20, New York.

46. Daniel Kahneman (2011). *Thinking Fast and Slow.* Farrar, Strauss and Giroux, New York.

47. Lee McIntyre (2006). *Dark Ages.* MIT Press, Boston, MA.

48. Sherwin B. Nuland (2003). *The Doctors' Plague: Germs, Childbed Fever, and the Strange Story of Ignac Semmelweis.* W. W. Norton, New York.

49. Christopher Hitchens (2007). *God is Not Great.* Twelve, Hachette Book Group, p. 150, New York.

50. Mike Alder (2004). https://philosophynow.org/issues/46/Newtons_Flaming_Laser_Sword.

51. Dwight Macdonald (1957). The bomb, in: *The Responsibility of Peoples and Other Essays in Political Criticism.* Victor Collantz, p. 113, London.

52. Harold Pinter (2005). *Nobel Lecture: Art, Truth and Politics.* "https://www.nobelprize.org/uploads/2018/06/pinter-lecture-e.pdf.

53. Stephen Jay Gould. *The Flamingo's Smile: Reflections in Natural History.* Norton, New York.

54. James A. Hijiya (2000). The Gita of J. Robert Oppenheimer. *Proceedings of the American Philosophical Society,* 144, No. 2, Jun. 2000.

55. http://www.faktoider.nu/oppenheimer_eng.html.

56. https://www.osti.gov/opennet/manhattan-project-history/Resources/order_drop.htm.

57. Dwight Macdonald (1957). *The Responsibility of Peoples and Other Essays in Political Criticism.* Victor Collantz, p. 105, London.

58. Silvan Schweber (2000). *In the Shadow of the Bomb.* Princeton University Press, Princeton, NJ.

59. https://hypertextbook.com/eworld/einstein/#first.

60. USGCRP, 2018: *Impacts, Risks, and Adaptation in the United States: Fourth National Climate Assessment,* Volume II [Reidmiller, D.R., C.W. Avery, D.R. Easterling, K.E. Kunkel, K.L.M. Lewis, T.K. Maycock, and B.C. Stewart (eds.)]. U.S. Global Change Research Program, Washington, DC, USA. doi: 10.7930/NCA4.2018. https://nca2018.globalchange.gov/chapter/front-matter-about/.

61. Expressed by President Donald Trump as a response to the US report USGCRP, 2018: Impacts, Risks, and Adaptation in the United States: Fourth National Climate Assessment, Volume II [Reidmiller, D.R., C.W. Avery, D.R. Easterling, K.E. Kunkel, K.L.M. Lewis, T.K. Maycock, and B.C. Stewart (eds.)]. U.S. Global Change Research Program, Washington, DC, USA. doi: 10.7930/NCA4.2018. https://nca2018.globalchange.gov/chapter/front-matter-about.

62. Expressed by senator Rick Santorum, https://www.huffingtonpost.com/entry/rick-santorum-cnn-climate_us_5bfaf22de4b0771fb6b9f767.

63. Nicholas Wade (2021). The origin of COVID: Did people or nature open Pandora's box at Wuhan? *Bulletin of the Atomic Scientists,* May 5, 2021.

64. Michael J. Selgelid (2016). Gain-of-function research: Ethical analysis. *Science and Engineering Ethics*, 22(4), 923–64. Published online 2016, August 8. doi: 10.1007/s11948-016-9810-1 (consulted Jan 15 2021).

65. Jennifer A. Doudna, and Samuel H. Sternberg (2018). *A Crack in Creation: Gene Editing and the Unthinkable Power to Control Evolution*. Vintage, New York.

66. Ainissa Ramirez (2021). Shining a light on the impacts of our innovations. *Issues in Science and Technology*, XXXVII(3), Spring 2021.

67. Bertrand Russell (1955). Science and human life. In *What is Science?* James Newman, Simon & Schuster, New York.

68. Richard P. Feynman. *What Do You Care What Other People Think?: Further Adventures of a Curious Character*. W. W. Norton, New York.

69. https://www.washingtonpost.com/politics/trump-slams-fed-chair-questions-climate-change-and-threatens-to-cancel-putin-meeting-in-wide-ranging-interview-with-the-post/2018/11/27/4362fae8-f26c-11e8-aeea-b85fd44449f5_story.html?utm_term=.9e1ed67fe70b.

70. Noam Chomsky (1967). The Responsibility of Intellectuals. *The New York Times Review of Books*. https://chomsky.info/19670223/

71. Karl Popper (1994). The moral responsibility of scientists. In: *The Myth of the Framework: In Defense of Science and Rationality* (first published in 1968). Routledge, Oxfordshire, England.

72. Charles P. Snow (1959). *The Two Cultures*. Cambridge University Press (Canto 1998), p. 60, and 98, Cambridge, England.

73. Shawn Lawrence Otto (2011). Decline and Fall. *New Scientist*, October 29, 2011, p. 42.

74. Noam Chomsky (2016). *Who Rules the World?* Metropolitan Books, p. 21, New York.

75. Natalie Obiko Pearson (October 26, 2016). Donald Trump's grandfather Friedrich Trump ran a restaurant, bar, and brothel in British Columbia. Bloomberg. https://www.bloomberg.com/features/2016-trump-familyfortune

76. http://sites.nationalacademies.org/BasedOnScience/climate-change-humans-are-causing-global-warming/index.htm?_ga=2.73167441.355346005.1565010278-978273653.1546549226.

77. José Ortega y Gasset (1932). *The Revolt of the Masses*. W. W. Norton & Company (revised edition 1884), New York.

78. Shawn Lawrence Otto (2016). *The War on Science: Who's Waging It, Why It Matters, What We Can Do About It*. Milkweed Editions, p. 423, Minneapolis, MN.

79. Bertrand Russell (1950). *Unpopular Essays*. Routledge Classics (2009), p. 93, Oxfordshire, England.

80. Penelope Lively (1997). *Moon Tiger*. Grove Press, p. 11, New York.

81. Kahneman, D., Slovic, P., and Tversky, A (eds.) (1982). *Judgment under Uncertainty: Heuristics and Biases*. New York, Cambridge University Press, pp. 249–67, Cambridge, England.

82. Leon Festinger, Henry Riecken and Stanley Schachter (1956). *When Prophecy Fails*. London, Pinter and Martin (2008), p. 3, London.

83. David Frum (November 4, 2018). The Real Lesson of my Debate with Steve Bannon. The Atlantic https://www.theatlantic.com/ideas/archive/2018/11/bannon-frum-munk-debate-what-really-happened/574867/.

84. Daniel Kahneman, Olivier Sibony, and Cass R. Sunstein (2021). *Noise: A Flaw in Human Judgment*. New York, Little, Brown Spark.

85. Irving Copi and Charles Cohen (2009). *Introduction to Logic* (13th edn.). Taylor & Francis, London.

86. Jamie Whyte (2004). *Crimes Against Logic*. McGraw Hill, New York.

87. Philip Ward (2012). *The Book of Common Fallacies: Falsehoods, Misconceptions, Flawed Facts, and Half-Truths That Are Ruining Your Life*. Skyhorse, New York.

88. Peter B. Medawar (1967). *The Art of the Soluble*, Methuen, London.

89. Paul Kurtz (1992). *The New Skepticism*. Prometheus, p. 201, Amherst, NY.

90. Terza Lettera Del Sig, Galileo Galilei Al Sig, Marco Velseri Delle Macchie Solari (1612). *L'autorità dell'opinione di mille nelle scienze non val per una scintilla di ragione di un solo.* https://www.liberliber.it/mediateca/libri/g/galilei/lettere/html/lett08c.htm (And not as is often cited in his "Dialogue on the two major systems of the world").

91. Bertrand Russell (1929). *Marriage and Morals*. Liveright, p. 57, New York.

92. Jamie Whyte (2004). *Crimes Against Logic*. McGraw Hill, p. 148, New York.

93. Simpson, Edward H (1951). The interpretation of interaction in contingency tables. *Journal of the Royal Statistical Society*, Ser. B 13, 238–41.

94. T. Edward Damer (2008). *Attacking Faulty Reasoning*. Cengage Learning. Boston, MA.

95. David Grimes (2019). *Good Thinking*. The Experiment Publishing, New York.

Chapter 7

1. Giordano Bruno (1591). *De Triplici Minimo et Mensura. as quoted in Giordano Bruno and Renaissance Science by Hilary Gatti*, 1998, Cornell University Press, p. 4, Ithaca, NY.
2. William R. Catton (1980). *Overshoot.* University of Illinois Press, Urbana-Champaign, IL.
3. Galileo Galilei, Académico Linceo (1615). *Letter to the Grand Duchess Christina of Tuscany.*
4. Rolfe Humphries (translator, 1968). *De Rerum Natura of Titus Lucretius Carus.* Indiana University Press, p. 58, Bloomington, IN.
5. Giorgio de Santillana (1955). *The Crime of Galileo.* University of Chicago Press, Chicago, IL.
6. Dava Sobel (1999). *Galileo's Daughter.* Walker Publishing Company, New York.
7. Giorgio de Santillana (1955). *The Crime of Galileo.* University of Chicago Press, p. 4, Chicago, IL.
8. Maurice A. Finocchiaro (2014). *The Trial of Galileo: Essential Documents.* Hackett Classics, p. 146, Cambridge, MA.
9. Galileo (1615). *Letter to the Grand Duchess Christina of Tuscany.* https://sourcebooks.fordham.edu/mod/galileo-tuscany.asp.
10. http://inters.org/Bellarmino-Letter-Foscarini. Consulted April 1, 2018.
11. Galileo Galilei (1632). *Dialogue Concerning the Two Chief World Systems.* Translated by Stillman Drake (2001). Modern Library, new edition, p. 381, New York.
12. William A. Wallace (2017). *Domingo De Soto and the Early Galileo: Essays on Intellectual History.* Taylor & Francis, New York.
13. Galileo (1615). *Letter to the Grand Duchess Christina of Tuscany.* https://sourcebooks.fordham.edu/mod/galileo-tuscany.asp.
14. Thomas Campanella (1994). *A Defense of Galileo.* IN. University of Notre Dame Press, p. 29, Notre Dame.
15. Rolfe Humphries (translator, 1968). *De Rerum Natura of Titus Lucretius Carus.* Indiana University Press, p. 142, Bloomington, IN.
16. Erasmus Darwin (1803). *Zoonomia.* Sect 39.4.8 of Generation. https://www.gutenberg.org/files/15707/15707-h/15707-h.htm.
17. Charles Robert Darwin (1859). *On the Origin of Species by Means of Natural Selection: Or, The Preservation of Favoured Races in the Struggle for Life.*

18. Charles Darwin (1958). *Autobiography*, Nora Barlow (ed.). Norton, p. 87, New York.

19. Francis Darwin (1887). *The Life and Letters of Charles Darwin* (Vol. 3). D. Appleton, p. 17, London.

20. Jacques Monod (1972). *Chance and Necessity: An Essay on the Natural Philosophy of Modern Biology*. Knopf, New York.

21. Pope John Paul II. Message to the pontifical academy of sciences: On evolution. https://humanorigins.si.edu/sites/default/files/MESSAGE%20TO%20THE%20PONTIFICAL%20ACADEMY%20OF%20SCIENCES%20(Pope%20John%20Paul%20II).pdf. Consulted April 1, 2018.

22. Paul Broks (2003). *Into the Silent Land: Travels in Neuropsychology*. Atlantic Monthly Press, p. 91, New York.

23. https://digitaldante.columbia.edu/text/library/the-convivio/book-02/#02.

24. Richard E. Green, *et al.* (May 7, 2010). A draft sequence of the Neandertal genome. *Science*, 328(5979), 710–22.

25. Theodosius Dobzhansky (March 1973). Nothing in biology makes sense except in the light of evolution. *American Biology Teacher*, 35.

26. Isaac Asimov (1980). Attributed to remarks to the National Coalition Against Censorship (NCAC).

27. Daniel Dennett (1996). *Darwin's Dangerous Idea: Evolution and the Meanings of Life*. Simon & Schuster, p. 46, New York.

28. William Paley (1802). *Natural Theology*.

29. Gordy Slack (2007). *The Battle Over the Meaning of Everything. Evolution, Intelligent Design, and a School Board in Dover PA*. John Wiley, New York.

30. Judge Jones (2018). *Memorandum Opinion*. https://law.justia.com/cases/federal/district-courts/FSupp2/400/707/2414073/ Consulted on April 12, 2018.

31. "God Is Not Threatened by Our Scientific Adventures", interview by Laura Sheahen, Beliefnet (undated).

32. Millikan, R. A. (1923). Science and society. *Science*, 58(1503), 293.

33. Daniel Dennett (1996). *Darwin's Dangerous Idea: Evolution and the Meanings of Life*. Simon & Schuster, New York.

34. Steven Jay Gould (1989). *Wonderful Life: The Burgess Shale and the Nature of History*. W. W. Norton & Company, p. 14, New York.

35. Charles Darwin (1859). *Letter to Charles Lyell. Darwin Correspondence Project*. www.darwinproject.ac.uk/entry-2503. Accessed April 16, 2018.

36. Gregory Blue (1999). Gobineau on China: Race theory, the "Yellow Peril" and the critique of modernity. *Journal of World History*, 10(1).

37. James W. Loewen (2018). *Sundown Towns. A Hidden Dimension of American Racism*. The New Press, New York.

38. Benjamin Ferencz (2020). *Parting Words. 9 Lessons for a Remarkable Life*. Sphere, London.

39. Rolfe Humphries (translator, 1968). *De Rerum Natura of Titus Lucretius Carus*. Indiana University Press, p. 102, Bloomington, IN.

40. David Eagleman (2020). *Livewired: The Inside Story of the Ever-Changing Brain*. Pantheon, p. 8, New York.

41. Anthony Gottlieb. The correspondence of René Descartes and Princess Elisabeth of Bohemia — a debate about mind, soul, and immortality. *The Ghost and the Princess*. https://www.laphamsquarterly.org/states-mind/ghost-and-princess.

42. Roger W. Sperry (1974). Lateral specialization in the surgically separated hemispheres. In F. Schmitt and F. Worden (eds.), *Neurosciences Third Study Program* (Ch. I, vol. 3, pp. 5–19). MIT Press, Cambridge, MA.

43. Francis Crick (1994). *Astonishing Hypothesis: The Scientific Search for the Soul*. Scribner, New York.

44. Hank Davis (2009). *Caveman Logic: The Persistence of Primitive Thinking in a Modern World*. Prometheus Books, Amherst, NY.

45. Patricia Smith Churchland (2013). *Touching a Nerve: The Self as Brain*. W. W. Norton, New York.

46. Bart D. Ehrman (2016). *Jesus Before the Gospels*. Harper One, New York.

47. Thomas Nagel (1974). *What is it like to be a bat? in Mortal Questions (1987)*. Cambridge Canto, Cambridge, England.

48. Paul Broks (2003). *Into the Silent Land: Travels in Neuropsychology*. Atlantic Monthly Press, p. 91.

49. Paul Churchland (2007). *Neurophilosophy at Work*. Cambridge University Press, p. 238, Cambridge, England.

50. Hannah Arendt (1964). *Eichmann in Jerusalem: A Report on the Banality of Evil*. Viking, p. 252, New York.

Chapter 8

1. William K. Clifford (1877). *The Ethics of Belief and Other Essays*. Prometheus Books (1999), p. 77, Amherst, NY.

2. Bertrand Russell (1928). *Skeptical Essays*. London: Routledge (2001), p. 14, Oxfordshire, England.

3. Charles S. Pierce (2009). *Idiot America. How Stupidity Became a Virtue in the Land of the Free*. Doubleday, New York.

4. Michael Shermer (1997). *Why People Believe Weird Things: Pseudoscience, Superstition, and Other Confusions of Our Time*. W H Freeman & Co, New York.

5. Kevin MacKay (2017). *Radical Transformation, Oligarchy, Collapse, and the Crisis of Civilization*. Between the Lines, p. 125, Auburn, ME.

6. Originally published in *Contemporary Review*, 1877. Reprinted in Lectures and Essays (1879). Presently in print in: *The Ethics of Belief and Other Essays*. Prometheus Books, 1999. Amherst, NY.

7. H. H. Price (1969). *Belief*. George Allen and Unwin, London.

8. Teofilo F. Ruiz (2011). *The Terror of History*. Princeton University Press, p. 36, Princeton, NJ.

9. Cornel West (2004). *Democracy Matters: Winning the Fight Against Imperialism*. Penguin Press, New York.

10. Charles Freeman (2003). *The Closing of the Western Mind: The Rise of Faith and the Fall of Reason*. Alfred A Knopf, New York.

11. Peter Boghossian (2013). *A Manual for Creating Atheists*. Pichstone Publishing, Durham, NC.

12. José Ortega y Gasset (1940). *Ideas y creencias*. Alianza Editorial (1999). Madrid.

13. Daniel Kahneman (2011). *Thinking Fast and Slow*. Farrar, Strauss and Giroux, p. 62, New York.

14. Carl Sagan (1997). *The Demon-Haunted World: Science as a Candle in the Dark*. Ballantine Books, p. 241, New York.

15. Ernst Hiemer (1938). *Der Giftpilz*. Stürmer-Verlag, Nürnberg.

16. Daniel Goldhagen (2002). *A Moral Reckoning. The Role of the Catholic Church in the Holocaust and its Unfulfilled Duty of Repair*. A. Knopf, New York.

17. Daniel Goldhagen (1996). *Hitler's Willing Executioners: Ordinary Germans and the Holocaust*. A. Knopf, New York.

18. James A. Connor (2004). *Kepler's Witch: An Astronomer's Discovery of Cosmic Order Amid Religious War, Political Intrigue, and the Heresy Trial of His Mother*. HarperOne, reprint edition (May 10, 2005), New York.

19. Brian P. Levack (1995). *The Witch-Hunt in Early Modern Europe*. Longman, London.

20. Heinrich Institoris (Kraemer) y Johann Sprenger (1487). *Maellus maleficarum* (The witch hammer), Part III, Second Head, Question XIV. http://www.malleusmaleficarum.org/files/MalleusAcrobat.pdf.

21. https://www.justice.gov/sites/default/files/olc/legacy/2010/08/05/memo-gonzales-aug2002.pdf.

22. Norman Cohn (1993). *Europe's Inner Demons*, University of Chicago Press, p. 233, Chicago, IL.

23. Michael S. Shermer (2002). *Why People Believe Weird Things*. Henry Holt and Co, New York.

24. Alan Sokal (2008). *Beyond the Hoax, Science Philosophy and Culture*. Oxford University Press, Oxford, England.

25. George Polya (1954). *Mathematics and Plausible Reasoning, Volume 1: Induction and Analogy in Mathematics*. Princeton University Press; Reprint edition (August 3, 1990), p. 8, Princeton, NJ.

26. Stephen E. Toulmin (1976). *Knowing and Acting. An Invitation to Philosophy*. Macmillan, p. 89, New York.

27. Isaac Newton (1700). Cited in David Castillejo (1981). *The Expanding Force in Newton's Cosmos, as Shown in His Unpublished Papers*. Ediciones de Arte y Bibliofilia, p. 116, Madrid.

28. Jerry Fodor (1981). *Representations: Philosophical Essays on the Foundations of Cognitive Science*. MIT Press, p. 316, Cambridge, MA.

29. Persi Diaconis and Frederick Mosteller (1989). Methods for studying coincidences. *Journal of the American Statistical Association*, 84(408).

30. David J. Hand (2014). The Improbability Principle. Why Coincidences, Miracles, and Rare Events Happen Every Day. *Scientific American*, New York.

31. Nassim Taleb (2007). *The Black Swan. The Impact of the Highly Improbable*. Random House, New York.

32. David Hume (1748). *An Enquiry Concerning Human Understanding*. Prometheus Book, p. 105, Amherst, NY.

33. https://www.vatican.va/roman_curia/congregations/cfaith/documents/rc_con_cfaith_doc_20000626_message-fatima_en.html.

34. Isaac Newton (1700). Cited in David Castillejo (1981). *The Expanding Force in Newton's Cosmos, as Shown in His Unpublished Papers*. Ediciones de Arte y Bibliofilia, p. 116, Madrid.

35. http://www.perseus.tufts.edu/hopper/text?doc=Perseus%3Atext%3A1999.02.0131%3Abook%3D3%3Acard%3D417.

36. Henry Louis Mencken (1949). *A Mencken Chrestomathy*. Vintage (1982), p. 88, New York.

37. The Bible KJV.

38. John Loftus (ed.) (2010). *The Christian Delusion: Why Faith Fails*. Prometheus Books, Amherst, NY.

39. https://www.catholic.org/encyclopedia/view.php?id=10963.

40. George B. Vetter (1973). *Magic and Religion, Their Psychological Nature, Origin, and Function*. Philosophical Library, p. 130, New York.

41. http://www.prb.org/Publications/Articles/2002/HowManyPeopleHa veEverLivedonEarth.aspx.

42. Charles Freeman (2003). *The Closing of the Western Mind: The Rise of Faith and the Fall of Reason*. Alfred A. Knopf, p. 322, New York.

43. Richard Dawkins. *The God Delusion*. Mariner Books; Reprint edition (2008). Boston, MA.

44. Daniel Dennett (2006). *Breaking the Spell: Religion as a Natural Phenomenon*. Viking Adult, New York.

45. Melvin Konner (1982). *The Tangled Wing: Biological Constraints on the Human Spirit*. Harper, p. 354 (revised edition 2002, Owl Books), New York.

Chapter 9

1. Paul B. Macready (1999). *An ambivalent Luddite at a Technological Feast*. Designfax. http://maccready.library.caltech.edu/islandora/object/ pbm%3A27832#P./6/mode/2up. [Accessed March 10, 2018].

2. Eisenhower's Farewell Address to the Nation, January 17, 1961.

3. Yevgeny Zamyatin (1921). *We* (Natasha Randall translator). The Modern Library (2006), p. 50, New York.

4. http://www.footprintnetwork.org/en/index.php/GFN/.

5. Georg Borgstrom (1969). *Too Many*. Macmillan, New York.

6. Russell Mokhiber and Robert Weissman (2005). *On the Rampage, Corporate Predators, and the Destruction of Democracy*. Common Courage Press, Monroe, ME.

7. Michael T. Klare (2012). *The Race for What's Left*. Picador, New York.

8. Paul Ehrlich (1968). *The Population Bomb, Population Control or Race to Oblivion*. Ballantine, New York.

9. Katha Pollitt (2021). We Owe It to the World's Children to Slow Population Growth. *The Nation*, July 12/19, 2021.

10. Nick Turse (2015). *Tomorrow's Battlefields. US Proxy Wars and Secret Ops in Africa*. Haymarket Books, Chicago, IL.

Chapter 10

1. Bill McKibben (2019). *Falter, Has the Human Game Begun to Play Itself Out?* Henry Holt, p. 70, New York.
2. Robin Hanson (2008). *Catastrophe, Social Collapse and Human Extinction in: Global Catastrophic Risks*, Nick Bostrom and Milan Ćirković (eds.). Oxford University Press, Oxford, England.
3. Varum Sivaram (2018). *Taming the Sun: Innovations to Harness Solar Energy and Power the Planet*. MIT Press, p. 59, Cambridge, MA.
4. Stolarski, R.S, and Cicerone, R.J. (1974). Stratospheric chlorine: A possible sink for ozone. *Can. J. Chem.* 52, 1610.
5. Molina, M. J., and Rowland, F. S (1974). Stratospheric sink for chlorofluoromethanes: Chlorine atom-catalysed destruction of ozone. *Nature*, 249(5460), 810.
6. Farman, J. C., Gardiner, B. G., and Shanklin, J. D. (1985). Large losses of total ozone in Antarctica reveal seasonal ClO_x/NO_x interaction. *Nature*, 315.
7. Press Release: The 1995 Nobel Prize in Chemistry. Nobelprize.org. https://www.nobelprize.org/prizes/chemistry/1995/summary/.
8. Lydia Dotto (1978). *The Ozone War*. Doubleday, New York.
9. Paul Crutzen (1995). My Life with O_3, NO_x and Other YZO_{xs}, Nobel Lecture. http://www.nobel.se/chemistry/laureates/1995/crutzen-lecture.html.
10. http://unfccc.int/cop4/conv-1.html.
11. Fidel Castro's speech at the Earth Summit, Rio, 1992.
12. Stephen A. Montzka, *et al.* (2018). An unexpected and persistent increase in global emissions of ozone-depleting CFC-11. *Nature*, 557, 413–417.
13. Svante Arrhenius (1895). On the influence of carbonic acid in the air upon the temperature of the ground. *The London Edinburgh, and Dublin Philosophical Magazine and Journal of Science*, 42, 252.
14. David Judt. Britain's long road to a just transition. *The Nation*, October 5/12, 2020.

15. Joseph D. Ortiz, and Roland Jackson (2020). Understanding Eunice Foote's 1856 experiments: heat absorption by atmospheric gases. *The Royal Society Journal of the History of Science*. https://doi.org/10.1098/rsnr.2020.0031.

16. Svante Arrhenius (1895). On the influence of carbonic acid in the air upon the temperature of the ground. *The London Edinburgh, and Dublin Philosophical Magazine and Journal of Science*, 42, 252.

17. Remarkable weather of 1911. *Popular Mechanics*. March 1912.

18. Waldemar Kaempffert, Science in Review; Warmer climate on the Earth may be due to more carbon dioxide in the air, *The New York Times*, October 28, 1956, p. 191.

19. James R. Garvey (August 1966). Air Pollution, and the Coal Industry. *Mining Congress Journal*, 18, 8.

20. https://sciencemetro.com/environment/one-oil-company-expertly-predicted-this-weeks-co2-milestone-almost-40-years-ago/.

21. James Hansen (2009). *Storms of My Grandchildren*. Bloombury, London.

22. Shawn Lawrence Otto (2016). *The War on Science: Who's Waging It, Why It Matters, What We Can Do About It*. Milkweed Editions, Minneapolis, MN.

23. Brad Plumer and Nadja Popovich (June 17, 2020). Emissions Are Surging Back as Countries and States Reopen. *The New York Times*.

24. Bill McKibben (2019). *Falter, Has the Human Game Begun to Play Itself Out?* Henry Holt, p. 74, New York.

25. Masson-Delmotte, V., et al. (eds.) (2019). Global Warming of 1.5°C. *An IPCC Special Report on the impacts of global warming of 1.5°C above pre-industrial levels and related global greenhouse gas emission pathways, in the context of strengthening the global response to the threat of climate change, sustainable development, and efforts to eradicate poverty.*

26. https://www.ipcc.ch/sr15/chapter/spm/ [Accessed January 2021].

27. Bill McKibben (2019). *Falter: Has the Human Game Begun to Play Itself Out?* Henry Holt, p. 59, New York.

28. Shawn Lawrence Otto (2011). *Fool Me Twice: Fighting the Assault on Science in America*. Rodale, p. 207, New York.

29. Laurence C. Smith, *et al.* (2017). Direct measurements of Greenland meltwater runoff. *Proceedings of the National Academy of Sciences*, 114(50), E10622–31; DOI: 10.1073/pnas.1707743114.

30. Jeff Goodell (2017). *The Water Will Come: Rising Seas, Sinking Cities and the Remaking of the Modern World.* Little, Brown, and Company. Boston, MA.

31. Raymond B. Huey and Peter D. Ward (2005). Hypoxia, Global Warming, and Terrestrial Late Permian Extinctions. *Science*, 308(5720), 398–401. Justin L. Penn, Curtis Deutsch, Jonathan L. Payne, and Erik, A. Sperling (2018). Temperature-dependent hypoxia explains biogeography and severity of end-Permian marine mass extinction. *Science.* DOI: 10.1126/science. aat1327.

32. https://skepticalscience.com/From-eMail-Bag-Carbon-Isotopes-Part-1.html.

33. Michel E. Mann and Tom Toles (2016). *The Madhouse Effect.* Columbia University Press, New York.

34. Dale Jamieson (2015). *Reason in a Dark Time.* Oxford University Press, Oxford, England.

35. BP Statistical Review of World Energy (June 2017). https://www. bp.com/content/dam/bp/en/corporate/pdf/energy-economics/statistical-review-2017/bp-statistical-review-of-world-energy-2017-full-report.pdf.

36. Michael Klare (2012). *The Race for What's Left. The Global Scramble for the World's Last Resources.* Metropolitan Books, New York.

37. National Research Council (2015). *Arctic Matters: The Global Connection to Changes in the Arctic.* The National Academies Press, Washington, DC, https://doi.org/10.17226/21717.

38. http://www.bbc.com/earth/story/20170504-there-are-diseases-hidden-in-ice-and-they-are-waking-up.

39. https://www.washingtonpost.com/graphics/2019/national/climate-environment/climate-change-siberia/.

40. https://www.reuters.com/article/us-climate-change-permafrost/scientists-amazed-as-canadian-permafrost-thaws-70-years-early-idUSKCN1TJ1XN?utm_medium=Social&utm_source=Facebook.

41. Committee on Abrupt Climate Change (2002), Abrupt Climate Change: Inevitable Surprises. National Research Council. National Academies Press. https://www.nap.edu/catalog/10136/abrupt-climate-change-inevitable-surprises.

42. Will Steffen, *et al.* (2018). Trajectories of the Earth System in the Anthropocene. *PNAS* August 9, 2018. https://doi.org/10.1073/pnas.1810141115.

43. Timothy M. Lenton, et al., Climate tipping points — too risky to bet against. *Nature*, 27 November 2019.

44. https://www.acq.osd.mil/eie/Downloads/CCARprint_wForward_e.pdf. consulted March 13, 2018.

45. Christian Parenti (2011). *Tropic of Chaos. Climate Change and the New Geography of Violence.* Nation Books, New York.

46. US global change research program climate science special report (CSSR). https://assets.documentcloud.org/documents/3920195/Final-Draft-of-the-Climate-Science-Special-Report.pdf

47. Jason Stanley (2019). *How Fascism Works.* Random House, New York.

48. https://news.yahoo.com/gop-rep-louie-gohmert-asks-190912530.html?gucc ounter=1&guce_referrer=aHR0cHM6Ly93d3cuYmluZy5jb20v&guce_referr er_sig=AQAAAA5gEN27Wg54HQMm1ALEAnuTd4Lcy2inW-L2KZ F3vLS6OASU-08iJ6v8U3aRQhSlGpwOGJkEgMcntO7OKWYcIY-5IUo WYycBVvFFm6kxy494GsiIlM2-glNB_eyt4m4H3MlSeBKffoHVZNSGuB2p-vHZJlI46gUcE5IBRsuK7uan.

49. Charles C. Mann (1994). Betting the Planet. In Peter Menzel (ed.), *Material World.* Sierra Club. San Francisco, CA.

50. Raymond Pierrehumbert (2019). There is no Plan B for dealing with the climate crisis. *Bulletin of the Atomic Scientists*, 75(5), 215–221.

51. Ken Caldeira1, Atul K. Jain, and Martin I. Hoffert (2003). Climate Sensitivity Uncertainty and the Need for Energy Without CO_2 Emission. *Science*, 299(5615), 2052–2054.

52. Alex Steffen (@AlexSteffen, Aug 28, 2017) Tweet.

53. Shawn Lawrence Otto (2016). *The War on Science: Who's Waging It, Why It Matters, What We Can Do About It.* Milkweed Editions, p. 8, Minneapolis, MN.

54. Michael Mann and Tom Toles (2016). *The Madhouse Effect.* Columbia University Press, New York.

55. Amitav Ghosh (2016). *The Great Derangement.* University of Chicago Press, p. 110, Chicago, IL.

56. Both quotations from wikiquote https://en.wikiquote.org/wiki/Mahatma_ Gandhi consulted April 15, 2018.

57. Shawn Lawrence Otto (2016). *The War on Science: Who's Waging It, Why It Matters, What We Can Do About It.* Milkweed Editions, Minneapolis, MN.

58. Robert Scheer. The Heck with that Global Warming Stuff. https://www. thenation.com/article/heck-global-warming-stuff/ Consulted March 13, 2018.

59. Sarah Churchwell (2018). *Behold, America: A History of America First and the American Dream.* Bloomsbury, London.

60. Peter Singer (2002). *One World. The Ethics of Globalization.* Yale University Press, New Haven, CT.

61. Les Roberts *et al.* (2004). Mortality before and after the 2003 invasion of Iraq: Cluster sample survey. *Lancet*, 364, 1857–1864. Published online October 29, 2004 https://www.thelancet.com/pdfs/journals/lancet/PIIS0140-6736(04)17441-2.pdf.

62. Michael Mann (2012). *The Hockey Stick and the Climate Wars: Dispatches from the Front lines.* Columbia University Press, New York.

63. David Wallace-Wells (2019). *The Uninhabitable Earth.* Tim Duggan Books, New York.

Chapter 11

1. Rolfe Humphries (1968). *De Rerum Natura of Titus Lucretius Carus.* Indiana University Press, p. 82, Bloomington, IN.

2. Ben Guarino (June 21, 2017). Stephen Hawking Calls for a Return to the Moon as Earth's Clock Runs Out. *The Washington Post.* https://www.washingtonpost. com/news/speaking-of-science/wp/2017/06/21/stephen-hawking-calls-for-a-return-to-the-moon-as-earths-clock-runs-out/.

3. Giordano Bruno. *On the Infinite Universe and Worlds.* (De L'infinito Universo et Mondi) (Third Dialog).

4. Fernando J. Ballesteros (2010). *E.T. Talk: How Will We Communicate with Intelligent Life on Other Worlds?* Springer, New York.

5. Isaac Asimov (Fall 1989). The relativity of wrong. *The Skeptical Inquirer*, 14(1), 35.

6. H.G Wells (1896). *The Island of Dr. Moreau.*

7. Jacques Monod (1972). *Chance and Necessity.* Collins, p. 160, Cork, Ireland.

8. Thomas Payne (1794). *The Age of Reason.* Gramercy Books (1993), p. 55, Bexley, OH.

9. Douglas Adams (2002). *The Ultimate Hitchhiker's Guide to the Galaxy.* p. 170, Del Rey-Ballantine, New York.
10. Michel Mayor and Didier Queloz (1995). A Jupiter-mass companion to a solar-type star. *Nature,* 378(6555), 355–9.
11. https://tess.mit.edu/.
12. Blaise Pascal (1669). *Thoughts.* http://oll.libertyfund.org/titles/pascal-the-thoughts-of-blaise-pascal.
13. Teofilo F. Ruiz (2011). *The Terror of History: On the Uncertainties of Life in Western Civilization.* Princeton University Press, Princeton, NJ.
14. John D. Barrow and Frank J. Tipler (1988). *The Anthropic Cosmological Principle.* Oxford University Press, Oxford.
15. Paul S. Braterman (2007). *Darwinism and Unbelief in the New Encyclopedia of Unbelief,* Tom Flynn (ed.), Prometheus Books, p. 500, Amherst, NY.

Chapter 12

1. Alan Calaprice (2011). *The Ultimate Quotable Einstein.* Princeton University Press, p. 274, Princeton, NJ.
2. Lewis Mumford (1934). *Technics and Civilization.* University of Chicago Press (2010 edition), p. 311, Chicago, IL.
3. Paul Tillich (1962). *The Power of Self-Destruction.* In: *God and the H-Bomb.* Macfadden Books, New York.
4. Günther Anders (1956). *Die Antiquiertheit des Menschen I.* C.H. Beck, Munich.
5. Chalmers Johnson (2000). *Blowback. The Cost and Consequences or American Empire.* Metropolitan Books, p. 9, New York.
6. https://www.youtube.com/watch?v=X2CE0fyz4ys.
7. Chalmers Johnson (2000). *Blowback: The Costs and Consequences of American Empire.* Metropolitan Books, p. 33, New York.
8. Sven Lindqvist (1992). *Exterminate All the Brutes.* The New York Press, New York.
9. Adam Hochschild (1998). *King Leopold's Ghost: A Story of Greed, Terror, and Heroism in Colonial Africa.* Pan Macmillan, Hampshire, England.
10. Ruben Andersson (2019). *No Go World.* University of California Press, Berkeley, CA.
11. Hans M. Kristensen Robert S. Norris (2017). Worldwide deployments of nuclear weapons. *Bulletin of the Atomic Scientists,* 73(5), 2017.

12. Elisabeth Eaves (2021). Why is America getting a new $100 billion nuclear weapon? https://thebulletin.org/

13. http://www.trumanlibrary.org/publicpapers/index.php?pid=104&st=&st1 — Radio Report to the American People on the Potsdam Conference.

14. Robert Lifton and Greg Mitchell (1995). *Hiroshima in America. Fifty Years of Denial*. Gosset-Putnam, New York.

15. Mills, Michael J., *et al.* (2008): Massive global ozone loss predicted following regional nuclear conflict. *PNAS*, doi:10.1073/pnas.0710058105

16. Robock, Alan, Luke Oman, and Georgiy L. Stenchikov (2007). Nuclear winter revisited with a modern climate model and current nuclear arsenals: Still catastrophic consequences. *Journal of Geophysical Research*, 112, D13107, doi:10.1029/2006JD008235.

17. Toon, Owen B., *et al.* (2007). Consequences of regional-scale nuclear conflicts. *Science*, 315, 1224–5.

18. Kingston Reif and Mandy Smithberger (2021). 500,000,000,000 reasons to scrutinize the US plan for nuclear weapons. *Bulletin of the Atomic Scientists*.

19. Lloyd Dumas (2010). *The Technology Trap: Where Human Error and Malevolence Meet Powerful Technologies*. Praeger, p. 103, Westport, CT.

20. Daniel Ellsberg (2017). *The Doomsday Machine: Confessions of a Nuclear War Planner*. Bloomsbury, London.

21. Suvrat Raju (2015). The unlearned lessons of 1945. *Bulletin of the Atomic Scientists*. https://thebulletin.org/roundtable_entry/the-unlearned-lessons-of-1945/.

22. Eric Hobsbawm (2007). *Globalization, Democracy and Terrorism*. Little Brown & Co., p. 167, New York.

23. Frantz Fanon (1963). *The Wretched of the Earth*. Grove Press, p. 253, New York.

24. James Baldwin (1998). The devil finds work. In *Collected Essays*, Library of America, p. 489, New York.

25. Roberto Sirvent and Danny Haiphong (2019). *American Exceptionalism and American Innocence: A People's History of Fake News: From the Revolutionary War to the War on Terror*. Skyhorse, p. 11, New York.

26. Andrew Bacevich (2021). *After the Apocalypse*. Metropolitan Books, p. 4, New York.

27. Roberto Sirvent and Danny Haiphong (2019). *American Exceptionalism and American Innocence: A People's History of Fake News: From the Revolutionary War to the War on Terror.* Skyhorse, p. 11, New York.

28. Stephen Bronner (2005). *Blood in the Sand: Imperial Fantasies, Right-Wing Ambitions, and the Erosion of American Democracy.* University Press of Kentucky, Lexington, KY.

29. Andrew Bacevich (2021). *After the Apocalypse.* Metropolitan Books, New York.

30. Fareed Zakaria (2011). *The Post American World 2.0.* Norton, New York.

31. Noam Chomsky and Edward Herman (1979, 2014). *The Washington Connection and Third World Fascism.* Haymarket Books, Chicago, IL.

32. William Blum (2008). *Killing Hope. US Military and CIA Interventions since World War II.* Common Courage Press, Monroe, ME.

33. Fintan O'Toole (2020). Donald Trump has destroyed the country he promised to make great again. April 25, 2020. https://www.irishtimes. com/opinion.

34. Christopher Simpson (1988). *Blowback: The First Full Account of America's Recruitment of Nazis and its Disastrous Effect on our Domestic and Foreign Policy.* Weidenfeld and Nicolson, London.

35. James Baker and Linda Zall (2020). The Medea Program. *Oceanography,* 33(1).

36. Nick Turse (2008). *The Complex: How the Military Invades our Everyday Lives.* Metropolitan Books, New York.

37. Stephen Kinzer (2006). *Overthrow: America's Century of Regime Change from Hawaii to Iraq.* Henry Holt, New York.

38. William Blum (2008). *Killing Hope. US Military and CIA Interventions since World War II.* Common Courage Press, Monroe, ME.

39. Andrew Bacevich (2021). *After the Apocalypse.* Metropolitan Books, p. 7.4, New York.

40. William Blum (2004). *Killing Hope U. S. Military and CIA Interventions since World War II.* Common Courage Press, Monroe, ME.

41. Jimmy Carter. State of the Union Address 1980, January 23, 1980.

42. Eric Hobsbawm (2007). *Globalization, Democracy and Terrorism.* Little Brown & Co., p. 158, New York.

43. Joseph Stiglitz (2017). *Globalization and Its Discontents Revisited.* Norton, p. 377, New York.

44. Andrew Bacevich (2010). *Washington Rules: America's Path to Permanent War*. Metropolitan Books, New York.

45. Max Blumenthal (2019). *The Management of Savagery*. Verso, p. 198, New York.

46. William Blum (2004). *Killing Hope U. S. Military and CIA Interventions since World War II*. Common Courage Press, p. 200, Monroe, ME.

47. A. J. Langguth (1978). *Hidden Terrors: The Truth about US Police Operations in Latin America*. Pantheon, New York.

48. https://www.wsj.com/articles/the-legal-case-for-striking-north-korea-first-1519862374. Consulted March 2018.

49. Andrew J. Bacevich (2020). *The Age of Illusions: How America Squandered Its Cold War Victory*. Metropolitan Books, p. 37, New York.

50. Kornbluh, Peter (2003). *The Pinochet File: A Declassified Dossier on Atrocity and Accountability*, The New Press, New York.

51. Secretary of State Colin L. Powell, Interview on Black Entertainment Television's Youth Town Hall, February 20, 2003. https://fas.org/irp/news/2003/02/dos022003.html. Consulted April 2, 2018.

52. A. J. Langguth (1978). *Hidden Terrors: The Truth about US Police Operations in Latin America*. Pantheon, New York.

53. John Dinges (2004). *The Condor Years, How Pinochet and his Allies Brought Terrorism to Three Continents*, The New Press, New York.

54. Seymour M. Hersh (1982). The Price of Power: Kissinger, Nixon, and Chile. *The Atlantic*, December 1982 Issue. https://www.theatlantic.com/magazine/archive/1982/12/the-price-of-power/376309/.

55. https://www.theguardian.com/culture/2000/jul/31/artsfeatures1.

56. Aldous Huxley (1958). *Brave New World Revisited*. Harper Perennial Modern Classics, New York.

57. George Orwell (1938). *Homage to Catalonia*. Bibliotech Press, p. 47.

58. Noam Chomsky (2003). *Hegemony or Survival, Americas Quest for Global Dominance*. Metropolitan Books, New York.

59. Kathy Gilsinian (2020). The War Machine Is Run on Contracts. *The Atlantic*.

60. Noam Chomsky. Remembering Fascism: Learning from the Past. Tuesday, April 20, 2010. http://www.truthout.org/remembering-fascism-learning-from-past58724.

61. Rob Riemen (2018). *To Fight Against This Age*. Norton, p. 2, New York.

62. Jason Stanley (2018). *How Fascism Works*. Random House, New York.

63. Brandy Lee (ed.) (2017). *The Dangerous Case of Donald Trump*. Thomas Dunne Books, p. 298, New York.

64. Chalmers Johnson (2004). *The Sorrows of Empire: Militarism, Secrecy, and the End of the Republic*. Metropolitan Books, New York.

65. David Cay Johnston (2019). *It's Even Worse Than You Think: What the Trump Administration is Doing to America*. Simon and Shuster, p. 6, New York.

66. Matthew White (2012). *The Great Big Book of Horrible Things*. Norton, p. xvii, New York.

67. Alice Calaprice (2011). *The Ultimate Quotable Einstein*. Princeton University Press, p. 280, Princeton, NJ.

68. Barbara MacMillan (2020). *War: How Conflict Shaped Us*. Random House, New York.

69. John Ralston Saul (1992). *Voltaire's Bastards*. Free Press, p. 188, New York.

70. David Swanson (2016). *War is a Lie*. Just World Books, Washington, DC.

71. Michael Zezima (2000). *Saving Private Power*. Soft Skull Press, New York.

72. Craig Whitlock (2021). *The Afghanistan Papers*. Simon & Schuster, New York.

73. https://watson.brown.edu/costsofwar/figures/2019/budgetary-costs-post-911-wars-through-fy2020-64-trillion.

74. https://worldbeyondwar.org/.

75. Robert L. Heilbroner (1991). *An Inquiry into the Human Prospect*. Norton, New York.

76. Bertrand Russell (1943). *An Outline of Intellectual Rubbish*. Haldeman-Julius.

77. Nazi leader Hermann Goering, interviewed by Gustave Gilbert during the Easter recess of the Nuremberg trials, April 18, 1946, quoted in Gilbert's book, *Nuremberg Diary*, Da Capo Press, Reprint edition (August 22, 1995), p. 279, Cambridge, MA.

78. Sheldon S. Wolin (2008). *Democracy Inc. Managed Democracy and the Specter of Inverted Totalitarianism*. Princeton University Press, Princeton, NJ.

79. Hannah Arendt (1951). *The Origins of Totalitarianism*. Schocken Books (2004), New York.

80. Philip Gourevitch (1998). *We Wish to Inform You that Tomorrow We Will be Killed with Our Families*. Farrar, Straus and Giroux, New York.

81. https://www.independent.co.uk/news/world/americas/bush-god-told-me-to-invade-iraq-6262644.html.

82. https://www.independent.co.uk/news/world/americas/bush-god-told-me-to-invade-iraq-6262644.html.

83. https://davidswanson.org/warlist/.

84. https://www.sipri.org/sites/default/files/2021-04/fs_2104_milex_0.pdf.

85. Tom Engelhardt (2018). *A Nation Unmade by War*. Haymarket Books, Chicago, IL.

86. Chris Hedges (2016). *Unspeakable. Talks with David Talbot*. Hot Books. p. 58, Abingdon, VA.

87. Smedley Butler (1935). *War is a Racket*. CreateSpace Independent Publishing, 2016, p. 7, Scotts Valley, CA.

88. Barbara MacMillan (2020). *War: How Conflict Shaped us*. Random House, p. 267, New York.

89. United States Space Command (1997). Vision for 2020. US Space Command, Peterson AFB, CO. Available at https://archive.org/details/pdfy-j6U3MFw1cGmC-yob.

90. https://www.sipri.org/media/press-release/2021/world-military-spending-rises-almost-2-trillion-2020.

91. Andrew J. Bacevich. http://www.bu.edu/pardeeschool/2014/06/28/bacevich-on-iraq-isis-and-more/.

92. Tom Engelhardt (2018). *A Nation Unmade by War*. Haymarket Books, p. 21, Chicago, IL.

93. Graham Allison (2017). *Destined for War: Can America and China Escape Thucydides' Trap?* Houghton Mifflin Harcourt, p. 29, New York.

94. Max Weber (1967/1922). *On Law in Economy and Society*. Touchstone, p. 335, Woodland Park, CO.

95. Frantz Fanon (1963). *The Wretched of the Earth*. Grove Press, p. 77, New York.

96. Karl Popper (1945). *The Open Society and Its Enemies*. Routledge, Oxfordshire, England.

97. James Baldwin (1998). Letter to My Nephew. In *Collected Essays*, Library of America, p. 292, New York.

98. World Inequality Report (2022). https://wid.world/news-article/world-inequality-report-2022/.

99. Joseph Stiglitz (2013). *The Price of Inequality: How Today's Divided Society Endangers Our Future*. W. W. Norton & Company, p. 361, New York.

100. Matt Taibbi (2010). *Griftopia*. Spiegel and Grau, New York.
101. Mark Charles and Soong-Chan Rah (2019). *Unsettling Truths: The Ongoing, Dehumanizing Legacy of the Doctrine of Discovery*. IVP Books, p. 163, Westmont, IL.
102. https://www.forbes.com/real-time-billionaires/#24ec69f23d78.
103. Louis Menard (2021). *The Free World*. Farrar, Strauss and Giroux, New York.
104. José María Tortosa (2001). *El Juego Global — Maldesarrollo y pobreza en el capitalismo mundial*. Icaria Antrazyt, España.
105. Phillip K. Dick. *The Philip K. Dick Reader*. Citadel; Reprint edition (January 26, 2016), New York.

Chapter 13

1. Norman Cohn (1996). *Warrant for Genocide: The Myth of the Jewish World-Conspiracy and the Protocols of the Elders of Zion*. Serif, London.
2. Piere Paolo Pasolini (1975). Abiura dalla Trilogia della vita. *Corriere della Sera*, July 18, 1975.
3. Nancy L. Rosenblum and Russell Muirhead (2020). *A Lot of People Are Saying: The New Conspiracism and the Assault on Democracy*. Princeton University Press, Princeton, NJ.
4. Peter Katel (2009). Conspiracy Theories, do they threaten democracy? *CQ Researcher*, 19(37), 885–908. www.cqresearcher.com, http://www.maxwell.syr.edu/uploadedFiles/news/Conspiracy%20Theories.pdf.
5. http://patriotsquestion911.com/professors.html#Margulis,
6. David Ray Griffin. *The New Pearl Harbor: The Disturbing Questions about the Bush Administration and 9/11*. Interlink Books, Northampton, MA.
7. Nancy L. Rosenblum and Russell Muirhead (2020). *A Lot of People Are Saying: The New Conspiracism and the Assault on Democracy*. Princeton University Press, Princeton, NJ.
8. Adrienne LaFrance (2020). How QAnonn is warping reality and discrediting science. *The Atlantic*, June 2020, p. 27.
9. Brian L. Keeley (March, 1999). Of conspiracy theories. *The Journal of Philosophy*, 96(3), 109–26.
10. Seth Kalichman (2009). *Denying AIDS: Conspiracy Theories, Pseudoscience, and Human Tragedy*. Springer, New York.

11. Kary Mullis (1998). *Dancing Naked in the Mind Field*. Pantheon, New York.
12. Seth C. Kalichman (2009). *Denying AIDS: Conspiracy Theories, Pseudoscience, and Human Tragedy*. Springer Copernicus, p. 56, New York.
13. https://motherjones.com/wp-content/uploads/LuntzResearch_environment. pdf.
14. David Michaels (2008). *Doubt is Their Product*. Oxford University Press, Oxford, England.
15. Adolf Hitler, 30 January 1939 Reichstag speech, in Domarus (1990). *Hitler Speeches and Proclamations, 1932–1945* (Vols. 1–4), p. 1449.
16. Stephen Eric Bronner (2000). *A Rumor About the Jews*. St Martin's Press, p. 17, New York.
17. Adam Hochschild (2018). *Lessons from a Dark Time and Other Essays*. University of California Press, Berkeley, CA.
18. Saul Friedlander (1997). *Nazi Germany and the Jews* (Vol. 1). Harper Collins, New York.
19. John Dominic Crossan (1996). *Who killed Jesus? Exposing the Roots of Anti-Semitism in the Gospel Story of the Death of Jesus*. Harper Collins, New York.
20. http://www.mercaba.org/MAGISTERIO/etsi_multa.htm.
21. http://www.humanist.de/kriminalmuseum/kirche.htm [Accessed March 6, 2018].
22. Will Eisner (2005). *The Plot: The Secret Story of The Protocols of the Elders of Zion*. Norton, New York.
23. Norman Cohn (1996). *Warrant for Genocide*. Serif, p. 157, London.
24. Karl Brammer. *Das Politische Ergebnis des Rathenauprozesses (2012). Auf Grund des Amtlichen Stenogramms Bearbeitet*. Verlag fuer Sozialwissenschaft, 1922, Nabu Press, p. 14, Berlin.
25. Ernst Klee, Willi Dresen and Volker Riess (eds.) (1988). *The Good Old Days*. Konecky & Konecky, Old Saybrook, CT.
26. Peter Longerich (2006). *Davon haben wir nichts gewusst! (¡About that we knew nothing!) Die Deutschen und die Judenverfolgung 1933–1945*. Siedler. Munich, Germany.
27. Daniel Goldhagen (1996). *Hitler's Willing Executioners: Ordinary Germans and the Holocaust*. Alfred Knopf, New York.
28. Hannah Arendt (1964). *Eichmann in Jerusalem: A Report on the Banality of Evil*. Viking, p. 276, New York.
29. Raul Hilberg (1985). *The Destruction of the European Jews*. Homes and Meyer, p. 293, Teaneck, NJ.

30. Raul Hilberg, Stanislaw Staron and Josef Kermisz (eds.) (1979). *The Warsaw Diary of Adam Czerniakow*. Stein and Day, Lanham, MD.
31. John Cornwell (1999). *Hitler's Pope: The Secret History of Pius XII*, Penguin, New York.
32. José M Sánchez (2002). *Pius XII and the Holocaust*. The Catholic University of America Press, Washington, DC.
33. Tom Heneghan (2020). Pope Pius XII, accused of silence during the Holocaust, knew Jews were being killed, researcher says. *The Washington Post*, April 29, 2020.
34. Gerald Steinacher (2010). *Nazis auf der Flucht. Wie Kriegsverbrecher Über Italien nach Übersee Entkamen*. Fischer, Frankfurt, Germany.
35. Michael C. Bender (2021). *Frankly, We Did Win This Election: The Inside Story of How Trump Lost*. Twelve, p. 132, New York.
36. Stephen Eric Bronner (2000). *A Rumor About the Jews*. St Martin's Press, p. 130, New York.
37. Oliver Schröm and Andrea Röpke (2001). *Stille Hile fur Braune Kameraden*. Links Verlag, Berlin.
38. Jorge Semprun (1997). *Literature or Life*. Viking, New York.
39. Maurice Halbwachs (1941). *On Collective Memory*. University of Chicago Press (1992), Chicago, Il.

Chapter 14

1. Isaac Asimov, A Cult of Ignorance. *Newsweek*, January 21, 1980. p. 19.
2. Lewis Thomas (1995). *Late Night Thoughts on Listening to Mahler's Ninth Symphony*. Penguin Books, New York.
3. Sven Lindqvist (1992). *Exterminate all the Brutes*. The New York Press, New York.
4. Daniel Ellsberg (2002). *Secrets: A Memoir of Vietnam and the Pentagon Papers*. Viking, p. 7, New York.
5. Lloyd J. Dumas (1999). *Lethal Arrogance*. St. Martin's Press, p. 91, New York.
6. Robin Fox (1997). Moral Sense and Utopian Sensibility. In *Conjecture and Confrontations*. Transaction Publishers, p. 123, New Brunswick, NJ.
7. Bill McKibben. Winning Slowly Is the Same as Losing. *Rolling Stone*, December 1, 2017.

8. Neil Postman (1985). *Amusing Ourselves to Death*. Penguin Books, New York.

9. Susan Ertz (1943). *Anger in the Sky*. Harper & Brothers. (http://www.nerleelef.com/books/Quotes.pdf)

10. Chris Hedges (2016). *Unspeakable: Talks with David Talbot*. Hot Books, p. 26, Abingdon, VA.

11. H. L. Mencken (1930, 1973). *Treatise on the Gods*. The Johns Hopkins University Press, p. 4, Baltimore, Maryland.

12. Richard Langsworth (2008). *Churchill by Himself*. Public Affairs Publisher, p. 583.

13. Richard Muller (2008). *Physics for Future Presidents: The Science Behind the Headlines*. W. W. Norton, New York.

14. Neil Postman (1985). *Amusing Ourselves to Death*. Penguin Books, New York.

15. Noam Chomsky interview at https://www.brainpickings.org/?s=chomsky.

16. Tom Nichols (2019). *The Death of Expertise*. Oxford University Press, p. 3.

17. Aldous Huxley (1932). *Brave New World*. HarperCollins (1998).

18. Caitlin Johnstone (2021). The Weird, Creepy Media Blackout On Recent Assange Revelations. https://caitlinjohnstone.substack.com/p/the-weird-creepy-media-blackout-on?token=eyJ1c2VyX2lkIjoxMjI0MTE5NiwicG9zdF9pZCI6MzgxNjI2MTksIl8iOiJpTWFDaCIsImlhdCI6MTYyNDk4MjYzMSwiZXhwIjoxNjI0OTg2MjcxLCJpc3MiOiJwdWItODIxMjQiLCJzdWIiOiJwb3N0LXJlYWN0aW9uIn0.gSnZ7nF8sO5TOxtBmXr4NYrM-1srnoidURJvF6oG3DI (retrieved July 2021).

19. Cited by Nuccio Ordine (2017). *The Usefulness of the Useless*. Paul Dry Books, p. 15, Philadelphia, PA.

20. Pierre Bourdieu (1999). *On Television*. New Press, New York.

21. https://www.facebook.com/notes/mark-zuckerberg/a-blueprint-for-content-governance-and-enforcement/10156443129621634/.

22. Roy Scranton (2015). *Learning to Die in the Anthropocene: Reflections on the End of a Civilization*. City Lights Open Media, p. 60, San Francisco, CA.

23. Shoshana Zuboff (2019). *The Age of Surveillance Capitalism: The Fight for a Human Future at the New Frontier of Power*. PublicAffairs, New York.

24. Edward Snowden (2019). *Permanent Record*. Metropolitan Books, New York.

25. https://www.democracynow.org/2021/6/28/julian_assange_extradition_case.

26. Daniel Ellsberg. (2014). Snowden would not get a fair trial — and Kerry is wrong. *The Guardian*.

27. Daniel Ellsberg (2002). *Secrets: A Memoir of Vietnam and the Pentagon Papers*. Viking, p. 192, New York.

28. Glenn Greenwald (2015). *No Place to Hide: Edward Snowden, the NSA, and the U.S. Surveillance State*. Metropolitan Books, p. 94, New York.

29. Byung-Chul Han (2014). *Psychopolitik, Neoliberalismus und die neuen Machttechniken*. Fisher, Frankfurt, Germany.

30. Clara Usiskin (2019). *America's Covert War in East Africa: Surveillance, Rendition, Assassination*. Oxford University Press, p. 199, Oxford, England.

31. Samuel D. Warren and Louis D. Brandeis (1890). The Right to Privacy. *Harvard Law Review*, 4(5), 193–220. https://www.jstor.org/stable/pdf/1321160. pdf?refreqid=excelsior%3A130a09946e1785e832645c01fd2a885d

32. Naomi Oreskes and Erick Conway (2010). *Merchants of Doubt: How a Handful of Scientists Obscured the Truth on Issues from Tobacco Smoke to Global Warming*. Bloomsbury, London.

33. Please see this amazing TED video: https://www.youtube.com/ watch?v=o2DDU4g0PRo.

34. David Michaels (2020). *The Triumph of Doubt: Dark Money and the Science of Deception*. Oxford University Press, Oxford, England.

35. Jim Sciutto (2019). *The Shadow War: Inside Russia's and China's Secret Operations to Defeat America*. Harper, New York.

36. Jaron Lanier (2019). *Ten Arguments for Deleting Your Social Media Accounts Right Now*. Picador, New York.

37. Maryanne Wolf (2018). *Reader, Come Home: The Reading Brain in a Digital World*. Harper, New York.

38. Neil Postman (1992). *Technopolis. The Surrender of Culture to Technology*. Knopf, p. 58, New York.

39. Caitlin Johnstone (2021). *Notes from the Edge of the Narrative Matrix*. Lightning Source, UK.

40. Francis Bacon (1620). *Novum Organum*, Joseph Devey, M.A. (ed.). P. F. Collier (1902), New York. Available at http://oll.libertyfund.org/. Consulted December 2016.

41. Edward Louis Bernays (1928). *Propaganda*. Horace Liveright, New York.

42. Jonathan Haidt (2012). *The Righteous Mind: Why Good People Are Divided by Politics and Religion*. Vintage, p. 125, New York.

43. Johann Wolfgang von Goethe (1749–1832). Es ist nichts schrecklicher als eine tätige *Unwissenheit*. *Goethe's World View Presented in His Reflections and Maxims*, Frederick Ungar (ed.), pp. 58–9 (1963).

44. Neil Postman (1985). *Amusing Ourselves to Death*. Penguin Books, New York.

Chapter 15

1. Robert B. Reich (2020). *The System. Who Rigged It, How We Fix It.* Knopf, p. 73, New York.
2. Crispin Tickell (2002). Address given before a conference on "The Earth Our Destiny"', at Portsmouth Cathedral.
3. Helen Dukas (1981). *Albert Einstein: The Human Side.* Princeton University Press, p. 95 (letter to a Brooklyn minister, November 20, 1950), Princeton, NJ.
4. Jonathan Haidt (2012). *The Righteous Mind: Why Good People Are Divided by Politics and Religion.* Vintage, p. 371, New York.
5. Carl Sandburg (1936). *The People, Yes* (1990), p. 43, Harvest, New York.
6. Edgar Morín (2006). *Breve historia de la barbarie en Occidente.* Paidós, Buenos Aires.
7. Jorge Riechmann (2005). *Un Mundo vulnerable: Ensayos sobre ecología, ética y tecnociencia.* Libros de la Catarata, Madrid.
8. Kevin MacKay (2017). *Radical Transformation, Oligarchy, Collapse, and the Crisis of Civilization.* Between the Lines, Canada.
9. Rachel Carson (1953). Letter to the Editor. *Washington Post.*
10. Peter Singer (2001). *Writings on an Ethical Life.* Harper Perennial, New York.
11. Chris Hedges (2015). *Wages of Rebellion.* Nation Books (book flap). New York.
12. Chris Hedges (2016). *Unspeakable.* Hot Books, Abington, VA.
13. Neil Postman (1985). *Amusing Ourselves to Death.* Penguin, New York.
14. Steve Turner (2011). *The Band that Played On.* Thomas Nelson, Edinburgh, Scotland.
15. Disaster at last befalls Capt. Smith. *The New York Times*, April 16, 1912.
16. Naomi Klein (2008). *The Shock Doctrine: The Rise of Disaster Capitalism.* Picador, p. 5, New York.
17. Fernando Savater (2003). *El valor de elegir.* Editorial Ariel, Madrid.
18. Jennifer Washburn (2015). *University Inc. The Corporate Corruption of Higher Education.* Basic Books, New York.
19. https://www.washingtonpost.com/opinions/2021/04/19/cornel-west-howard-classics/.
20. Abraham Flexner (1939). The Usefulness of Useless Knowledge. *Harpers*, issue 179, June/November.
21. Erazim Kohak (1984). *The Embers and the Stars: Philosophical Inquiry into the Moral Sense of Nature.* University of Chicago Press, p. x, Chicago.
22. Chris Hedges (2015). *Wages of Rebellion: The Moral Imperative of Revolt.* Nation Books, New York.

23. Wassily Leontief (1982). Academic Economics. *Science*, 217, 217.

24. Herman E. Daly, John B. Cobb (1994). *For the Common Good — Redirecting the Economy Toward Community, the Environment, and a Sustainable Future.* Beacon Press, Boston.

25. Michael J. Sandel (2012). *What Money Can't Buy. The Moral Limits of Markets.* Farrar, Strauss, and Giroux. p. 7, New York.

26. The Winter 2017 issue of *New Humanist*. A citadel that stormed itself, p. 33.

27. Nanjala Nyabola. Vaccine Nationalism is Patently Unjust. *The Nation.* April 19/26, 2021.

28. Robert B. Reich (2020). *The System. Who Rigged It. How We Fix It.* Knopf. P. 17, New York.

29. Wendy Brown (2015). *Undoing the Demos. Neoliberalism's Stealth Revolution.* Zone Books, New York.

30. Caitlin Johnstone. My Experiments with Hacking Capitalism. https://caitlinjohnstone.com/2021/05/24/my-experiments-with-hacking-capitalism/.

31. Michael Hudson (2021). *Super Imperialism: The Economic Strategy of American Empire.* Dresden, Germany: Islet Verlag.

32. Richard D. Wolff (2016). *Capitalism's Crisis Deepens.* Haymarket Books, p. ix, Chicago, IL.

33. Roberto Sirvent and Danny Haiphong (2019). *American Exceptionalism and American Innocence: A People's History of Fake News: From the Revolutionary War to the War on Terror.* Skyhorse, p. 89, New York.

34. Jonathan Tepper and Denise Hearn (2019). *The Myth of Capitalism: Monopolies and the Death of Competition.* Wiley, New York.

35. Richard D. Wolff (2020). *Understanding Socialism.* Democracy at Work, New York.

36. Robert B. Reich (2020). *The System. Who Rigged It. How We Fix It.* Knopf, p. 43, New York.

37. Matt Taibbi (2010). *Griftopia.* Spiegel and Grau, p. 32, New York.

38. William Blum (2004). *Killing Hope: U. S. Military and CIA Interventions since World War II.* Common Courage Press, p. 20, Monroe, ME.

39. Bill McKibben (2019). *Falter, Has the Human Game Begun to Play Itself Out?* Henry Holt, p. 121, New York.

40. Robert L. Heilbroner (1991). *An Inquiry into the Human Prospect: Looked at Again for the 1990s.* Norton, New York.

41. Small, K.A. and Van Dender, K. (2005). The Effect of Improved Fuel Economy on Vehicle Miles Traveled: Estimating the Rebound Effect Using U.S. State Data, 1966–2001. UC Berkeley: University of California Energy Institute. Retrieved from https://escholarship.org/uc/item/1h6141nj.

42. Walter Weisskopf (1971). *Alienation and Economics*. Dutton, p. 187, New York.

43. Adam Smith (1776). *An Inquiry into the Nature and Causes of the Wealth of Nations*. Book 1, Chapter 8. http://geolib.com/smith.adam/won1-08.html [Accessed June 1, 2018].

Chapter 16

1. H.L. Mencken (1920), *Prejudices: Second Series*, Chapter 4: The Divine Afflatus. Alfred A. Knopf, New York.

2. Chris Hedges (2016). *Wages of Rebellion*. Norton, p. 43, New York.

3. William J. Ripple, *et al.* (2017). 15,364 scientist signatories from 184 countries, World Scientists' Warning to Humanity: A Second Notice. *BioScience*, 67(12), 1026–1028. https://doi.org/10.1093/biosci/bix125.

4. Thomas Homer-Dixon (2000). *The Ingenuity Gap: Facing the Economic, Environmental, and Other Challenges of an Increasingly Complex and Unpredictable World*. Vintage, p. 1, Canada.

5. H.L. Mencken (1921). Bayard vs. Lionheart. *Baltimore Evening Sun*, July 26, 1920 (reprinted in the book: *On Politics; A Carnival of Buncombe*).

6. Roger Hallam (2019). *Common Sense for the 21st Century*. Chelsea Green, White River Junction, VT.

7. Tzvetan Todorov (2014). In Defense of the Enlightenment. *The Atlantic*, pp. 149–151.

8. Carl Sagan (1989, 2013). *Cosmos*. Ballantine Books, p. 356, New York.

9. Bertrand Russell (1955). Science and Human Life. In: *The Basic Writings of Bertrand Russell*, Chapter 78 (2009), Routledge Classics, London.

10. Darrell Fasching (1993). *The Ethical Challenge of Auschwitz and Hiroshima: Apocalypse or Utopia?* SUNY, p. 318, New York.

11. Ivan Illich (1970). *Deschooling Society*. Harper and Row, New York.

12. Natalia Ginzburg (1962). *The Little Virtues*. Einaudi, Rome, Italy.

13. Rob Riemen (2018). *To Fight Against This Age*. Norton, pp. 75 and 86, New York.

14. Mark Pavlick and Caroline Luft (eds.) (2019). *The United States, Southeast Asia, and Historical Memory*. Haymarket Books, Chicago, IL.

15. Thomas Friedman (2016). *Thank You for Being Late*. Farrar, Straus and Giroux, p. 184, New York.

16. Laurie Garret (1994). *The Coming Plague*. Farrar, Strauss and Giroux, New York.

17. Thomas Homer Dixon (2006). *The Upside of Down: Catastrophe, Creativity, and the Renewal of Civilization*. Island Press, p. 308, Washington, DC.

18. Andrew J. Bacevich (2020). *The Age of Illusions: How America Squandered Its Cold War Victory*. Metropolitan Books, p. 195, New York.

19. Timothy C. Winegard (2019). *The Mosquito. A Human History of Our Deadliest Predator*. Dutton, New York.

20. Brian Fagan (2009). *The Long Summer*. Basic Books, p. 252, New York.

21. Roy Scranton (2015). *Learning to Die in the Anthropocene*. City Lights Books, p. 108, San Francisco, CA.

22. Daniel R. Altschuler (2002). *Children of the Stars*. Cambridge University Press, p. 208, Cambridge, England.

Chapter 17

1. Thomas Huxley (1863). *Man's Place in Nature*. Modern Library (2001), New York.

2. Rachel Carson (1962). *Silent Spring*. Houghton Mifflin, p. 5, Boston, MA.

3. Arthur Schopenhauer (1859). *The World as Will and Representation* (vol. 1), trans. E. F. J. Payne. Dover, 1969, p. 326, New York.

4. Peter Longerich (2001). *Der Ungeschriebene Befehl*. Piper Verlag, p. 191. *The Unwritten Order* (2016). (In a secret speech made in July 1944 to a gathering of senior SS officers), p. 128, The History Press, Cheltenham, England.

5. Jason Stanley (2018). *How Fascism Works*. Random House, p. 157, New York.

6. Walter Laqueur (1980). *The Terrible Secret. Suppression of the Truth about Hitler's Final Solution*. Little Brown and Co, Boston, MA.

7. Arthur Schopenhauer. *The Project Gutenberg eBook of The World as Will and Idea* (vol. 1 of 3) by Arthur Schopenhauer, p. 418. https://www.gutenberg.org/files/38427/38427-pdf.pdf.

8. Wendy Orent (2004). *Plague*. Free Press, New York.

9. James Howard Kunstler (2005). *The Long Emergency*. Atlantic Monthly Press, New York.

Appendix: Further Reading

Because I have included so many references to books, I have prepared a (very subjective) list of those I think you will benefit by reading them.

About Perspective

Yuval Harari (2015). *Sapiens: A Brief History of Humankind*. Harper. This book offers a wonderful trip to understand how we got to where we are.
Neil deGrasse Tyson and Donald Goldsmith (2004). *Origins: Fourteen Billion Years of Cosmic Evolution*. Norton. All you wish to know from the Big Bang to *Homo Sapiens*.
Daniel R. Altschuler (2002). *Children of the Stars*. Cambridge University Press. Why not?

About the Scientific Temper and Technology

Carl Sagan (1995). *The Demon-Haunted World: Science as a Candle in the Dark*. Ballantine Books. A classic book by the eminent Carl Sagan, easy and fun to read, a must read.
Peter Atkins (2003). *Galileo's Finger: The Ten Great Ideas of Science*. Oxford University Press. A well-written book by one of the best science writers.
Shawn Lawrence Otto (2016). *The War on Science: Who's Waging It, Why It Matters, What We Can Do About It*. Milkweed Editions. A thorough discussion of the growing anti-science sentiment and its consequences.
Lee McIntyre (2019). *The Scientific Attitude*. MIT Press. A very readable treatment about the scientific temper.
Clifford Conner (2020). *The Tragedy of American Science: From Truman to Trump*. Haymarket Books. A presentation of science gone wrong, going from the disinterested search for knowledge to another tool for the military-industrial complex.

Tom Nichols (2019). *The Death of Expertise.* Oxford University Press. Here you can read why expert knowledge is important, and how although we are exposed to a tsunami of information, the noxious idea that all voices are to be taken with equal seriousness and any claim to the contrary is elitist will end up destroying democracy.
Lloyd Dumas (2010). *The Technology Trap: Where Human Error and Malevolence Meet Powerful Technologies.* Praeger. A book that looks at the many ways human fallibility and the characteristics of modern technology can combine to cause us to threaten ourselves to make things go not just wrong but catastrophically wrong.

About Logic and Biases

Daniel Kahneman (2011). *Thinking Fast and Slow.* Farrar, Strauss and Giroux. Written in a readable style by one of the founders of cognitive science and Nobel laureate (in economics).
Hank Davis (2009). *Caveman Logic: The Persistence of Primitive Thinking in a Modern World.* Prometheus Books. To understand why we do think the way we think.
David Grimes (2019). *Good Thinking.* The Experiment Publishing. "A beautifully reasoned book about our own unreasonableness."
Jamie White (2004). *Crimes Against Logic.* McGraw Hill. A short, very readable exposition.

About the Catastrophic Convergence

A thorough and complete treatment of the problems we face, and the underlying science, can be found in: Milton W. Cole, Angela D. Lueking and David L. Goodstein (2018). *Science of the Earth, Climate and Energy.* Singapore: World Scientific.
Dale, Jamieson's, *Reason in a Dark Time* (Oxford University Press, 2014) tells us why we have not managed much progress in the face of the threat of Climate Change.
Thomas Homer Dixon, *The Upside of Down: Catastrophe, Creativity, and the Renewal of Civilization.* (Island Press, 2006), although different in tone and issues covered, and perhaps more "academic," is comparable in terms of the concern for the future of humanity. It provides a detailed discussion of the problems we face from a systems theory perspective, and although written 14 years ago, it is still current.
Another book that challenges current views and analyzes our predicament from a historical perspective is Donald Worster's *Shrinking the Earth* (Oxford University Press, 2016).

Although concentrated on Climate Change and the future havoc it entails, the book by David Wallace-Wells, *The Uninhabitable Earth* (Tim Duggan Books, 2019) covers some of what I have to say.
Brian Fagan (2004). *The Long Summer: How Climate Changed Civilization*. Basic Books.
And then there is Bill McKibben's, *Falter, Has the Human Game Begun to Play Itself Out?* (Henry Holt, 2019), a book written by a pioneer of the environmental movement, again discussing the future of humanity under the threat of Climate Change. As the one by Wallace-Wells it is a book you wish everyone would read.

About Beliefs

Justin L. Barrett (2004). *Why Would Anyone Believe in God?* Altamira Press. A lucid and pioneering explanation of our tendency to believe.
Charles Freeman (2003). *The Closing of the Western Mind: The Rise of Faith and the Fall of Reason*. New York, Alfred A Knopf. A thousand-year gap in our development.
George B. Vetter (1973). *Magic and Religion, Their Psychological Nature, Origin, and Function*. Philosophical Library, New York. A classic.
You will find an excellent and enjoyable discussion about morals in the book by Jonathan Haidt (2012). *The Righteous Mind: Why Good People Are Divided by Politics and Religion*. Vintage.

About Conspiracies

Nancy L. Rosenblum and Russell Muirhead (2020). *A Lot of People Are Saying: The New Conspiracism and the Assault on Democracy*. Princeton University Press. "A short book, treating conspiratorial thinking, including a new modality which is conspiracy without even a theory. Perpetuated by repetition, reaching new heights during the Trump administration, undermining democracy."
Stephen Eric Bronner (2003). *A Rumor About the Jews: Conspiracy, Anti-Semitism, and the Protocols of Zion* (Oxford University Press). Here you have a history of this notorious fabrication, probably the most influential work of antisemitism ever written — one which has renewed salience in a "post truth" society dominated by "fake news" and conspiracies — which explores its influence on right-wing movements throughout the twentieth century and the ongoing appeal of bigotry.

On Nazism

Benjamin Ferencz (2020). *Parting Words. 9 Lessons for a Remarkable Life*. Sphere. An inspiring autobiographical book written by the last surviving prosecutor at the Nuremberg Trial.

Daniel Goldhagen (1996). *Hitler's Willing Executioners: Ordinary Germans and the Holocaust*. A. Knopf.

Raul Hilberg (1985). *The Destruction of the European Jews*. Homes and Meyer. An abridgment of a monumental three-volume study by one of the preeminent scholars in the field.

Saul Friedlander (1997). *Nazi Germany and the Jews* (Volumes I and II). HarperCollins. Another thorough exposition of the great tragedy.

About Empire, War, and Peace

The story of how the US became the "leader of the free world" is superbly told by Louis Menard (2021). *The Free World*. Farrar, Strauss and Giroux.

Max Blumenthal (2019). *The Management of Savagery*. Verso. "Excavates the real story behind America's dealing with the world and shows how the extremists' horses that now threaten peace across the globe are the inevitable flowering of America's imperial designs."

Noam Chomsky and Edward S. Herman (1979, 2014). *The Washington Connection and Third World Fascism*. Haymarket Books. "A brilliant and convincing account of United States backed suppression of political and human rights in Latin America, Africa, and the role of the media in misreporting these policies."

Andrew Bacevich (2020). *The Age of Illusions: How America Squandered Its Cold War Victory*, and (2021). *After the Apocalypse*. Metropolitan Books.

Tom Engelhardt (2018). *A Nation Unmade by War*. Haymarket Books.

Michael Parenti (2011). *Face of Imperialism*. Pergamon.

Stephen Bronner (2005). *Blood in the Sand: Imperial Fantasies, Right-Wing Ambitions, and the Erosion of American Democracy*. University Press of Kentucky.

Clara Usiskin (2019). *America's Covert War in East Africa, Surveillance, Rendition, Assassination*. Hurst.

And If you are honestly interested in getting a story that is closer to the truth, then try Daniel A Sjursen (2021). *A True History of the United States*. Lebanon, NH: Steerforth Press.

To get insight into the shady world of PMC read: Jeremy Scahill (2007). *Blackwater: The Rise of the Most Powerful Mercenary Army*. Nation Books.

Three indispensable books by Chalmers Johnson: *Dismantling the Empire: America's Last Best Hope; Blowback; and Nemesis: The Last Days of the American Republic*, all from Metropolitan Books.

Steve Coll's book: *Ghost Wars: The Secret History of the CIA*, Afghanistan and Bin Laden, from the Soviet Invasion to September 10, 2001. The detail of this shady business is well laid out in this book.

About Economic Matters

Shoshana Zuboff (2019). *The Age of Surveillance Capitalism: The Fight for a Human Future at the New Frontier of Power*. Public Affairs. This is a rather long treatise dealing with the new reality of a new economic order hitherto not quite recognized, and that threatens our human condition. Urgent and well written.

Nicholas Georgescu-Roegen (1971). *The Entropy Law and the Economic Process*. Harvard University Press. A classic.

Michael Hudson (2021). *Super Imperialism: The Economic Strategy of American Empire*. ISLET. This is probably the most technical of the lot, but understandable and worth the effort. It will open your eyes.

Noam Chomsky (2016). *Who Rules the World?* Metropolitan Books. No need to elaborate.

Naomi Klein (2008). *The Shock Doctrine: The Rise of Disaster Capitalism*. Picador.

About the WWW and Narratives

Roberto Sirvent and Danny Haiphong (2019). *American Exceptionalism and American Innocence: A People's History of Fake News — From the Revolutionary War to the War on Terror*. Skyhorse.

Neil Postman (1985). *Amusing Ourselves to Death*. Penguin Books. This is a classic book that, although referring to TV (for obvious reasons), can be understood today as it applies to all our new communication technologies.

Daniel Ellsberg (2002). *Secrets: A Memoir of Vietnam and the Pentagon Papers*. Viking. A fascinating story about how governments lie and use official secrecy to keep their actions (often criminal) hidden from the public.

About Everything

Read one of the greatest (and slowly forgotten) thinkers of all time, and an important influence on me: Bertrand Russell. A good selection is to be found in: *The Basic Writings of Bertrand Russell* (Routlege Classics, 2009).

Journals

I might also suggest a subscription to one or two of the following journals, and their blogs: *Scientific American* (https://www.scientificamerican.com), *New Scientist* (https://www.newscientist.com), *Nautilus* (https://nautil.us), *The Bulletin of the Atomic Scientists* (https://thebulletin.org), *New Humanist* (https://newhumanist.org.uk), *The Nation* (https://www.thenation.com), and *Jacobin* (https://www.jacobinmag.com), all of them high quality publications.

Honour, justice, compassion, and freedom are ideas that have no converts. There are only people, without knowledge, understanding or feeling, who intoxicate themselves with words, repeat words, shout them out, imagining they believe them without believing in anything else but profit personal advantage, and their own satisfaction. Words fly away — and nothing remains do you see? Absolutely nothing, you man of good faith! Nothing at all. One moment, and nothing remains — except a lump of dirt, a cold, dead lump of dirt thrown out into black space, spinning round and extinguished sun. Nothing neither thought, sound, nor soul. Nothing.

— Joseph Conrad (1896)*

*Zdzislaw Najder (2007). *Joseph Conrad: A Life*. Camden House, p. 255.

Index